ORGANOCHALCOGEN COMPOUNDS

Advances in Green and Sustainable Chemistry

ORGANOCHALCOGEN COMPOUNDS

SYNTHESIS, CATALYSIS, AND NEW PROTOCOLS WITH GREENER PERSPECTIVES

Edited by

EDER JOÃO LENARDÃO
Laboratório de Síntese Orgânica Limpa—LASOL, CCQFA, Universidade Federal de Pelotas—UFPel, Pelotas, Brazil

CLAUDIO SANTI
Group of Catalysis, Synthesis and Organic Green Chemistry, Department of Pharmaceutical Sciences, University of Perugia, Perugia, Italy

GELSON PERIN
Laboratório de Síntese Orgânica Limpa—LASOL, CCQFA, Universidade Federal de Pelotas—UFPel, Pelotas, Brazil

DIEGO ALVES
Laboratório de Síntese Orgânica Limpa—LASOL, CCQFA, Universidade Federal de Pelotas—UFPel, Pelotas, Brazil

Series Editor

BÉLA TÖRÖK
TIMOTHY DRANSFIELD

ELSEVIER

Elsevier
Radarweg 29, PO Box 211, 1000 AE Amsterdam, Netherlands
The Boulevard, Langford Lane, Kidlington, Oxford OX5 1GB, United Kingdom
50 Hampshire Street, 5th Floor, Cambridge, MA 02139, United States

ISBN: 978-0-12-819449-2

For Information on all Elsevier publications
visit our website at https://www.elsevier.com/books-and-journals

Publisher: Susan Dennis
Acquisitions Editor: Charlotte Rowley
Editorial Project Manager: Emerald Li
Production Project Manager: Kumar Anbazhagan
Cover Designer: Greg Harris
Cover graphic designed by Dr. Micol Santi

Typeset by MPS Limited, Chennai, India

Contents

List of contributors

Laura Abenante Laboratório de Síntese Orgânica Limpa—LASOL—CCQFA—Universidade Federal de Pelotas—UFPel, Pelotas, Brazil

Diego Alves Laboratório de Síntese Orgânica Limpa—LASOL—CCQFA—Universidade Federal de Pelotas—UFPel, Pelotas, Brazil

Juliano B. Azeredo Department of Pharmacy, Federal University of Pampa, Uruguaiana, Rio Grande do Sul (RS), Brazil

Luana Bagnoli Department of Pharmaceutical Sciences (Group of Catalysis, Synthesis and Organic Green Chemistry), University of Perugia, Perugia, Italy

Angelita M. Barcellos Laboratório de Síntese Orgânica Limpa—LASOL—CCQFA—Universidade Federal de Pelotas—UFPel, Pelotas, Brazil

Antonio L. Braga Department of Chemistry, Federal University of Santa Catarina - UFSC, Florianópolis, Santa Catarina (SC), Brazil

Ítalo F.C. Dias Laboratório de Síntese Orgânica Limpa—LASOL—CCQFA—Universidade Federal de Pelotas—UFPel, Pelotas, Brazil; Group of Catalysis, Synthesis and Organic Green Chemistry, Department of Pharmaceutical Sciences, University of Perugia, Perugia, Italy

Daniela Hartwig Laboratório de Síntese Orgânica Limpa—LASOL—CCQFA—Universidade Federal de Pelotas—UFPel, Pelotas, Brazil

Michio Iwaoka Department of Chemistry, School of Science, Tokai University, Hiratsuka-shi, Kanagawa, Japan

Raquel G. Jacob Laboratório de Síntese Orgânica Limpa—LASOL—CCQFA—Universidade Federal de Pelotas—UFPel, Pelotas, Brazil

Eder J. Lenardão Laboratório de Síntese Orgânica Limpa—LASOL—CCQFA—Universidade Federal de Pelotas—UFPel, Pelotas, Brazil

Francesca Mangiavacchi Group of Catalysis, Synthesis and Organic Green Chemistry, Department of Pharmaceutical Sciences, University of Perugia, Perugia, Italy

Francesca Marini Department of Pharmaceutical Sciences (Group of Catalysis, Synthesis and Organic Green Chemistry), University of Perugia, Perugia, Italy

José E.R. Nascimento Laboratório de Síntese Orgânica Limpa—LASOL—CCQFA—Universidade Federal de Pelotas—UFPel, Pelotas, Brazil

Agata J. Pacuła-Miszewska Department of Organic Chemistry, Faculty of Chemistry, Nicolaus Copernicus University, Torun, Poland

Martina Palomba Fondazione Istituto Italiano di Tecnologia, Genova, Italy

Filipe Penteado Laboratório de Síntese Orgânica Limpa—LASOL—CCQFA—Universidade Federal de Pelotas—UFPel, Pelotas, Brazil

Gelson Perin Laboratório de Síntese Orgânica Limpa—LASOL—CCQFA—Universidade Federal de Pelotas—UFPel, Pelotas, Brazil

Jamal Rafique Institute of Chemistry (INQUI), Federal University of Mato Grosso do Sul - UFMS, Campo Grande, Mato Grosso do Sul (MS), Brazil; Institute of Chemistry (IQ), Federal University of Goiás - UFG, Goiânia, Goiás (GO), Brazil

Ornelio Rosati Group of Catalysys, Synthesis and Organic Green Chemistry, Department of Phamaceutical Sciences, University of Perugia, Perugia, Italy

Sumbal Saba Institute of Chemistry (IQ), Federal University of Goiás - UFG, Goiânia, Goiás (GO), Brazil

Manoela Sacramento Laboratório de Síntese Orgânica Limpa—LASOL—CCQFA—Universidade Federal de Pelotas—UFPel, Pelotas, Brazil

Luca Sancineto Group of Catalysis, Synthesis and Organic Green Chemistry, Department of Pharmaceutical Sciences, University of Perugia, Perugia, Italy

Claudio Santi Group of Catalysys, Synthesis and Organic Green Chemistry, Department of Phamaceutical Sciences, University of Perugia, Perugia, Italy

Carl H. Schiesser Seleno Therapeutics Pty. Ltd., Brighton East, VIC, Australia

Paulo H. Schneider Institute of Chemistry (IQ), Department of Organic Chemistry, Federal University of Rio Grande do Sul - UFRGS, Porto Alegre, Rio Grande do Sul (RS), Brazil

Shingo Shimodaira Department of Chemistry, School of Science, Tokai University, Hiratsuka-shi, Kanagawa, Japan

Márcio S. Silva Laboratório de Síntese Orgânica Limpa—LASOL—CCQFA—Universidade Federal de Pelotas—UFPel, Pelotas, Brazil

Liane K. Soares Laboratório de Síntese Orgânica Limpa—LASOL—CCQFA—Universidade Federal de Pelotas—UFPel, Pelotas, Brazil

Jacek Ścianowski Department of Organic Chemistry, Faculty of Chemistry, Nicolaus Copernicus University, Torun, Poland

PART I

Synthesis of Organochalcogens

CHAPTER

1

Multicomponent reactions in the synthesis of organochalcogen compounds

Angelita M. Barcellos[1], *Francesca Mangiavacchi*[2], *Laura Abenante*[1], *Ítalo F.C. Dias*[1,2] and *Manoela Sacramento*[1]

[1]Laboratório de Síntese Orgânica Limpa—LASOL—CCQFA—Universidade Federal de Pelotas—UFPel, Pelotas, Brazil, [2]Group of Catalysis, Synthesis and Organic Green Chemistry, Department of Pharmaceutical Sciences, University of Perugia, Perugia, Italy

1.1 Introduction

Nowadays, the focus of chemistry is reshaping the conventional methods to alternative sustainable variants, by the development of different synthetic strategies applying the *green chemistry* concepts.[1] In this way, multicomponent reactions (MCRs) are a strategy used in organic synthesis and medicinal chemistry to prepare elaborate and complex molecular scaffolds.[2] MCRs employ a minimum of three reactants that are mixed in a single reaction vessel, and most of the atoms from the starting materials are incorporated in the final product. This strategy has many green advantages over stepwise linear synthesis as—inter alia—atom, time, and energy economy, less waste generation, operational simplicity, efficiency, less human effort.[3] Consequently, MCRs are receiving continuous attention by chemists to prepare a wide range of natural products,[4] bioactive compounds,[5] and heterocycles.[6]

Organochalcogen Compounds
DOI: https://doi.org/10.1016/B978-0-12-819449-2.00002-1

The search for new potential therapeutic agents is a big area in chemistry; every day, researchers discover new molecules that offer multitarget pharmaceutical applications. In this sense, organochalcogen compounds, mainly those containing sulfur, selenium, and tellurium atoms, have been well documented in the last decade as presenting a diverse and interesting biological activity.[7] In parallel, the green concepts in modern chemistry have been widely explored in the synthesis of these compounds, such as nonconventional reaction media, application of alternative energy sources, among others.[8]

Herein we provide a brief overview of the development of fast and experimentally simple methods for the preparation of organochalcogen compounds by MCRs. Considering the high number of published scholarly works in this topic, for a better discussion, we selected some representative works, according to the chemical class of the synthesized organochalcogen compound to report here. This chapter is divided into six sections: (1) chalcogen-functionalized heterocycles; (2) heterocycles with chalcogen embedded in the ring; (3) selenoethers; (4) vinyl chalcogenides; (5) phosphorus derivatives; and (6) miscellaneous.

1.2 Chalcogen-functionalized heterocycles

Transition metal (TM)-catalyzed cross-coupling reactions are common transformations widely applied to access complex heterocycles.[9,10] Among others, copper salts, which are accessible, abundant, and easy to handle, have been demonstrated to be a versatile catalyst for the formation of new C–C and C–heteroatom bonds, as well as for C–H functionalization.[9] Recently, C–Se bond formation research has been focused on TM-catalyzed cross-coupling methodologies, which can be differentiated on the basis of the source of the selenium.

In 2015, Alves and coworkers[11] developed a synthetic strategy for the direct synthesis of 4-(organylselanyl)-1*H*-pyrazoles (**1**) variously decorated, through a one-pot copper-catalyzed cyclocondensation and C–H activation, the latter of which allows the selenylation reaction using diaryl diselenides (**2**) as the source of selenium (Scheme 1.1, Eq. (1)). The optimal conditions involve stirring a mixture of arylhydrazines (**3**), 2,4-diones (**4**), diselenide (**2**), CuBr (10 mol%), and bpy (10 mol%) in dimethyl sulfoxide (DMSO) at 100°C for 24 hours in an open flask. The scope of the reaction was widely explored using variously substituted hydrazines (**3**) (with both electron-donating group [EDG] and electron-withdrawing group [EWG] substituents in the aromatic ring), diverse diketones (**4**) (including hindered and asymmetric ones), and diaryl diselenides (**2**) with several different substituents. The reaction was successfully applied

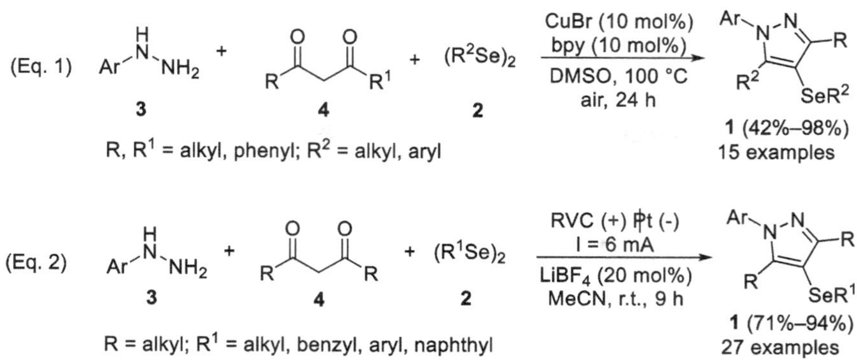

SCHEME 1.1 Synthesis of 4-(organylselanyl)-1*H*-pyrazoles.

for the synthesis of 15 examples in yields ranging from good (42%) to excellent (98%).

A further improvement in the development of a sustainable protocol for the synthesis of chalcogen-functionalized pyrazoles was demonstrated by Wu et al.[12] They reported an electrochemical multicomponent synthesis of 4-selanylpyrazoles (**1**) in agreement with 5 of the 12 principles of green chemistry. They avoided the use of metal catalysts and chemical oxidant; instead, electrons were used as the driving force for the multicomponent transformation (Scheme 1.1, Eq. (2)). In the model reaction, arylhydrazines (**3**), reacted with 2,4-pentadiones (**4**) and diorganyl diselenide (**2**). The optimum reaction conditions selected a constant current of 6 mA, reticulated vitreous carbon as the anode, and platinum plate as the cathode, in acetonitrile. The addition of electrolyte enhanced the reaction performance, and thus, $LiBF_4$ (20 mol%) was selected as such agent. The scope of the reaction was largely explored by producing 24 variously decorated 4-selanylpyrazoles (**1**). The efficiency of the protocol was also demonstrated through a fivefold successful scale-up.

Recently, Jacob et al. reported a copper-catalyzed tandem synthesis of sulfur- and selenium-functionalized azonines (**5**), nine-membered *N*-heterocycles (Scheme 1.2).[13] After careful evaluation of the reaction conditions, the best result was the use of *o*-phenylenediamine (**6**), 2-azidobenzaldehyde (**7**), propargylselenides (**8**), CuI (10 mol%) as a catalyst and Et_3N (2.0 equivalents) as base, in 1,4-dioxane as solvent at 100°C for 24 hours under inert atmosphere. The synthetic protocol was successfully extended to different substituted propargylselenides and the homologs sulfur derivatives, affording a small library of 12 new arylselanyl- or arylsulphenyl-alkyl-1,2,3-triazolo-1,3,6-triazonines (**5**). A plausible mechanism of the MCR copper(I)-catalyzed reaction was described by the authors. First, the condensation reaction between *o*-phenylendiamine (**6**) and 2-azidobenzaldehyde

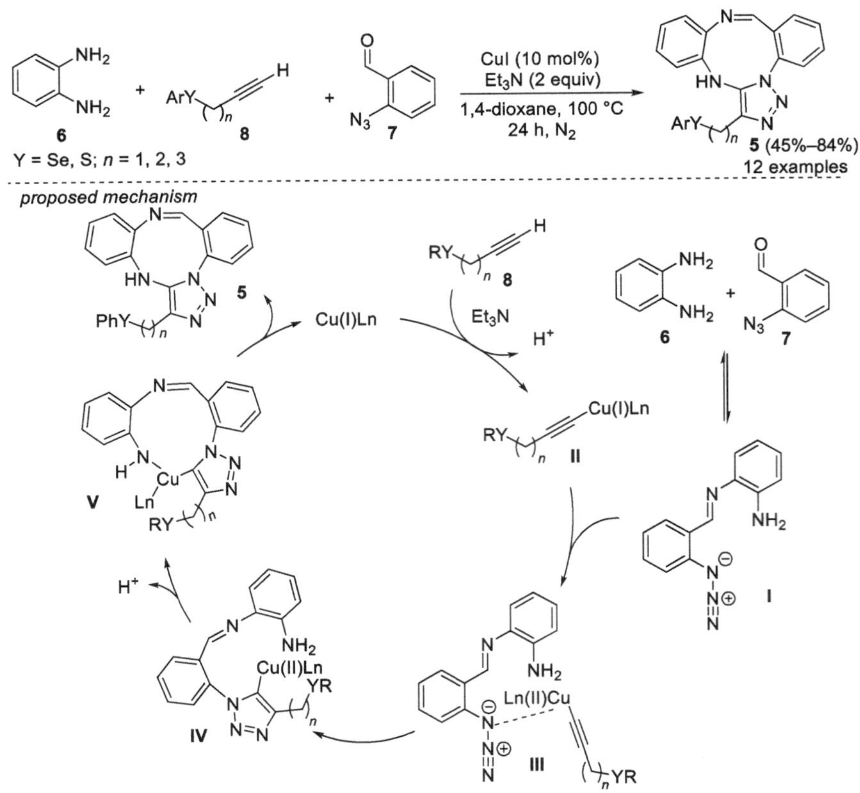

SCHEME 1.2 Synthesis of sulfur- and selenium-functionalized azonines.

(7) affords the imine-adduct intermediate (**I**), while the reaction between the copper salt and the alkyne (**8**) gives the complex (**II**). Then, the reaction between intermediates (**I**) and (**II**) gives the intermediate (**III**). The latter, after a copper(I)-catalyzed azide-alkyne cycloaddition (CuAAC) reaction forms the reactive intermediate (**IV**). The lattermost undergoes an Ulmann-type intramolecular coupling, giving intermediate (**V**), which after reductive elimination affords the target azonine (**5**) (Scheme 1.2).

A similar synthetic strategy, involving a CuAAC reaction, was adopted by Wu et al.[14] for the synthesis of various 5-thio- or 5-selenotriazoles in a one-pot MCR using elemental sulfur or selenium, respectively. The authors reported the synthesis of more than 50 examples in yields of up to 91%. The successful synthesis of five-, six-, or eight-membered rings fused with the triazole unity (**9–14**) is worth mentioning (Fig. 1.1).

The use of elemental selenium as a selenium source in selenylation reactions has several advantages and has been largely explored. Selenium powder is readily accessible, commercially available, affordable, easy to

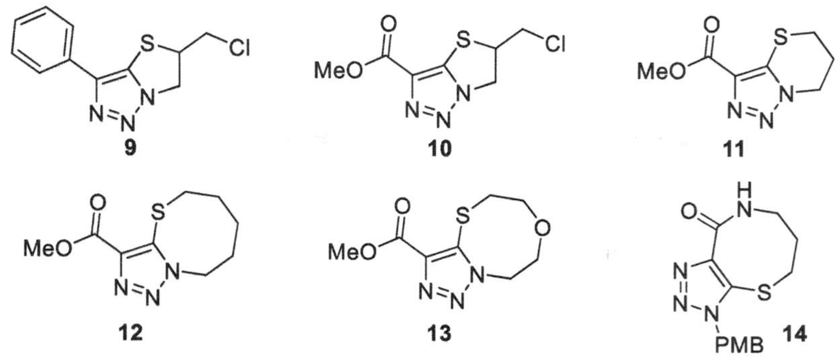

FIGURE 1.1 Fused bicyclic thiotriazoles.[14]

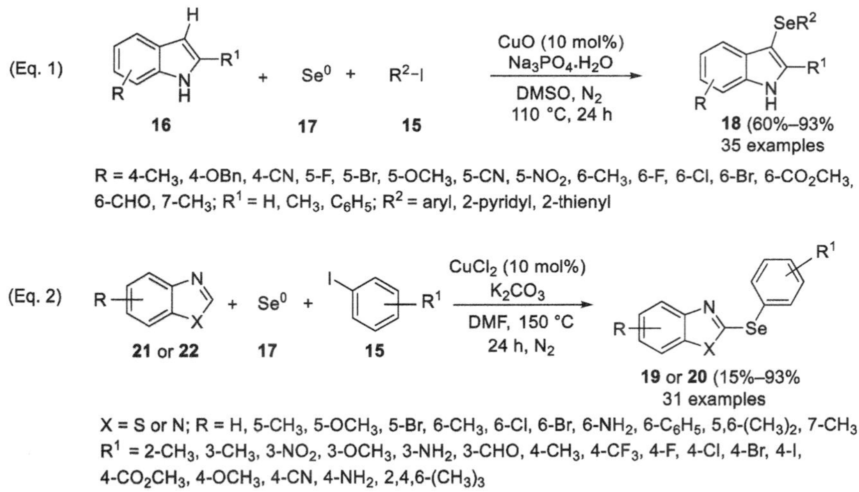

SCHEME 1.3 Synthesis of selanyl-azoles.

handle, and moisture-stable. In 2016, Wu and coworkers[15] envisioned the strategy of a single selenium atom bridging two cross-coupling partners under copper-catalysis for the C_3-selenofunctionalization of indoles. The model reaction involved the use of iodoarene (**15**), indole (**16**), and elemental selenium (**17**) (Scheme 1.3, Eq. (1)). To optimize the yield of selanylindoles (**18**), a screening of the reaction conditions was performed. Among a series of investigated copper salts, CuO resulted in the best performance associated to $Na_3PO_4 \cdot H_2O$ as base and DMSO as the solvent at 110°C for 24 hours. The efficiency and versatility of the optimal conditions were tested on both substituted aryl iodides and indoles. EDGs, as well as EWGs, were well tolerated when included in the iodobenzene, achieving moderate to excellent yields (14 examples, 51%–92% yield).

Variously decorated indoles were also good substrates for the reaction, and no influence of the substituents was observed in the reactivity. The reaction was observed to have high regioselectivity, giving in all the cases the C_3-substitution product. The free *N*–H group is pivotal for the transformation, and thus, in the case of 1-methylindole and electron-poor derivatives (such as *N*-Ac, *N*-Ts), the target product (**18**) was not observed. Furthermore, the authors reported a successful scale-up and the application of the synthetic protocol for the synthesis of a tubulin polymerization inhibitor, namely, [3-(3,4,5-trimethoxyphenylselenyl)-methyl]-indole, through a one-pot two-step reaction.

More recently, the same research group reported a copper-catalyzed cascade reaction to access dissymmetric diaryl selenides, such as 2-arylselanyl-azoles (**19**) and (**20**), under ligand-free conditions (Scheme 1.3, Eq. (2)).[16] The model reaction involved the use of benzothiazole (**21**) or benzoimidazole (**22**), iodoarene (**15**), and elemental selenium (**17**). After a careful screening of the best reaction conditions, K_2CO_3 was selected as a base, $CuCl_2$ (10 mol%) as a catalyst in dimethylformamide (DMF) as solvent at 150°C for 24 hours under inert atmosphere. The synthetic protocol showed a broad scope, with the synthesis of 31 compounds in up to 92% yield. Moreover, the efficient gram scale-up confirmed the utility of the synthetic protocol. The hypothesized reactive species in this reaction is PhSeCu.

The functionalization of indoles has been explored by several research groups. Li and coworkers[17] reported a Cu-catalyzed MCR for the selenylation of imidazoheterocycles (**23**) and indoles (**16**), using elemental selenium (**17**) as the selenium source and iodoarenes (**15**) (Scheme 1.4). The selected optimized conditions use $Cu(OAc)_2$ as a catalyst and KOH as a base in DMF or NMP at 120°C for 12 hours to prepare compounds (**18**) and (**24**). The protocol was successfully applied in a broad substrate scope, to both *N*–H- and *N*-substituted indoles (**18**). The prepared compounds were screened for their in vitro antiproliferative activity in five human cancer cell lines, and most of them presented a comparative activity respect to 5-fluorouracil, used as control. Three of the prepared

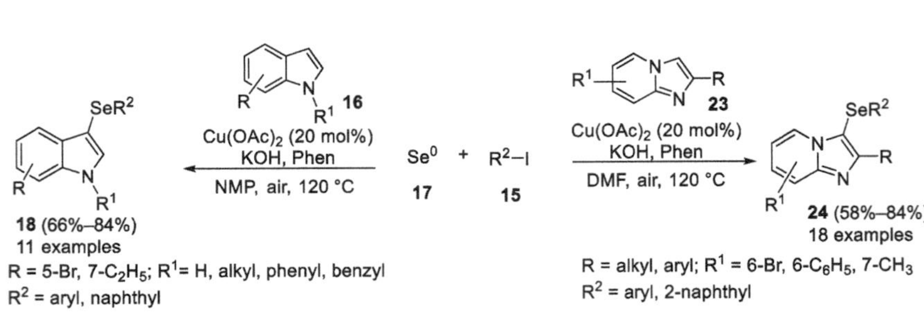

SCHEME 1.4 Selenylation of imidazoheterocycles and indoles.

compounds showed a potent antiproliferative activity against lung (PC-9 and H1975) and breast carcinomas (MCF-7).

An interesting example of metal-free C_3-selenium-functionalization in indoles was reported by Ji et al.[18] They used selenium powder (**17**) as the selenium source, indole (**16**), arylisocianide (**25**), and diamine (**26**), in the presence of TEMPO (2,2,6,6-tetramethylpiperidinooxy) as a catalyst, and oxygen as a green oxidant (Scheme 1.5). The reaction proceeded under mild conditions and with a broad substrate scope, affording 38 examples of selenylated compound (**27**) in 53%–98% yield. The mechanism of the reaction was investigated through electron spin resonance experiments, which evidenced a nitrogen-centered radical. The proposed mechanism is depicted in Scheme 1.5. First, the basic media favors the reaction between isocyanide and selenium, giving the intermediate isoselenocianide (**I**), which reacts with the amine affording the selenolate (**II**), which is in equilibrium with the nitrogen anion (**III**). Then, intermediate (**III**) is oxidized by O_2/TEMPO giving the nitrogen-centered radical (**IV**), which moves to the more active selenium radical (**V**). In the meantime, the base promotes the deprotonation of indole, giving the corresponding nitrogen anion (**VI**) and carbon anion (**VII**) intermediates. The subsequent cross-coupling of radicals (**V**) and (**VII**) gives the radical cation (**VIII**), which undergoes a single electron transfer, affording the target compound (**27**).

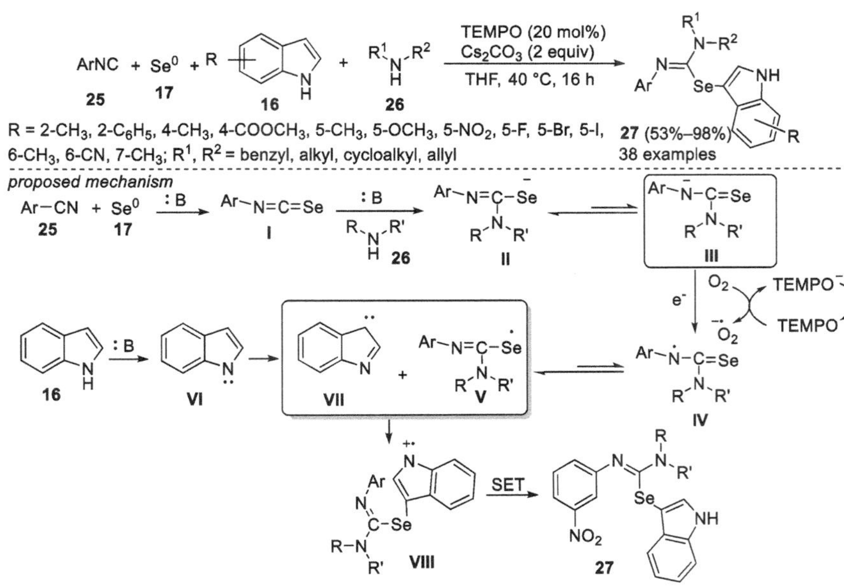

SCHEME 1.5 Synthesis of 3-selenylindole derivatives.

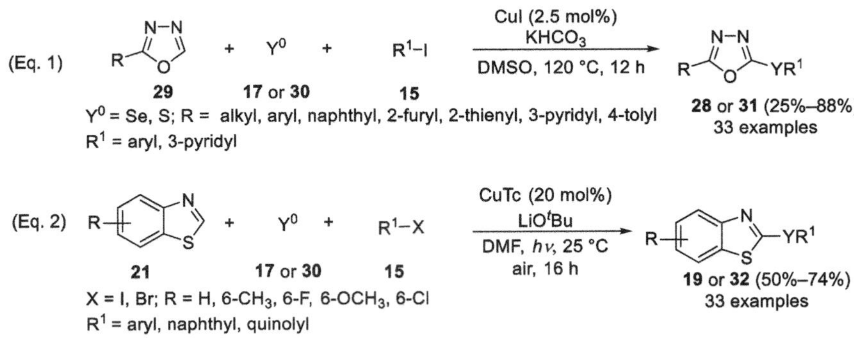

SCHEME 1.6 Synthesis of seleno/thio-oxadiazoles and thio/seleno-substituted benzothiazoles.

In 2018, Braga and coworkers[19] described a copper-catalyzed three-component reaction for the synthesis of the pharmaceutical relevant scaffold seleno-oxadiazoles (**28**) using selenium powder (**17**) as the selenium source (Scheme 1.6, Eq. (1)). The screening to find the optimal reaction conditions selected copper(I) iodide (2.5 mol%) as a catalyst and $KHCO_3$ as a base, in DMSO at 120°C for 12 hours. The versatility of the protocol was evaluated by extending the substrate scope to different iodoarenes (**15**), diverse oxadiazole (**29**, ODZ), and elemental sulfur (**30**) as the chalcogen source, affording 33 selanyloxadiazoles (**28**) and thiooxadiazoles (**31**).

A more sustainable copper-catalyzed protocol was recently reported by Ackerman and coworkers exploiting the photo-induction.[20] Photocatalysis is a powerful tool in organic synthesis,[21,22] and it has been largely explored for the C–H functionalization of a diversity of substrates. In the attempt to improve the sustainability of copper-catalyzed protocols with respect to the general harsh temperature that is required, Akerman's group was able to perform a photo-induced copper-catalyzed C–H chalcogenylation at room temperature (Scheme 1.6, Eq. (2)). The model reaction involved the use of elemental chalcogen (**17**) or (**30**), benzothiazole (**21**), haloarenes (**15**), and copper(I) thiophene-2-carboxylate (CuTc) as catalyst, in DMF at room temperature for 16 hours. The scope of the reaction was largely explored affording 33 thio- (**32**) and seleno-substituted benzothiazoles (**19**) in moderate to good yields.

Elemental selenium has also been used in the three-component reaction to access 4-organoselenylisoxazoles (**33**)[23] and selenylated isochromenones (**34**),[24] as reported by Wu and coworkers (Scheme 1.7). Both protocols involve silver-mediated catalysis and boronic acid derivatives (**35**) in addition to the appropriate substrate (**36**) or (**37**) and elemental selenium (**17**). The control experiments for the mechanism elucidation evidenced a radical pathway in which Ag(II)/oxidant system reacts with the boronic acid

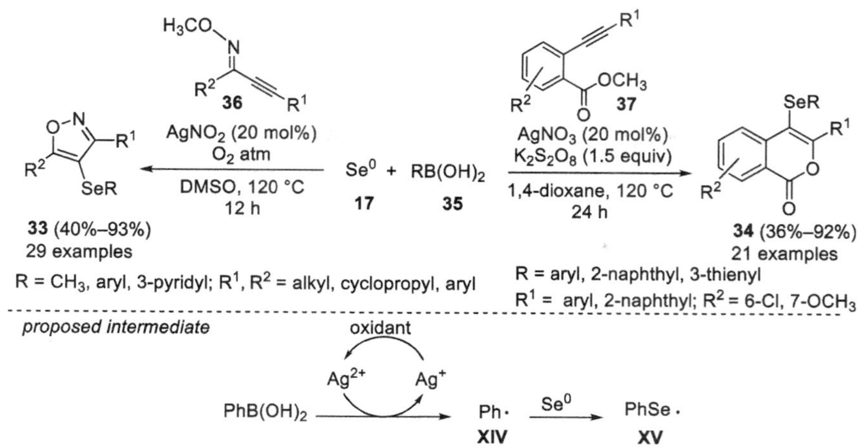

SCHEME 1.7 Synthesis of 4-organoselenylisoxazoles, and selenylated isochromenones.

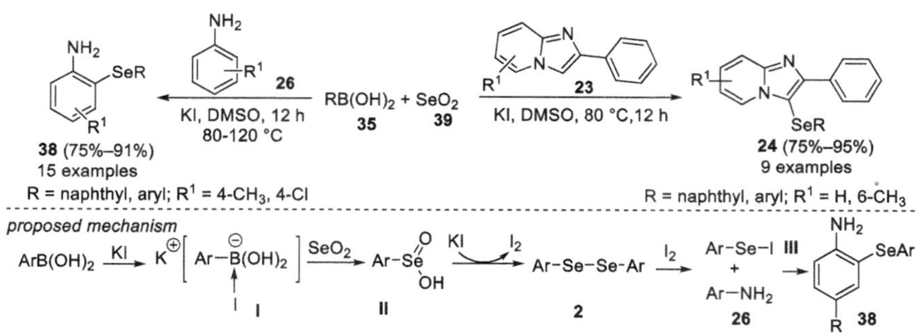

SCHEME 1.8 Synthesis of arylselenium-substituted aniline and imidazo-[1,2-*a*]-pyridines.

generating the phenyl radical intermediate (**XIV**), which then reacts with elemental selenium (**17**) to give the selenium-centered radical (**XV**).

An interesting example of using selenium oxide as the selenium source to generate arylselenium-substituted compounds was described by Zhou and coworkers.[25] SeO_2 as selenium source is an attractive (large availability and stability, easy to handle, and safe) and challenging compound, generally used as a catalyst or oxidant. With the aim to develop new selenium-based synthetic protocols, the authors were able to synthetize arylselenium-substituted aniline (**38**) and imidazo-[1,2-*a*]-pyridines (**24**) (Scheme 1.8). In the model reaction it was used SeO_2 (**39**), arylboronic acid derivatives (**35**), and the appropriate substrate (**23**) or (**26**). The optimized conditions were selected adding KI (1.0 equivalent) as substoichiometric reagent, in DMSO at 80°C for 12 hours. A broad reaction scope was described using different 4-substituted anilines (**26**) (both EWG and

EDG were well tolerated). The unsubstituted anilines, in which a competitive substitution between *ortho-* and *para-* positions was possible, gave rise to the *para*-substituted anilines as the main product. The reaction showed satisfactory results in the case of pyridine (C_5-substitution) and indole (C_3-substitution) as well. A good performance was observed in the case of variously substituted imidazo-[1,2-*a*]-pyridines (**23**), while pyridine boronic acid and aliphatic boronic acid were less reactive (<5% yield). Scheme 1.8 describes a plausible mechanism that starts by the reaction between the arylboronic acid and KI, affording the intermediate (**I**). The interaction of (**I**) with SeO_2 (**39**) forms the arylseleninic acid (**II**), which is subsequently reduced to diselenide (**2**). The diselenide then reacts with iodine, formed in the previous step, giving the corresponding electrophilic ArSeI derivative (**III**). The reaction of (**III**) with the nucleophilic amine gives the target product (**38**).

1.3 Heterocycles with chalcogen embedded in the ring

Using microwave as an alternative energy source, Choudhury and coworkers[26] have prepared trisubstituted 1,3-thiazoles (**40**), through a domino reaction (Scheme 1.9). In this method, equivalent amounts of arylglyoxals

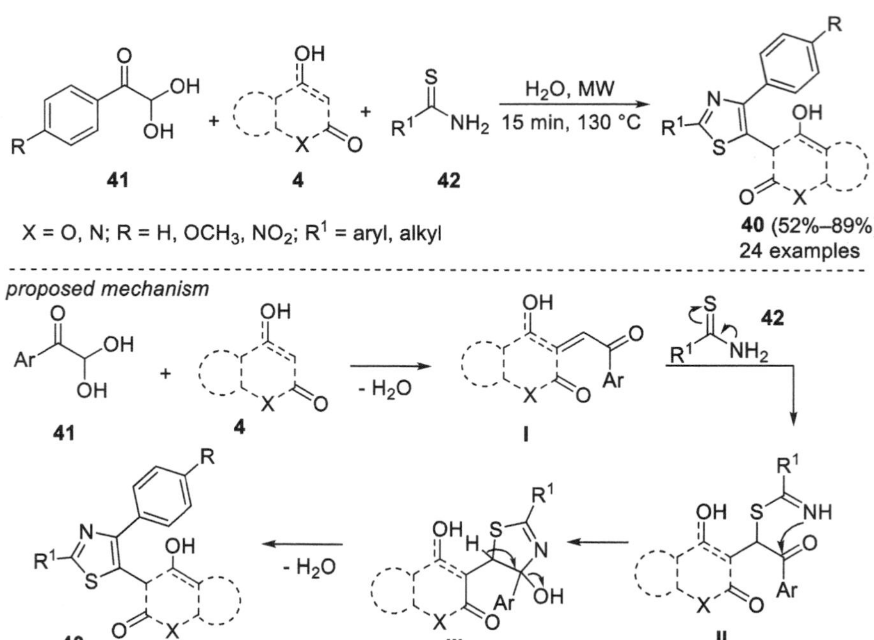

SCHEME 1.9 Synthesis of trisubstituted 1,3-thiazoles.

(**41**), cyclic 1,3-dicarbonyls (**4**) and thioamides (**42**) reacted in aqueous media for 15 minutes at 130°C under microwave irradiation. This protocol was applied in the synthesis of 24 differently substituted 1,3-thiazoles, in yields ranging from 52% to 89%. The authors also performed the reaction with thiobenzamides, including 4-Cl and 4-OCH_3-substituted ones, providing the respective products (**40**) in moderate yields. In addition, cyclic 1,3-dicarbonyls (**4**) containing 4-OCH_3 and 4-NO_2 groups attached to the aromatic ring were also tested. The method proved to be efficient against the tested substituents, obtaining 24 examples. The proposed mechanism for the formation of trisubstituted 1,3-thiazoles initiates by the Knoevenagel-type reaction of (**41**) with (**4**) to afford the alkene intermediate (**I**). The thioamide (**42**) then reacts with the intermediate (**I**) through a thia-Michael addition, providing the intermediate (**II**). Once formed, the intermediate (**II**) undergoes a cyclization, losing water and giving the product (**40**) (Scheme 1.9).

In 2017, Cheng and coworkers[27] described the synthesis of 3,5-diaryl 1,2,4-thiadiazoles (**43**) from arylmethyl bromides (**44**), arylamidines (**45**) and elemental sulfur (**30**) in the presence of lithium *tert*-butoxide (4 equivalents) in toluene at 140°C, under nitrogen atmosphere for 12 hours (Scheme 1.10, Eq. (1)). After the study of optimization of the reaction conditions, the authors tested the efficiency of the method against electron-donating and electron-withdrawing substituents bonded to the aromatic ring of both the arylamidines (**45**) and the arylmethyl bromides (**44**). In both cases, the 3,5-diaryl 1,2,4-thiadiazoles (**43**) were obtained in good yields (35%–70%). Interestingly, the method was suitable to heteroarylamidines (**45**) (R^1 = 3- and 4-pyridyl), by using

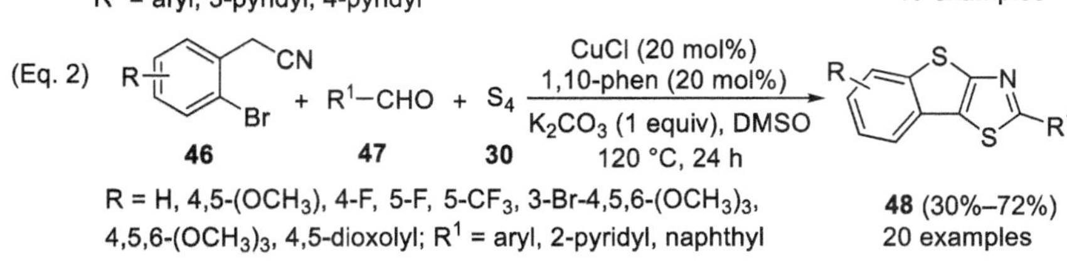

SCHEME 1.10 Synthesis of 3,5-diaryl 1,2,4-thiadiazoles and 2-arylbenzo[4,5]thieno[2,3-*d*]thiazole fused with thiophene.

DMSO as solvent instead of toluene, and the respective products were obtained in yields of 45% and 47%, respectively.

A series of 2-arylbenzo [4.5]thieno[2,3-*d*]thiazole fused with thiophene (**48**) were prepared by Du and coworkers[28] through the reaction of 2-(2-bromophenyl)acetonitrile (**46**) with various aromatic aldehydes (**47**) and elemental sulfur (**30**) (Scheme 1.10, Eq. (2)). In this method, CuCl (20 mol%) was used as catalyst in the presence of 1,10-phenanthroline (20 mol%), K_2CO_3 (1 equivalent), and DMSO as solvent at 120°C for 24 hours. The efficiency and generality of this protocol was evaluated in reactions using EDG and EWG attached to the aromatic ring of both the aldehyde and the 2-(2-bromophenyl)acetonitrile. In both cases the products (**48**) were obtained in good yields (30%–72%) providing a total of 20 examples. When the trifluoromethyl group was attached to the 2-(2-bromophenyl)acetonitrile; however, the desired product was not obtained.

The synthesis of benzoselenophenes (**49**) from acetophenones (**50**), indoles (**16**), and elemental selenium (**17**) was described in 2019 by Deng and coworkers (Scheme 1.11).[29] In this method, the authors used iodine bromide (1.2 equivalents), *N*-methyl-2-pyrrolidone as a solvent at 140°C for 8 hours. In this work, different 1-methyl-2-substituted-1*H*-indoles (**16**) were tested, as 2-aryl, 2-alkyl, 2-silyl, 2-ester, and 2-alkynyl substituents. Among the tested indoles, 1-methyl-2-phenyl-1*H*-indole provided the respective product in 83% yield, while 1-methyl-2-(thiophen-2-yl)-1*H*-indole afforded the expected benzoselenophene (**49**) in 57% yield. Different ketones containing EDG and EWG in the aromatic ring, as well as 1-(anthracen-2-yl)ethanone, 1-(phenanthren-3-yl)ethenone, and 1-(naphthalen-2-yl)ethenone were also evaluated, generating the desired products in good yields (34%–63%). In this work, 39 compounds were prepared in yields ranging

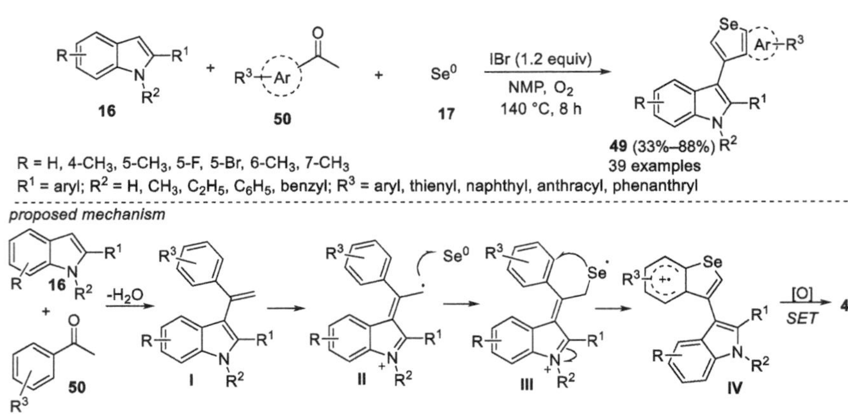

SCHEME 1.11 Synthesis of benzoselenophenes.

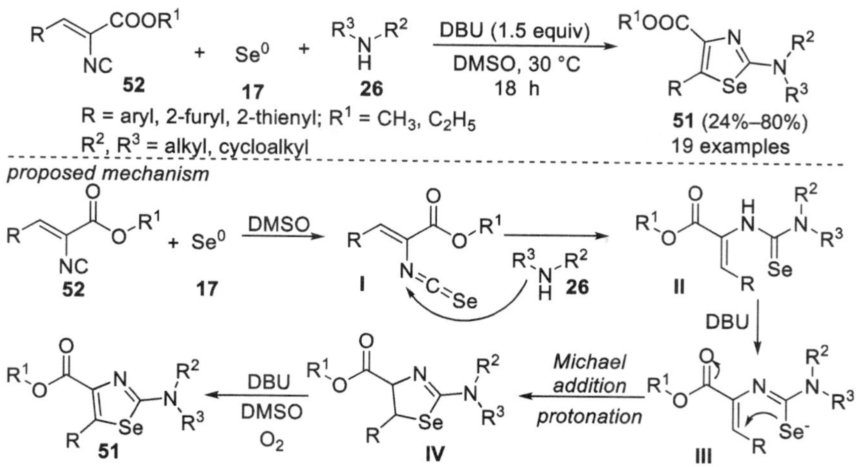

SCHEME 1.12 Synthesis of 2-amino-1,3-selenazole derivatives.

from 33% to 88%, demonstrating the suitability of the method to different groups. The mechanism of this MCR starts by the formation of the intermediate (**I**), through the dehydrative condensation of indole (**16**) with acetophenone (**50**). Through a single-electron transfer (SET) process, the intermediate (**I**) generates (**II**), which reacts with elemental selenium to form the radical cation (**III**). In the sequence, an intramolecular cyclization leads to the formation of the intermediate (**IV**), which after a new SET reaction, is oxidized to the desired benzoselenophene (**49**) (Scheme 1.11).

In 2020, Ji and coworkers[30] described the synthesis of 2-amino-1,3-selenazole derivatives (**51**) under TM-free conditions (Scheme 1.12). After a long study of the optimization step, the best reaction condition was determined as stirring a mixture of α,β-unsaturated isocyanides (**52**), elemental selenium (**17**) (1.5 equivalents), secondary amines (**26**) (1.5 equivalent), and 1,8-diazabicyclo[5.4.0]undec-7-ene (DBU) as base (1.5 equivalents) in DMSO as solvent at 30°C for 18 hours. The authors tested isocyanates with *E* and *Z* configurations, noting that the *Z*-isomer provided the desired product in higher yields. Thus, α,β-unsaturated isocyanides of *Z*-configuration (**52**) containing EDG and EWG at the *ortho, meta,* and *para*-positions of the aromatic ring, giving the respective products (**51**) in good yields. When heteroaryl α,β-unsaturated isocyanides (2-thienyl and 2-furyl) were used, only 2-thienyl provided the respective product (57% yield). Different secondary amines were also investigated by the authors, leading to the formation of the products of interest in good yields. When an amine containing the CN group was used, a considerable decrease in yield was observed (25%). Pyrrolidine, morpholine, piperidine, and 5,6,7,8-tetrahydroisoquinoline were suitable substrates, affording the respective products in 42%, 35%, 48%, and 50% yield, respectively. In a general way, acyclic amines were observed to provide better results than the cyclic ones, while no product was formed starting from primary amines. The developed approach was applied in the synthesis of Fluoxetine, a drug used for depression treatment, which was obtained in 53% yield. In the proposed mechanism, isocyanide (**52**) reacts with elemental selenium (**17**) in the presence of DMSO, forming the intermediate (**I**) (Scheme 1.12). Then, this intermediate undergoes a nucleophilic attack by amine (**26**), generating the intermediate (**II**), which is deprotonated by DBU to give the intermediate (**III**). In the sequence, a Michael addition followed by a protonation affords the intermediate (**IV**), which in the presence of O_2 and DMSO is oxidized to the desired product (**51**).

A simple and efficient method to obtain 2-phenylnaphtho[2,1-*d*] selenazoles (**53**) was described in 2021, by Liang and coworkers[31] (Scheme 1.13). The authors used naphthalen-2-amine (**26**), aldehydes (**47**), and selenium powder (**17**) in a medium containing I_2 (20 mol%) in DMSO as solvent, at 140°C for 6 hours. By this procedure, a total of 29 selenazoles were prepared in 22% to 86% yield. Aldehydes containing EDG and EWG attached to the aromatic ring were tested, providing the desired products in good to excellent yields. In addition, the protocol tolerates alkyl, naphthyl, heteroaromatic and vinyl aldehydes. However, when benzaldehyde was tested with different anilines, and it was not possible to obtain the product of interest. Furthermore, the authors investigated the influence of the position of the amino group and observed that the reactivity of naphthalen-2-amine is higher than that of naphthalen-1-amine. This can be explained by the fact that the cyclization to obtain the final product is an electrophilic process, since the position 1 of 2-naphthylamine has a higher electron cloud density than the position 2 of 1-naphthylamine.

In 2021, Bavanthula and coworkers[32] described the synthesis of 1,3,4-selenediazoles (**54**) through a cascade reaction between hydrazine (**3**), arylaldehydes (**47**) and elemental selenium (**17**) in the presence of iodine (Scheme 1.14). The developed method does not use metal; it is simple and is tolerant to several functional groups. The best condition employs I_2 as a catalyst (20 mol%) and DMSO as a solvent at 150°C for 4 hours.

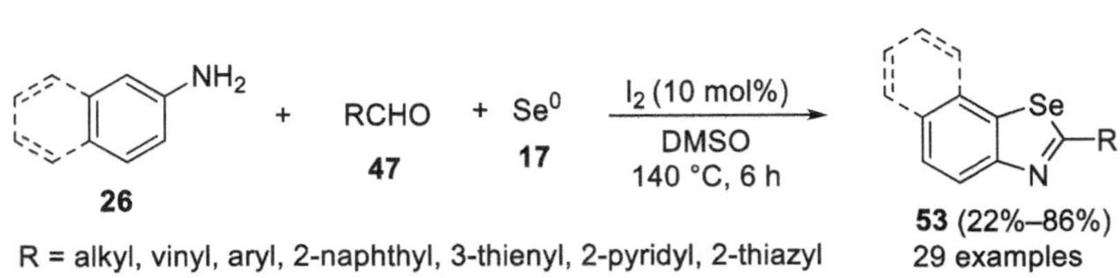

SCHEME 1.13 Synthesis of 2-phenylnaphtho[2,1-*d*]selenazoles.

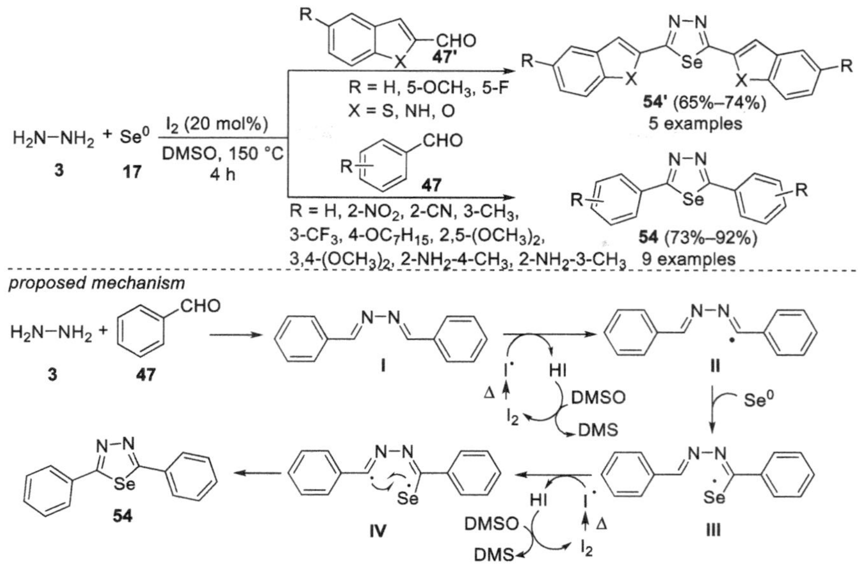

SCHEME 1.14 Synthesis of 1,3,4-selenediazoles.

The authors performed the synthesis of different 1,2,3-selenediazoles from benzaldehydes substituted with neutral, EDG and EWG at the *ortho*, *meta*, and *para* positions of the aromatic ring, affording the products (**54**) in 73%–92% yield. In addition to benzaldehyde, benzothiophene, and benzofuran, indole-substituted aldehydes (**47′**) were successfully used, generating the bis-indolyl selenadiazoles (**54′**) in good yields (65%–74%). To demonstrate the applicability of the method, the reaction was carried out in scales of 5 and 10 mmol. The mechanism for the formation of 1,2,3-benzoselazols (**54**) initiates with the reaction between the hydrazine (**3**) and the aldehyde (**47**) to form the di-imine intermediate (**I**). Then, the intermediate (**I**) reacts with radical iodine, formed after heating the molecular iodine, leading to the intermediate (**II**) and HI, which after reacting with DMSO returns to the catalytic cycle as I_2. In the sequence, the radical intermediate (**II**) reacts with elemental selenium (**17**), forming the Se-centered radical intermediate (**III**). This intermediate reacts with radical iodine leading to the bis-radical (**IV**), which is converted to the product (**54**) after an intramolecular cyclization.

1.4 Selenoethers

In 2011, Huang and coworkers[33] developed a novel three-component one-pot stereoselective synthesis of (*E*)-4-alkene-1,3-diketo selenoethers

(**55**), through the reaction between lithium selenolates (**56**), 1-(1-alkynyl) cyclopropyl ketones (**57**), and aldehydes (**47**) (Scheme 1.15). The reaction was carried out under nitrogen atmosphere in 1,4 dioxane as solvent for 20 hours at room temperature. The desired functionalized selenoethers (**55**) were obtained in moderate to good yields (43%–85%), and the protocol was able to tolerate many functionalities in the substrates.

In 2018, Perin and coworkers[34] proposed an alternative protocol to obtain *β*-methoxy- (**58**) and *β*-hydroxyselenides (**59**) under mild reaction conditions, from unactivated alkenes (**60**) using Oxone as a stoichiometric oxidant (Scheme 1.16). The products (**58**) were achieved in moderate to excellent yields (35%–94%) at room temperature after 3.5–20 hours, using CH_3OH as both nucleophile and solvent. The hydroxyselenides

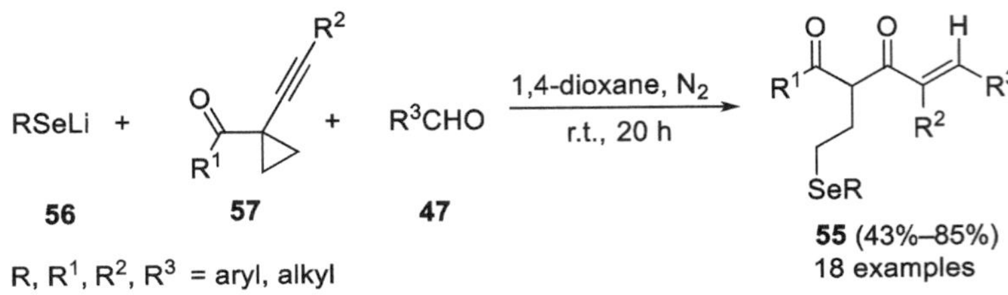

SCHEME 1.15 Synthesis of (*E*)-4-alkene-1,3-diketo selenoethers.

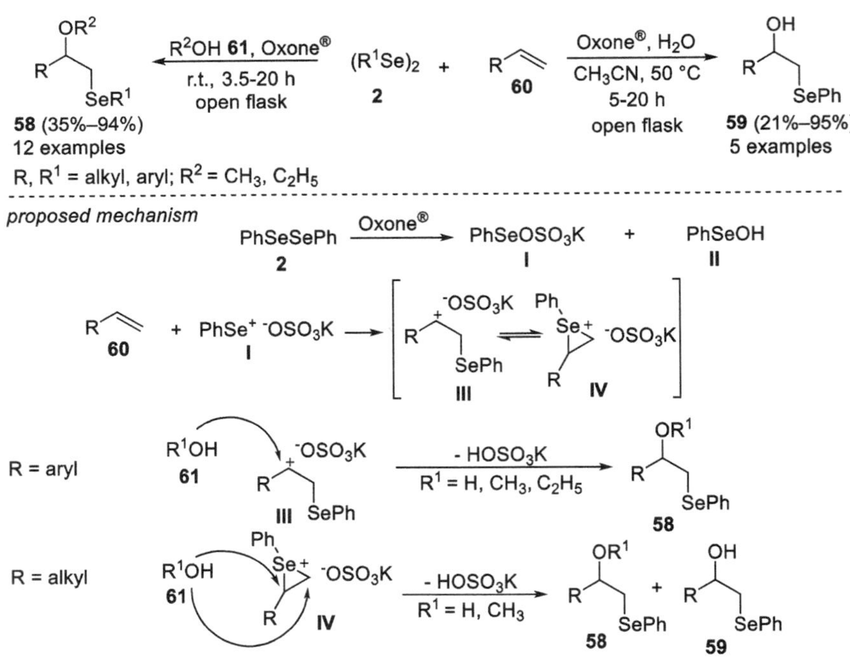

SCHEME 1.16 Synthesis of *β*-methoxy- and *β*-hydroxyselenides.

(**59**) instead, were reached after 5–20 hours of reaction, at 50°C, in 21%–95% yield using a mixture of H_2O/CH_3CN as solvent. The method was successfully applied to aromatic and aliphatic alkenes (**60**) and diselenides (**2**). In the proposed mechanism, Oxone and diselenide (**2**) react to generate in situ the intermediates (**I**) and (**II**). Subsequently, the species (**I**) reacts with alkene (**60**) generating (**III**). This last can be stabilized through seleniranium species (**IV**). In the presence of aryl alkenes (**60**), the intermediate (**III**) suffers a nucleophilic attack by the alcohol or water to form the products (**58**) or (**59**). Instead, when alkyl alkenes (**60**) were used, the seleniranium (**IV**) is favored, and it can be opened by the nucleophile, producing the products (**58**) or (**59**) (Scheme 1.16).

In 2020, Huang et al.[35] described a new three-component iron-catalyzed aminoselenation of alkenes (**60**) with diselenides (**2**) and amines (**26**), promoted by visible-light irradiation, and without using ligand and base (Scheme 1.17). The reaction was carried out under air atmosphere, at room temperature using 10 mol% of $FeBr_3$, under blue LED for 24 hours, and the respective β-aminoselenides (**62**) were obtained in 28%–99% yield. Different alkenes (**60**), aromatic and aliphatic diselenides (**2**), and primary and secondary amines (**26**) were well tolerated, showing the robustness of the method. As depicted in the Scheme 1.17, the reaction

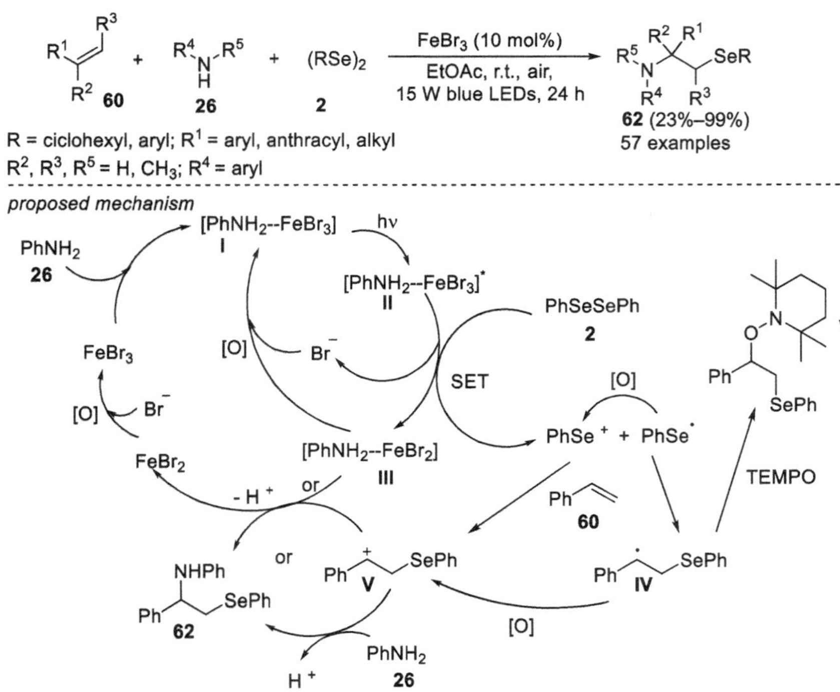

SCHEME 1.17 Synthesis of β-aminoselenides.

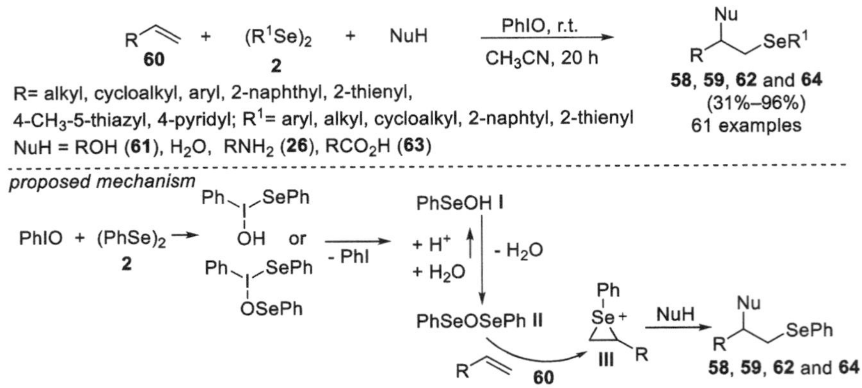

SCHEME 1.18 Selenofunctionalization of olefins.

mechanism starts with the interaction of amine (**26**) with Fe(III) to form the complex (**I**), which is excited to state (**II**). Then, diselenide (**2**) reacts with (**II**) through an SET, forming a selenium radical and a cation species of selenium. Subsequently, these two selenium intermediates add to the alkene (**60**), generating species (**IV**) and (**V**). Moreover, it was observed that in the presence of TEMPO, a radical scavenger, the intermediate (**VI**) was reached. Finally, the intermediate (**V**) reacts with the compound (**26**) to achieve the product (**62**) or the complex (**III**), releasing Fe(II).

Similarly, in 2021, Ling and coworkers[36] reported an alternative three-component selenofunctionalization of olefins (**60**), using diselenide (**2**) and amines (**26**), water, alcohol (**61**), or carboxylic acid (**63**), in the presence of hypervalent iodine(III), the iodosylbenzene (PhIO) as an oxidant (Scheme 1.18). The reaction was performed in CH_3CN or in mixtures with H_2O or alcohols (3:1), in open flask, at room temperature for 20 hours. In this study, a total of 61 products (**58**) (Nu = OR), (**59**) (Nu = OH), (**62**) (Nu = NH), and (**64**) (Nu = RCO_2) were obtained in 31%–95% yield and with high diastereoselectivity. Moreover, the protocol demonstrated to have excellent functional group tolerance. In the proposed mechanism, first, the diselenide (**2**) suffers an oxidation to form the species (**I**) and/or (**II**). The lattermost is the electrophile that can react with the alkene (**60**) to form the intermediate (**III**). Finally, the seleniranium intermediate (**III**) reacts with different nucleophiles to form the respective products (**58**), (**59**), (**62**), and (**64**).

1.5 Vinyl chalcogenides

In 2018, Sun et al.[37] developed a new multicomponent cascade selenosulphonation of alkynes (**67**) to obtain β-(seleno)vinyl sulfones (**65**)

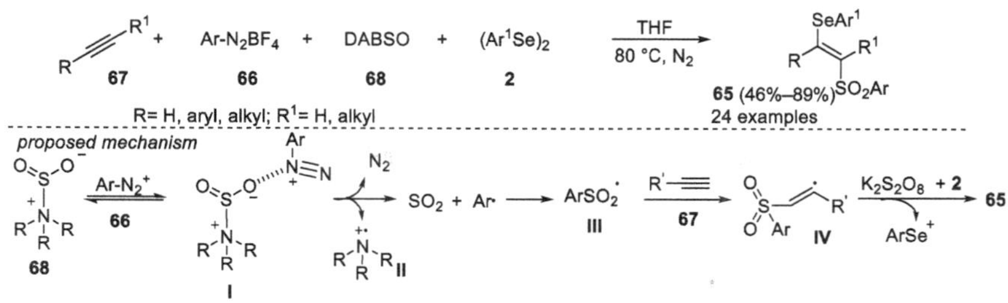

SCHEME 1.19 Selenosulphonation of alkynes.

(Scheme 1.19). The reaction was carried out in the presence of phenyldiazonium or 4-methylphenyldiazonium tetrafluoroβorate (**66**), alkynes (**67**), 1,4-diazobicylo[2.2.2]octane bis (sulfur dioxide) [DABSO] (**68**), and diselenide (**2**) in tetrahydrofuran (THF) at 80°C, under argon atmosphere. The method demonstrated good functional-group tolerance, and the products (**65**) were achieved in yields of 46%–89%, with good regio- and stereoselectivity. Moreover, good result was obtained using diphenyl ditelluride instead of diselenide, and the corresponding vinyl telluride was formed in 63% yield. The authors proposed a radical mechanism in which the complex (**I**) is formed from the aryldiazonium cation and DABSO. Once formed, complex (**I**) decomposes in sulfur dioxide, nitrogen, aryl radical, and radical cation (**II**). Then, sulfur dioxide and the aryl radical combines to form the radical (**III**), which reacts with alkyne (**67**) giving the vinyl radical (**IV**). Finally, arylselenyl radical formed from diselenide (**2**) couples with (**IV**), affording the respective (*E*)-*β*-(seleno)vinyl sulfone (**65**).

In 2019, Chen et al.[38] reported a new protocol to synthesize *β*-sulphonylvinylselanes (**69**), containing two vinylselanyl units, through an MCR of alkynes (**67**), diselenides (**2**), and DABSO (**68**), promoted by visible light (Scheme 1.20). The desired compounds (**69**) were obtained from poor to good yields (3%–71%) after 3–6 hours of reaction, under blue LED irradiation in CH_3CN as solvent under air atmosphere at room temperature. Different alkynes (**67**), aryl- and alkyl-substituted, were investigated together with and aryl and alkyl diselenides (**2**) and the products (**69**) were prepared with high selectivity for *E* configuration. Also, disulfide was tested, but unfortunately the corresponding product formed only in trace. The proposed mechanism involves first the formation of $PhSe^{\bullet}$ by the irradiation of diselenide (**2**) with blue light. Then, this radical reacts with (**67**) to generate the intermediate (**I**), which is in equilibrium with the isomer (**II**). In the sequence, the vinyl radical reacts with DABSO, affording the sulphonyl radical (**III**), which reacts with a new equivalent of alkyne (**67**) to form the bis-vinyl intermediate (**IV**). Finally, diselenide (**2**) reacts with

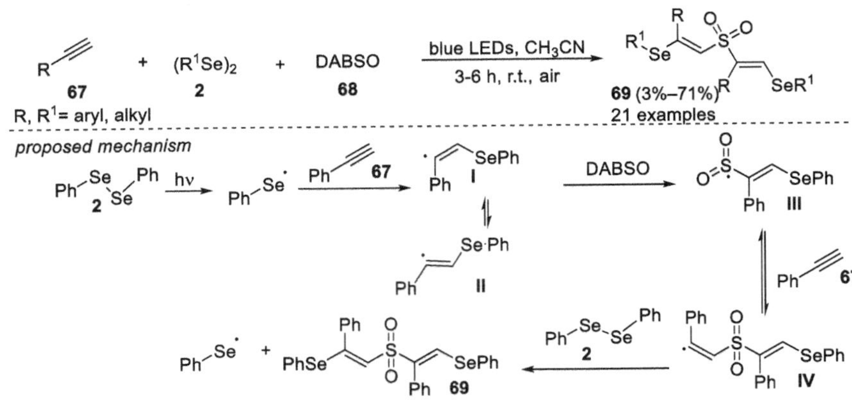

SCHEME 1.20 Synthesis of β-sulphonylvinylselanes.

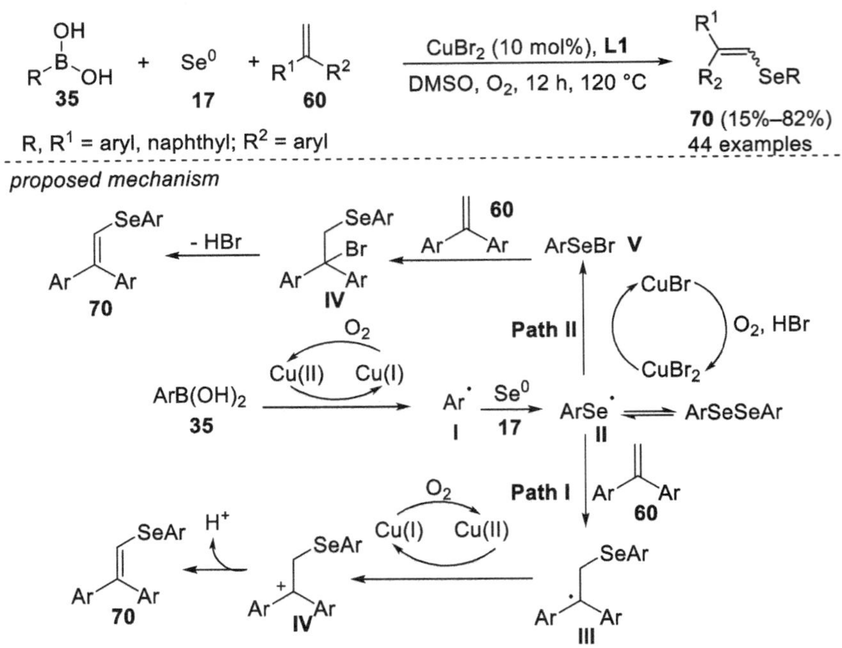

SCHEME 1.21 Three-component reaction to obtain vinyl selenides.

(IV) to give the product **(69)**, with the formation of a new selenium radical, which can be recycled (Scheme 1.20).

In 2020, Yang et al.[39] described a three-component reaction to obtain vinyl selenides **(70)**, through Cu catalysis, using arylboronic acids **(35)**, Se^0 powder **(17)**, and diarylethylenes **(60)** (Scheme 1.21). The products **(70)** were achieved in 15%–82% yield after 12 hours of reaction, at 120°C, under oxygen atmosphere in DMSO as solvent, and tricyclohexylphosphonium tetrafluoroborate (**L1**) as ligand. Various substituted arylboronic acids, bearing EWG and EDG, at *meta-*, *ortho-*, and *para-* positions, were studied, and all of them worked well in the reaction. Moreover, different diarylethylenes (**60**), with halogen, methyl, and methylthio substitutents were used, and maleimides were also tested, and all of them demonstrated to be good substrates under optimal conditions. The mechanism starts with the formation of the radical (**I**), generated in situ, which activates Se^0 (**17**), forming the radical intermediate (**II**). The radical (**II**) adds to diarylethylene (**60**) giving the intermediate (**III**), which reacts by a SET with Cu(II) to give the species (**IV**). Finally, through a deprotonation of (**IV**) is formed the product (**70**) (Scheme 1.21, path I). Alternatively, the species (**II**) can react with $CuBr_2$ to form ArSeBr (**V**), and this electrophile adds to (**60**) to produce the cation intermediate (**IV**), which produces (**70**) after losing a proton to form HBr (Scheme 1.21, path II).

1.6 Phosphorus derivatives

Selenophosphinic derivates are a new class of selenium compounds with enormous biological potential; however, only a few multicomponent protocols to prepare them are available in comparison with the sulfur analogs. In 2013, Opratina and coworkers[40] reported the synthesis of thioselenophosphinic *S*- and *Se*-esters (**71**), using elemental selenium (**17**), vinyl ether (**72**) and diorganylphosphine sulfide (**73′**) as starting materials (Scheme 1.22). The thiol derivative (**73**) is the phosphorus(III) active species, a tautomer of (**73′**). The drawback of using phosphine sulfides (**73′**) and elemental selenium is the formation of a mixture of thioselenophosphinic *S*- and *Se*-esters. The predominance of *S*-ester isomer in the product mixture was attributed to the molecular structures of the intermediate thioselenophosphinic acids, and to the vinyl ether (**72**) employed as starting material.

To avoid the problem of the formation of isomeric mixture, which resulted from tautomerism of the reaction intermediates, in 2011, Gusarova and coworkers[41] reported the MCR of elemental selenium, alkenes and secondary phosphanes, giving exclusively diselenophosphinic esters, with

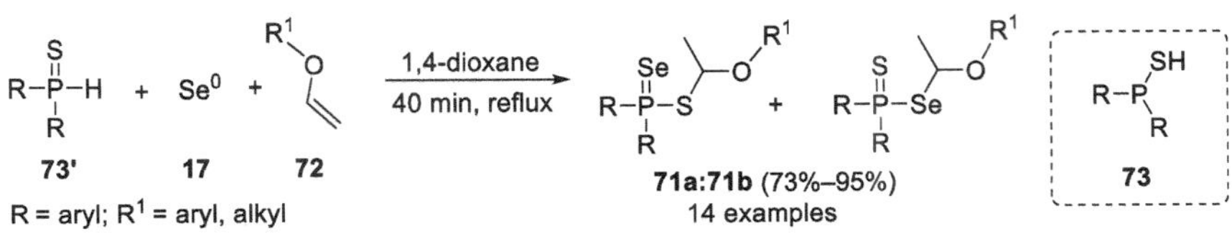

SCHEME 1.22 Synthesis of thioselenophosphinic S- and Se-esters.

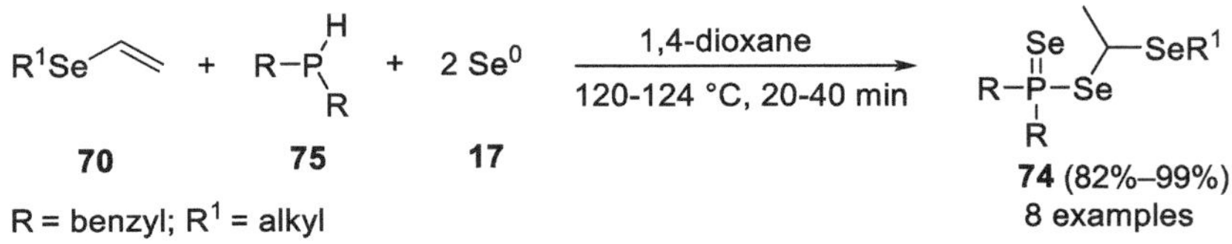

SCHEME 1.23 Synthesis of diselenophosphinic esters.

two selenium atoms bonded at the phosphorous one. Three years later, the same group[42] extended the protocol to the synthesis of diselenophosphinic esters (**74**), by starting from vinyl selenides (**70**) instead of alkenes, in the reaction with elemental selenium (**17**) and secondary phosphanes (**75**) (Scheme 1.23). The use of vinyl selenides (**70**) allowed the authors to obtain eight different diselenophosphinic *Se*-esters (**74**), via the formation of diselenophosphinic acid $R_2P(Se)SeH$ as intermediate.

1.7 Miscellaneous

The property of selenium and sulfur species to interact with copper salts to form stable complexes with chalcogen clusters has been explored in coupling reactions. In this sense, Zhang and coworkers[43] described the synthesis of trifuoromethyl alkanes (**76**) and trifuoromethyl alkenes (**77**) in low to good yields, through the fluoromethylthiolation/fluoromethylselenation of unsaturated hydrocarbons. The authors took advantage of the easy formation of the chalcogen–copper complex (**78**), which was used in the reaction with alkynes (**67**) or alkenes (**60**) and ethyl iodofluoroacetate (**79**), in the presence of CsF as base and dimethoxyethane (DME) as solvent at 100°C. The mechanism studies suggested that the presence of the complex chalcogen–copper is vital to obtain the key radical intermediate (**II**) (Scheme 1.24). In the proposed mechanism, ethyl difluoroiodoacetate (**79**) is reduced by bpyCu(I)YCF_3 to form bpyCu(II)IYCF_3 and the difluoroalkyl radical intermediate (**I**). which reacts with the unsaturated hydrocarbon (**60**) or (**67**) to give the intermediate (**II**). Then, the intermediate (**II**) reacts with bpyCu(II)IYCF_3, leading to the Cu(III) species (**III**), which, after a reductive elimination, produces the desired products (**76**) or (**77**) and releases bpyCuI.

Other interesting use of copper complexes in chalcogen chemistry was described in 2017 by Liu and coworkers,[44] in the MCR of isonitriles (**25**), elemental selenium (**17**), amines (**26**), and iodoarenes (**15**) under basic conditions, to prepare carbamimidoselenoates (**80**) (Scheme 1.25). After a long optimization study, the best condition was established using excess of (**17**) and (**26**) (1.5 equivalents), and of (**15**) (2.0 equivalents), in the presence of CuI as catalyst, 1,10-phenanthroline as ligand, and Cs_2CO_3 as

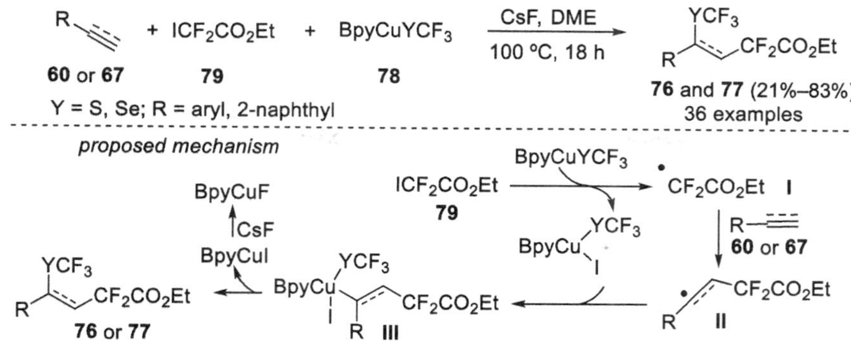

SCHEME 1.24 Synthesis of trifuoromethyl alkanes and trifuoromethyl alkenes.

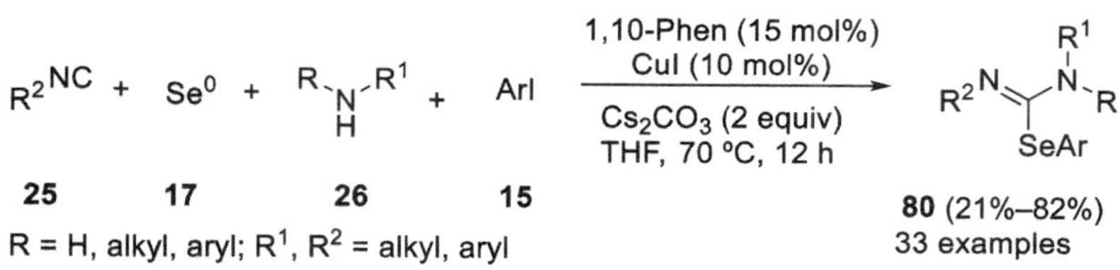

SCHEME 1.25 Synthesis of carbamimidoselenoates.

base, in THF as solvent at 70°C under an air atmosphere for 12 hours. Carbamimidoselenoates (**80**) bearing an acyclic isoselenourea skeleton were prepared by using isonitriles and amines containing alkyl and aryl groups, and these compounds were isolated in acceptable yields using four substrates in a single step.

Gaykar and coworkers[45] described the use of electrophilic species of selenium (**81**) to promote the *ortho*-aminoselenation of arynes. In this study, the first step is the in situ generation of the aryne (**I**) by the 1,2-elimination of (trimethylsilyl)aryl triflate (**82**) (Scheme 1.26). After the generation of the aryne, it suffers a nucleophilic addition by the tertiary amine (**26**), generating the 1,3-zwitterionic intermediate (**II**). The negative charge of the intermediate (**II**) attacks the electrophilic selenium species (**81**), leading to the formation of the *o*-selanyl quaternary amine (**III**). Finally, demethylation of (**III**) by a bromide anion affords the desired product (**83**) and MeBr.

A metal-free procedure to prepare alkynyl alkyl selenides (**84**) was reported in 2018, by Wu and coworkers.[46] The authors used elemental selenium (**17**), epoxides (**85**), and terminal alkynes (**67**) in the presence of tetra-*n*-butylammonium iodide as a phase transfer catalyst under basic conditions. The authors described that the elemental selenium (**17**) in basic conditions suffered a disproportionation generating selenite and selenide anion (**I**), which reacts with the epoxide (**85**), leading to

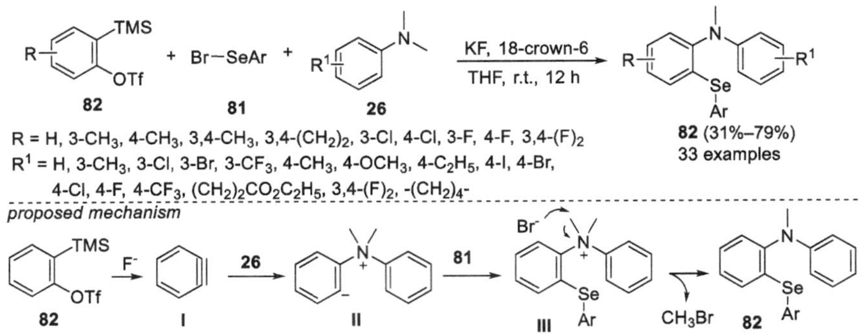

SCHEME 1.26 Ortho-aminoselenation of arynes.

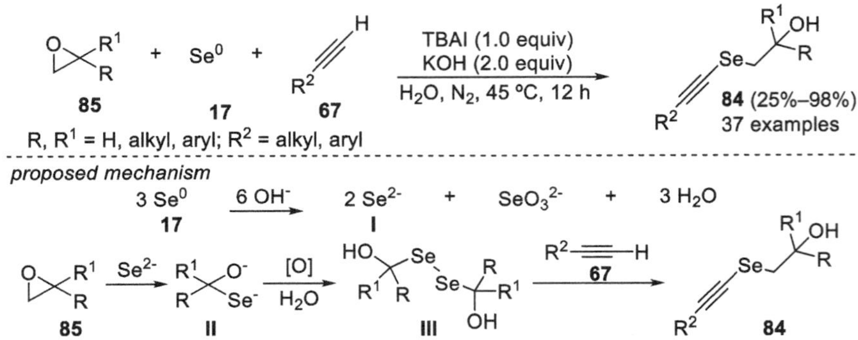

SCHEME 1.27 Synthesis of β-hydroxy-selenoacetylene.

the formation of the instable intermediate (**II**), which is immediately oxidized into the stable diselenide (**III**). Under the basic conditions, the diselenide (**III**) generated in situ couples with the terminal alkyne (**67**), giving the desired β-hydroxy-selenoacetylene (**84**) (Scheme 1.27).

1.8 Conclusions and perspectives

Over the last decades, developments in MCR-based strategies have revealed new insights and possibilities to prepare elaborate and complex molecular scaffolds. Specifically, in this chapter, we presented the panorama regarding the preparation of various classes of organochalcogen compounds, such as chalcogen-functionalized heterocycles, heterocycles with chalcogen embedded in the ring, selenoethers, vinyl chalcogenides, phosphorus derivatives, among others. Among the advantages of this approach for introducing a chalcogen atom in different molecules, is the availability of high atom-efficiency and low time-consuming protocols.

As demonstrated in this chapter, a variety of chalcogen sources have been efficiently used in these MCRs, including dichalcogenides, electrophilic species, elemental chalcogen, and other inorganic sources. Looking for ways to further green these methods, many authors have associated MCRs with electrochemistry, photochemistry, the use of catalytic reagents, among other techniques that are in agreement with the principles of green chemistry. There are many opportunities still to be explored for the future development of novel MCRs focusing on mild reaction conditions, short reaction times and good yields, and alternative feedstocks.

Acknowledgments

This study was financed in part by the Coordenação de Aperfeiçoamento de Pessoal de Nível Superior—Brasil (CAPES)—Finance Code 001. FAPERGS, CNPq and FINEP are acknowledged for financial support. This work has been carried out in the framework of the international network, "Selenium, Sulfur, Redox and Catalysis" (SeSRedCat).

Conflict of interest

There are no conflicts to declare.

Credit author statement

Francesca Mangiavacchi, Laura Abenante, Ítalo F.C. Dias, Manoela Sacramento: Methodology and Writing-Original draft preparation; Angelita M. Barcellos: Supervision and Writing- Reviewing and Editing.

References

1. (a) Erythropel, H. C.; Zimmerman, J. B.; De Winter, T. M.; Petitjean, L.; Melnikov, F.; Lam, C. H.; Lounsbury, A. W.; Mellor, K. E.; Janković, N. Z.; Tu, Q., et al. *Green Chem* **2018,** *20*, 1929–1961. Available from: https://doi.org/10.1039/c8gc00482j.
   (b) Cioc, R. C.; Ruijter, E.; Orru, R. V. A. *Green Chem* **2014,** *16*, 2958–2975. Available from: https://doi.org/10.1039/c4gc00013g.
2. (a) Younus, H. A.; Al-Rashida, M.; Hameed, A.; Uroos, M.; Salar, U.; Rana, S.; Khan, K. M. *Expert Opin. Ther. Pat.* **2021,** *31*, 267–289. Available from: https://doi.org/10.1080/13543776.2021.1858797.
   (b) Zhang, W.; Zhi, S.; Ma, X. *Org. Biomol. Chem.* **2019,** *17*, 7632–7650. Available from: https://doi.org/10.1039/c9ob00772e.
3. (a) Brauch, S.; van Berkel, S. S.; Westermann, B. *Chem. Soc. Rev.* **2013,** *42*, 4948–4962. Available from: https://doi.org/10.1039/c3cs35505e.
   (b) Gu, Y. *Green Chem* **2012,** *14*, 2091–2128. Available from: https://doi.org/10.1039/c2gc35635j.
   (c) Singh, M. S.; Chowdhury, S. *RSC Adv* **2012,** *2*, 4547–4592. Available from: https://doi.org/10.1039/c2ra01056a.

(d) Alvim, H. G. O.; Da Silva Júnior, E. N.; Neto, B. A. D. *RSC Adv* **2014,** *4*, 54282–54299. Available from: https://doi.org/10.1039/c4ra10651b.

(e) Ghasemzadeh, M. A.; Mirhosseini-Eshkevari, B.; Tavakoli, M.; Zamani, F. *Green Chem* **2020,** *22*, 7265–7300. Available from: https://doi.org/10.1039/d0gc01767a.

4. Touré, B. B.; Hall, D. G. *Chem. Rev.* **2009,** *109*, 4439–4486. Available from: https://doi.org/10.1021/cr800296p.
5. Graebin, C. S.; Ribeiro, F. V.; Rogério, K. R.; Kümmerle, A. E. *Curr. Org. Synth.* **2019,** *16*, 855–899. Available from: https://doi.org/10.2174/1570179416666190718153703.
6. (a) Fairoosa, J.; Neetha, M.; Anilkumar, G. *RSC Adv* **2021,** *11*, 3452–3469. Available from: https://doi.org/10.1039/d0ra10472h.

   (b) Neochoritis, C. G.; Zhao, T.; Dömling, A. *Chem. Rev.* **2019,** *119*, 1970–2042. Available from: https://doi.org/10.1021/acs.chemrev.8b00564.

   (c) Estévez, V.; Villacampa, M.; Menéndez, J. C. *Chem. Soc. Rev.* **2010,** *39*, 4402–4421. Available from: https://doi.org/10.1039/b917644f.

   (d) Rotstein, B. H.; Zaretsky, S.; Rai, V.; Yudin, A. K. *Chem. Rev.* **2014,** *114*, 8323–8359. Available from: https://doi.org/10.1021/cr400615v.
7. (a) Barcellos, A. M.; Abenante, L.; Sarro, M. T.; Leo, I. D.; Lenardão, E. J.; Perin, G.; Santi, C. *Curr. Org. Chem.* **2017,** *21*, 2044–2061. Available from: https://doi.org/10.2174/1385272820666161020162113.

   (b) Mugesh, G.; Du Mont, W.-W.; Sies, H. *Chem. Rev.* **2001,** *101*, 2125–2179. Available from: https://doi.org/10.1021/cr000426w.

   (c) Nogueira, C. W.; Rocha, J. B. T. *Arch. Toxicol* **2011,** *85*, 1313–1359. Available from: https://doi.org/10.1007/s00204-011-0720-3.
8. (a) Lenardão, E. J.; Santi, C.; Sancineto, L. *New Frontiers in Organoselenium Compounds, Chapter 1: Organoselenium Compounds as Reagents and Catalysts to Develop New Green Protocols;* Springer: Gewerbestrasse, Switzerland, 2018.

   (b) Penteado, F.; Monti, B.; Sancineto, L.; Perin, G.; Jacob, R. G.; Santi, C.; Lenardão, E. J. *Asian J. Org. Chem.* **2018,** *7*, 2368–2385. Available from: https://doi.org/10.1002/ajoc.201800477.

   (c) Santoro, S.; Azeredo, J. B.; Nascimento, V.; Sancineto, L.; Braga, A. L.; Santi, C. *RSC Adv* **2014,** *4*, 31521–31535. Available from: https://doi.org/10.1039/C4RA04493B.

   (d) Perin, G.; Alves, D.; Jacob, R. G.; Barcellos, A. M.; Soares, L. K.; Lenardão, E. J. *ChemistrySelect* **2016,** *2*, 205–258. Available from: https://doi.org/10.1002/slct.201500031.
9. Guo, X.-X.; Gu, D.-W.; Wu, Z.; Zhang, W. *Chem. Rev.* **2015,** *115*, 1622–1651. Available from: https://doi.org/10.1021/cr100198w.
10. Liu, C.; Zhang, H.; Shi, W.; Lei, A. *Chem. Rev.* **2011,** *111*, 1780–1824. Available from: https://doi.org/10.1021/cr100379j.
11. Oliveira, D. H.; Aquino, T. B.; Edmilson, J.; Nascimento, R.; Perin, G.; Jacob, R. G.; Alves, D. *Adv. Synth. Catal.* **2015,** *357*, 4041–4049. Available from: https://doi.org/10.1002/adsc.201500625.
12. Wu, Y.; Chen, J.-Y.; Ning, J.; Jiang, X.; Deng, J.; Deng, Y.; Xu, R.; He, W.-M. *Green. Chem.* **2021,** *23*, 3950–3954. Available from: https://doi.org/10.1039/D1GC00562F.
13. Aquino, T. B.; do Nascimento, J. E. R.; Dias, Í. F. C.; de Oliveira, D. H.; Barcellos, T.; Lenardão, E. J.; Perin, G.; Alves, D.; Jacob, R. G. *Tetrahedron Lett.* **2018,** *59*, 1080–1083. Available from: https://doi.org/10.1016/j.tetlet.2018.01.072.
14. Zhang, L.-L.; Li, Y.-T.; Gao, T.; Guo, S.-S.; Yang, B.; Meng, Z.-H.; Dai, Q.-P.; Xu, Z.-B.; Wu, Q.-P. *Synthesis* **2019,** *51*, 4170–4182. Available from: https://doi.org/10.1055/s-0039-1690618.
15. Luo, D.; Wu, G.; Yang, H.; Liu, M.; Gao, W.; Huang, X.; Chen, J.; Wu, H. *J. Org. Chem.* **2016,** *81*, 4485–4493. Available from: https://doi.org/10.1021/acs.joc.6b00229.

16. Gao, C.; Wu, G.; Min, L.; Liu, M.; Gao, W.; Ding, J.; Chen, J.; Huang, X.; Wu, H. *J. Org. Chem.* **2017,** *82*, 250–255. Available from: https://doi.org/10.1021/acs.joc.6b02388.
17. Guo, T.; Dong, Z.; Zhang, P.; Xing, W.; Li, L. *Tetrahedron Lett.* **2018,** *59*, 2554–2558. Available from: https://doi.org/10.1016/j.tetlet.2018.05.046.
18. Liu, H.; Fang, Y.; Wang, S.-Y.; Ji, S.-J. *Org. Lett.* **2018,** *20*, 930–933. Available from: https://doi.org/10.1021/acs.orglett.7b03783.
19. Peterle, M. M.; Scheide, M. R.; Silva, L. T.; Saba, S.; Rafique, J.; Braga, A. L. *ChemistrySelect* **2018,** *3*, 13191–13196. Available from: https://doi.org/10.1002/slct.201801213.
20. Gandeepan, P.; Moa, J.; Ackermann, L. *Chem. Commun.* **2017,** *53*, 5906–5909. Available from: https://doi.org/10.1039/C7CC03107F.
21. Shaw, M. H.; Twilton, J.; MacMillan, D. W. C. *J. Org. Chem.* **2016,** *81*, 6898–6926. Available from: https://doi.org/10.1021/acs.joc.6b01449.
22. Romero, N. A.; Nicewicz, D. A. *Chem. Rev.* **2016,** *116*, 10075–10166. Available from: https://doi.org/10.1021/acs.chemrev.6b00057.
23. Wu, J.; Yang, Y.-F.; Huang, X.-B.; Gao, W.-X.; Zhou, Y.-B.; Liu, M.-C.; Wu, H.-Y. *ACS Omega* **2020,** *5*, 23358–23363. Available from: https://doi.org/10.1021/acsomega.0c03254.
24. Jin, G.-Q.; Gao, W.-X.; Zhou, Y.-B.; Liu, M.-C.; Wu, H.-Y. *RSC Adv.* **2020,** *10*, 30439–30442. Available from: https://doi.org/10.1039/D0RA06016J.
25. Ren, Y.; Xu, B.; Zhong, Z.; Pittman, C. U., Jr.; Zhou, A. *Org. Chem. Front.* **2019,** *6*, 2023–2027. Available from: https://doi.org/10.1039/C9QO00299E.
26. Karamthulla, S.; Pal, S.; Khan, Md. N.; Choudhury, L. H. *RSC Adv.* **2014,** *4*, 37889–37899. Available from: https://doi.org/10.1039/c4ra06239f.
27. Zhou, Z.; Liu, M.; Sun, S.; Yao, E.; Liu, S.; Wu, Z.; Yu, J.-T.; Jiang, Y.; Cheng, J. *Tetrahedron Lett.* **2017,** *58*, 2571–2573. Available from: https://doi.org/10.1016/j.tetlet.2017.05.061.
28. Zhang, W.; Tao, S.; Ge, H.; Li, Q.; Ai, Z.; Li, X.; Zhang, B.; Sun, F.; Xu, X.; Du, Y. *Org. Lett.* **2020,** *22*, 448–452. Available from: https://doi.org/10.1021/acs.orglett.9b04206.
29. Ni, P.; Tan, J.; Zhao, W.; Huang, H.; Xiao, F.; Deng, G.-J. A. *Org. Lett.* **2019,** *21*, 3518–3522. Available from: https://doi.org/10.1021/acs.orglett.9b00739.
30. Liu, H.-W.; Fang, Y.; Wang, S.-Y.; Ji, S.-J. *J. Org. Chem.* **2020,** *85*, 3508–3516. Available from: https://doi.org/10.1021/acs.joc.9b03234.
31. Chen, W.; Zhu, X.; Wang, F.; Yang, Y.; Deng, G.; Liang, Y. *J. Org. Chem.* **2020,** *85*, 3349–3357. Available from: https://doi.org/10.1021/acs.joc.9b03154.
32. Bowroju, S. K.; Bavanthula, R. *RSC Adv.* **2021,** *11*, 5724–5728. Available from: https://doi.org/10.1039/d0ra10576g.
33. Xu, J.; Wu, L.; Huang, X. *J. Org. Chem.* **2011,** *76*, 5598–5605. Available from: https://doi.org/10.1021/jo2005439.
34. Perin, G.; Santoni, P.; Barcellos, A. M.; Nobre, P. C.; Jacob, R. G.; Lenardão, E. J.; Santi, C. *Eur. J. Org. Chem.* **2018,** *2018*, 1224–1229. Available from: https://doi.org/10.1002/ejoc.201701775.
35. Huang, B.; Li, Y.; Yang, C.; Xia, W. *Green. Chem.* **2020,** *22*, 2804–2809. Available from: https://doi.org/10.1039/C9GC04163J.
36. Liang, Z.-P.; Yi, W.; Wang, P.-F.; Liu, G.-Q.; Ling, Y. *J. Org. Chem.* **2021,** *86*, 5292–5304. Available from: https://doi.org/10.1021/acs.joc.1c00257.
37. Sun, K.; Shi, Z.; Liu, Z.; Luan, B.; Zhu, J.; Xue, Y. *Org. Lett.* **2018,** *20*, 6687–6690. Available from: https://doi.org/10.1021/acs.orglett.8b02733.
38. Chen, H.; Ding, R.; Tang, H.; Pan, Y.; Xu, Y.; Chen, Y. *Chem. Asian J.* **2019,** *14*, 3264–3268. Available from: https://doi.org/10.1002/asia.201901049.
39. Yang, Y.-F.; Li, C.-Y.; Leng, T.; Huang, X.-B.; Gao, W.-X.; Zhou, Y.-B.; Liu, M.-C.; Wu, H.-Y. *Adv. Synth. Catal.* **2020,** *362*, 2168–2172. Available from: https://doi.org/10.1002/adsc.202000015.

40. Oparina, L. A.; Artem'ev, A. V.; Vysotskaya, O. V.; Kolyvanov, N. A.; Bagryanskaya, I. Yu; Doronina, E. P.; Gusarova, N. K. *Tetrahedron* **2013,** *69,* 6185–6195. Available from: https://doi.org/10.1016/j.tet.2013.05.038.
41. Gusarova, N. K.; Artem'ev, A. V.; Malysheva, S. F.; Tarasova, O. A.; Trofimov, B. A. A. *Tetrahedron Lett.* **2011,** *52* (51), 6985–6987. Available from: https://doi.org/10.1016/j.tetlet.2011.10.094.
42. Artem'ev, A. V.; Chernysheva, N. A.; Gusarova, N. K.; Yas'ko, S. V.; Liao, J.-H.; Liu, C. W.; Albanov, A. I.; Trofimov, B. A. *Heteroat. Chem.* **2014,** *25,* 135–139. Available from: https://doi.org/10.1002/hc.21144.
43. Zhang, B.-S.; Gao, L.-Y.; Zhang, Z.; Wen, Y.-H.; Liang, Y.-M. *Chem. Commun.* **2018,** *54,* 1185–1188. Available from: https://doi.org/10.1039/C7CC09083H.
44. Liu, H.; Fang, Y.; Yin, L.; Wang, S.-Y.; Ji, S.-J. *J. Org. Chem.* **2017,** *82,* 10866–10874. Available from: https://doi.org/10.1021/acs.joc.7b01663.
45. Gaykar, R. N.; Guin, A.; Bhattacharjee, S.; Biju, A. T. *Org. Lett.* **2019,** *21,* 9613–9617. Available from: https://doi.org/10.1021/acs.orglett.9b03789.
46. Wu, G.; Min, L.; Li, H.; Gao, W.; Ding, J.; Huang, X.; Liu, M.; Wu, H. *Green. Chem.* **2018,** *20,* 1560–1563. Available from: https://doi.org/10.1039/C8GC00341F.

CHAPTER

2

Alternative energy source: synthesis of selenium compounds

Angelita M. Barcellos, Filipe Penteado, Gelson Perin, Diego Alves and Márcio S. Silva

Laboratório de Síntese Orgânica Limpa—LASOL—CCQFA—Universidade Federal de Pelotas—UFPel, Pelotas, Brazil

2.1 Introduction

Sustainability can be defined as the quality of a process of causing little or no damage to the environment and therefore able to continue for a long time. Based on this concept, green chemistry is a vast field, and it is used as a guideline to produce environmentally benign synthetic protocols and prudent management of chemical and energy raw materials.[1–5] Some of the green chemistry principles describe ideas as preventing waste, nonconventional energy source, less hazardous synthesis, renewable feedstock, and catalysis.[6–10] On the other hand, rigorous regulations are placed on the chemical industries, especially pharmaceutical factories, because of the amount of waste produced in the process (higher E factor),[11,12] directly affecting the companies and inspiring alternatives methods to produce the large number of compounds required by the society. Consequently, several newer protocols have appeared, and improvements regarding safety, security, and costs have been reached. Nevertheless, there are newer challenges facing the chemical industry settling constant innovations and waste reductions.[13,14]

Organoselenium compounds are an important and useful organic class, which have received considerable attention because of their widespread application, especially in medicinal chemistry.[15] This use can be explained because the element selenium plays a central role in the

Organochalcogen Compounds
DOI: https://doi.org/10.1016/B978-0-12-819449-2.00006-9

activity of several enzymes, such as glutathione peroxidase.[16,17,18] Additionally, organoselenium compounds demonstrate other important biological properties, including antiviral, antifungal, and antiprotozoal activities.[19,20,21,22] Although selenium substrates allow chemo- and regioselective reactions, forming complex organic compounds, under mild conditions,[23,24,25,26,27] alternative benign reaction methodologies have become an imperative part of organoselenium chemistry.[28,29] This synergic process employing organoselenium compounds can connect the innovative sustainable reactions with the reactivity modulation in different types of selenium species. For instance, solvents are responsible for up to 90% of mass utilization in the fine chemical and pharmaceutical industries, and thus, the use of unconventional solvents in organoselenium reactions is a valuable alternative to employ safe solvents (glycerol, ionic liquids, supercritical carbon dioxide, or water),[30,31] improving the yields with a wide substrate scope.

Among the green chemistry principles, the employing an alternative energy source, such as microwave (MW), ultrasound (US), and mechanochemical mixing (ball milling), has drastically improved reactivity and selectivity of a large number of reactions,[32,33,34] also involving organic chalcogen compounds.[35,36,37] Microwave is a consolidated energy source to provide organosulfur compounds, which can be seen in several literature reports.[38,39,40,41] Additionally, studies on the synthesis of organosulfur compounds employing mechanochemical mixing[42,43,44] and ultrasound irradiation[45,46] have demonstrated them as effective alternatives to conventional heating. Based on the organosulfur reports, it is possible to observe that a great amount of detail regarding substrate scope, stability, reaction conditions, time, and yields is required to provide effective and general alternative methodologies, such as thioureas and disulfides, which are applied as key reagents on the development of sustainable reactions.[40,42,44,45]

In this context, herein we provide a brief overview of the recent achievements and protocols created by alternative energy source (MW, US, and ball milling) to access organoselenium compounds, which minimize the generation of byproducts and/or side reactions and improve the reactivity and selectivity. Regarding the selenium species used as starting material, a number of protocols explore the reactivity of diselenide compounds. The intrinsic characteristics of diselenides compounds, attributable to its Se–Se weak bond, represent an attractive synthetic reagent to incorporate selenium element into complex molecules, mainly using an alternative energy source. Considering the variety of synthetic protocols and selenium compounds described, for a better discussion, this chapter is divided into five sections, according to the reaction type: addition reaction, substitution reaction, ring-opening reaction of epoxides, cyclization reaction, and miscellaneous synthetic methodologies.

2.2 Addition reactions

Vinyl chalcogenides are important chalcogen-based compounds, valuable from synthetic[27,47] and biological[48–50] aspects. In this sense, the reaction between in situ−prepared nucleophilic chalcogen-based species (selenium and tellurium) and alkynes, has been settling down as a remarkable tool to access these compounds under thermal conditions.[27,30,47] In order to improve efficiency, MW irradiation has been recently employed as a more efficient energy source for the formation of vinyl chalcogenides.

In 2007, Lenardão and coworkers[51] reported a simple and efficient method for the hydrochalcogenation of activated alkynes **1**, bearing electron-withdrawing groups (EWG = CO_2Me, COMe, CN), by employing as nucleophiles phenylchalcogenolate species, generated from the reductive cleavage of diphenyl dichalcogenide, promoted by $NaBH_4$. The methodology has demonstrated to be general, by employing several alkyne derivatives **1** under three different reaction conditions: (1) under room temperature (method **A**), (2) under MW irradiation (method **B**), and (3) under conventional heating (method **C**) (Scheme 2.1).

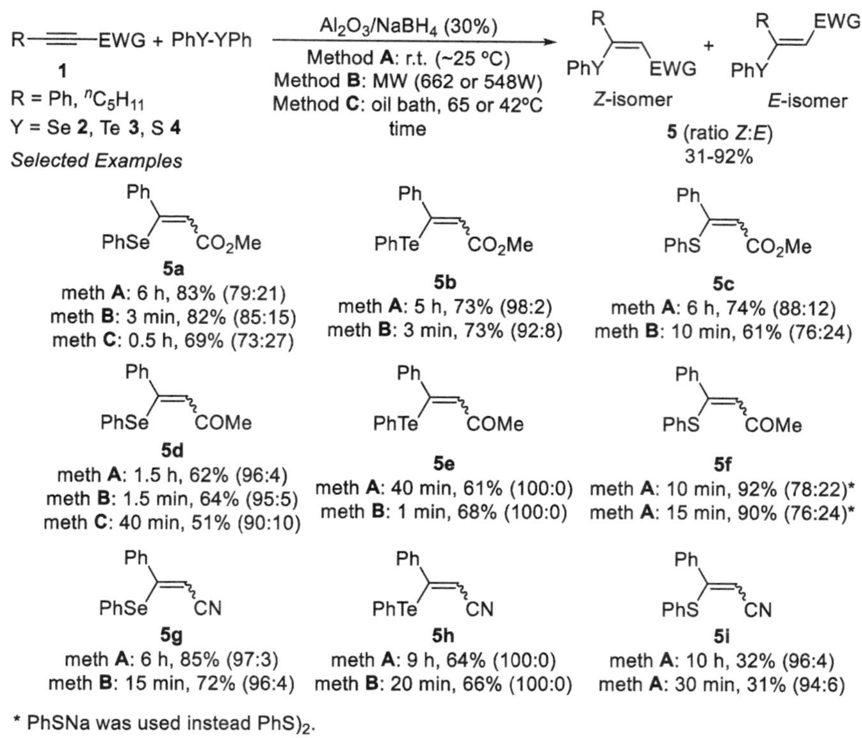

SCHEME 2.1 Hydrochalcogenation of activated alkynes.

In general, the reaction yields fluctuate from poor to excellent, however keeping a high selectivity toward the (Z)-isomer derivative **5**. Among the diphenyl dichalcogenide starting materials, the ditelluride derivatives demonstrated to positively affect the reaction selectivity, reaching 100% of the (Z)-isomer in most cases. On the other hand, MW irradiation was highly efficient in promoting the reactions quickly, affording the desired product **5** in just few minutes, while under room temperature and under a gently heating, long reaction times were necessary (Scheme 2.1).

In 2016, Perin and coworkers[52] reported a thermally controlled mono- and bis-chalcogenation of diynes, under conventional and MW irradiation conditions. In the optimization study, authors set that diynes **6** and diorganyl dichalcogenides, in the presence of $NaBH_4$ (2 equivalent) as a reductant agent and PEG-400 as solvent, under inert atmosphere (N_2), were able to provide the product **7** or **8**. The synthesis of the vinyl chalcogenides **7** and **8** can be modulated easily by a simple change in the reaction temperature from 30°C to 90°C, respectively. In general, MW irradiation was more efficient to promote the synthesis, requiring just few minutes, while conventional heating required long reaction times (Scheme 2.2).

In 2015, Braga and coworkers[53] reported a simple and efficient MW-assisted I_2-catalyzed alkoxy-chalcogenation of alkenes **9**, by employing diorganyl dichalcogenide as substrate. A brief reaction optimization study has defined the best reaction condition by using I_2 (20 mol%) and dimethyl sulfoxide (DMSO; 1 equivalent), under MW irradiation (100 W) at 50°C, for just 10 minutes. Under the optimized reaction condition, 25 chalcogen ether derivates (**10**) were prepared in poor to excellent yields. In general, a good substrate tolerance was achieved, with electron-rich and electron-deficient styrenes (**9**) and diphenyl diselenides (**2**) satisfactorily employed to access the products **10a–e**. On the other hand, *meta*-trifluoromethyl diphenyl diselenide (**2**) ($R^1 = 3\text{-}CF_3$) was not suitable to the methodology, giving the product **3f** in only 21% yield. Meanwhile, dibutyl diselenide has demonstrated to be a better substrate, affording the product **10g** in 57% yield (Scheme 2.3). Additionally, *i*-propyl and benzyl alcohols were also well employed, giving the products **10h** and **10i** in 80% and 95% yield, respectively, as well as 2-methylstyrene, which yielded the product **10j** in 63%. Different nucleophilic species could also be used, including acetic acid and inorganic azides, which afforded the products **10k** and **10l** in 42% and 21% yield, respectively (Scheme 2.3).

2.3 Substitution reactions

During the last decade, the use of in situ–formed electrophilic chalcogen-based species in electron-rich aromatic compounds has been emerging as a powerful tool to access chalcogen-containing substituted

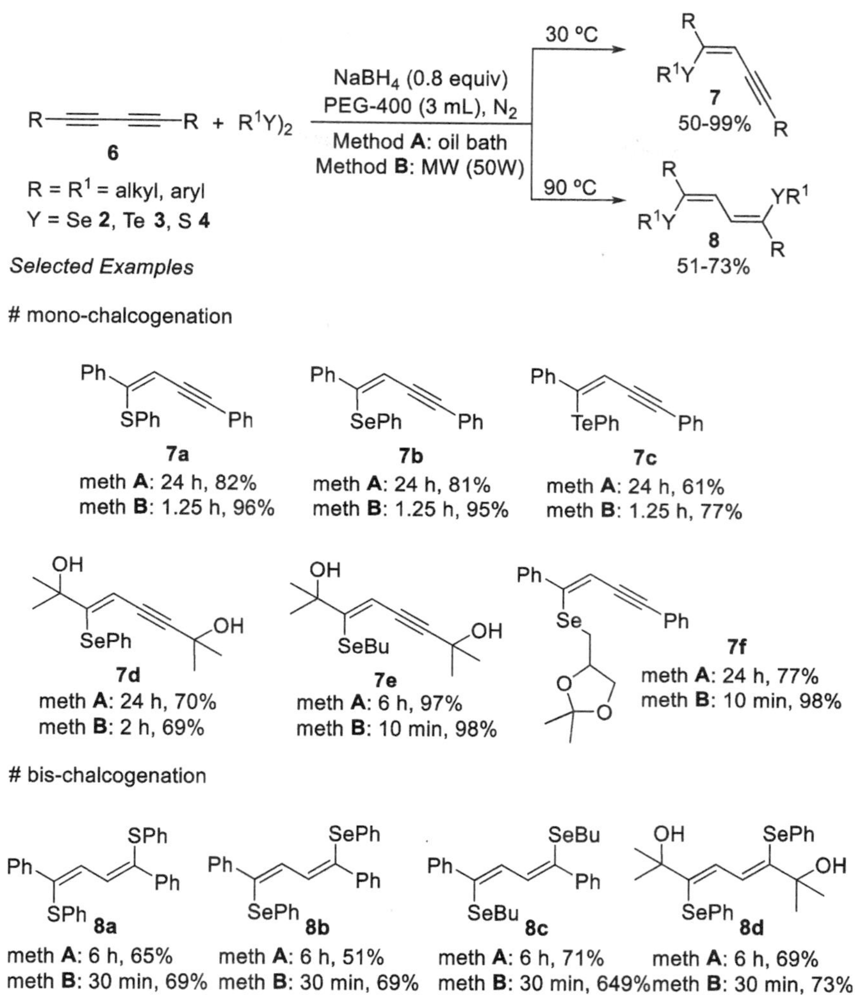

SCHEME 2.2 Thermally controlled mono- and bis-chalcogenation of diynes.

compounds. In this context, two main classes of nucleophiles have been employed, namely, activated arene and heteroaromatic compounds. Thus, we will present here a concise discussion about the use of alternative energy sources promoting electrophilic substitutions between electron-deficient chalcogen species and activated arenes and heteroarenes.

2.3.1 Substitution reactions of activated arenes

In 2016, Braga and coworkers[54] disclosed a robust metal- and solvent-free MW-mediated I_2-catalyzed regioselective synthesis of

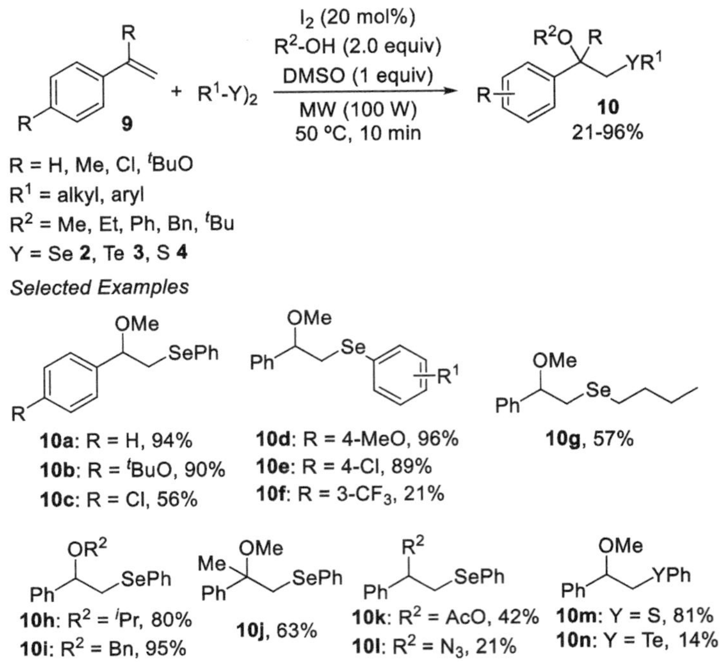

SCHEME 2.3 MW-assisted I_2-catalyzed alkoxy-chalcogenation of alkenes **9**.

unsymmetrical chalcogenides through oxidative C–Se/C–S bond formation via direct C–H bond functionalization of activated arenes. Based on a comprehensive optimization study, the authors set the best reaction condition by using the respective arenes (**11**), diorganyl dichalcogenides (**2**) and (**4**), molecular iodine (20 mol%), DMSO (3 equivalent), being heated for 10 minutes at 100°C, by using 100 W MW irradiation as an alternative energy source (Scheme 2.4).

Employing the optimized reaction condition, the authors afforded a large scope of products **12** and **13** in good-to-excellent yields, demonstrating the huge applicability of the protocol. Thus, by reacting diphenyl diselenide (**2**) with *N*-protected anilines (R^1 = Me, Et, and ^{n}Bu), the desired products **12a–c** were adequately reached in excellent yields. Similarly, *ortho* and *para* anilines (R^1 = H; R^2 = Me, Cl) reacted smoothly to give the products **12d–g** in 72%–87% yield. In addition, benzo[*d*][1,3]dioxol-5-amine and thiazol-2-amine heterocycles demonstrated to be suitable to the present protocol, reaching the products **12h** and **12i** in 82% and 75% yield, respectively (Scheme 2.4). Expanding the reaction scope, authors employed phenol and anisole as oxygen-activated arenes (**13**), which were able to give the respective products **12j** and **12k** in 82% and 88% yield, respectively (Scheme 2.4).

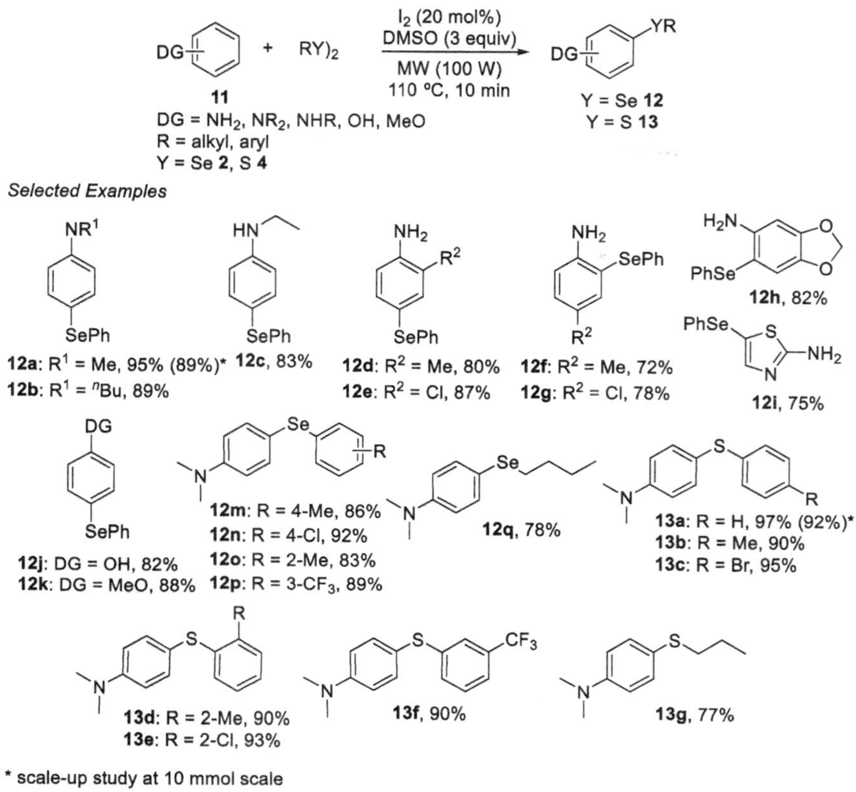

SCHEME 2.4 MW-activated I_2-catalyzed synthesis of unsymmetrical chalcogenide.

Thus, the scope of diorganyl dichalcogenides **2** and **4** was also studied, by employing several derivatives as substrate toward *N,N*-dimethylanilines. Thus, electron-rich and electron-deficient *para*-substituted diphenyl diselenides **2** (R = 4-Me and 4-Cl) were successfully employed to afford the products **12m** and **12n** in excellent yields of 86% and 92%, respectively. Satisfactorily, *ortho*-substituted derivative (R = 2-Me) was also suitable, giving the product **12o** in 83% yield, overcoming some possible steric hindrance effect, as well as 3-trifluoromethyl diphenyl diselenide, which afforded the product **12p** in 89% yield. Additionally, the diselenide scope was not limited to only aromatic substrates and dibutyl diselenide reacted smoothly to give the product **12q** in 78% yield (Scheme 2.4).

Hence, *N,N*-dimethylanilines were additionally used to amplify the reaction scope to unsymmetrical sulfides (**13**), by reaction with several alkyl and aryl disulfides (**4**). In general, any remarkable difference between electron-neutral, electron-rich, and electron-deficient *para*-substituted diphenyl disulfide **5** (R = H, Me, and Br) was observed, and the respective products **13a–c** were obtained in excellent yields. Similarly, *ortho*-substituted diphenyl disulfides (R = Me and Cl) were also satisfactorily employed as substrate, giving the desired products **13d** and **13e** in 90% and 93% yield, respectively, as well as 3-fluoromehyl diphenyl disulfide, which gave the product **13f** in 90% yield. Finally, dialkyl disulfide (R = Pr) was suitable to the protocol, giving the product **13g** in 77% yield. Thus, in order to demonstrate the potential of the present developed protocol, authors provided a reaction scale-up study, performing the reaction to reach compounds **12a** and **13a** in 10 mmol scale. The desired products were obtained in excellent yields, demonstrating the robust applicability of the developed protocol (Scheme 2.4).

Based on the obtained results and literature, the authors have proposed a plausible reaction mechanism, describing the main reaction pathways involved in the transformation (Scheme 2.5).

Initially, the selenium electrophilic specie **I** (PhSeI) is formed through an MW-mediated reaction between iodine and diphenyl diselenide (**2**). Next, the activated arene (**11**) attacks the reactive selenium specie, giving the positively charged intermediate **II**, which after a proton elimination affords the desired product (**12a**) and formation of HI. After this, DMSO

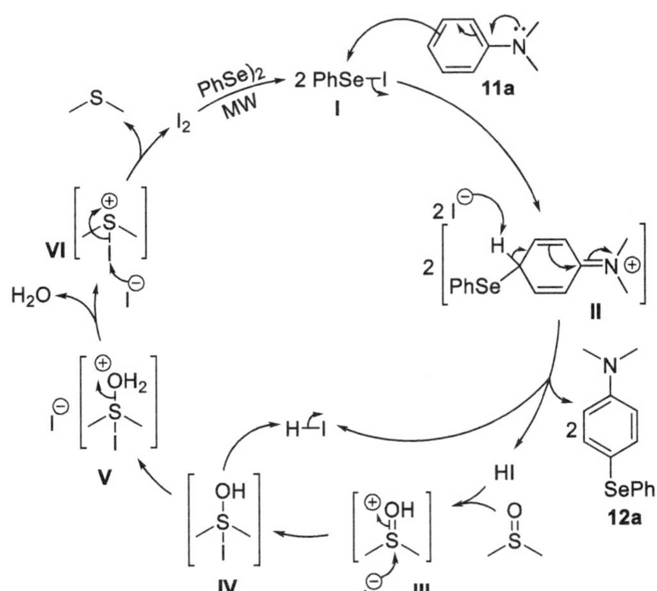

SCHEME 2.5 Reaction mechanism for the I_2-catalyzed synthesis of unsymmetrical chalcogenides.

undergoes a protonation by HI, leading to the formation of the intermediate **III**, which is subsequently attacked by the ion iodine, giving the intermediate **IV**. Over again, the intermediate **IV** is protonated by HI, giving the intermediate **V**, which is dehydrated to afford the intermediate **VI**. Finally, the intermediate **VI** undergoes a nucleophilic substitution of iodine anion, restoring the I_2 as catalyst to restart the reaction cycle and releasing dimethyl sulfide as a byproduct (Scheme 2.5).

In 2012, Ranu and coworkers[55] described a Zn-promoted synthesis of unsymmetrical diaryl chalcogenides, through the coupling between diphenyl dichalcogenides and aryl diazonium tetrafluoroborate salts (**14**), addressing the desired products in good-to-excellent yields. Through a detailed reaction optimization study, authors have set that the synergism between 50 mol% of Zn and dimethyl carbonate (DMC) was needed as a green and reusable solvent to reach the desired products satisfactorily. Additionally, the reaction could be employed under conventional conditions (oil bath, 80°C), as well as under MW irradiation (170 W), the last being much more efficient, reducing the reaction time drastically and keeping excellent reaction yields. Thus, it was employed as an extensive comparative study between conventional heating and MW irradiation as energy source (Scheme 2.6).

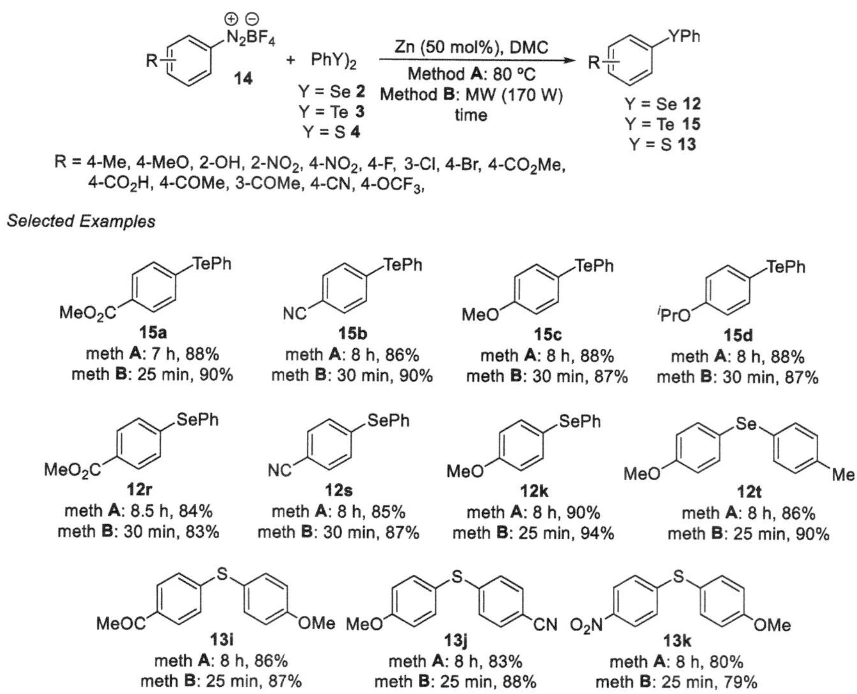

SCHEME 2.6 Comparative study in Zn-promoted synthesis of diaryl chalcogenides.

Initially, diaryl ditelluride (**3**) was submitted to the optimized reaction condition in the presence of several electron-rich and electron-deficient aryl diazonium fluoroborate salts **14** derivatives. Thus, regarding the products **15a–d**, it could be noticed that the electronic effect was not mandatory to access them satisfactorily; however, the reaction time was drastically reduced by performing the reaction under MW irradiation condition. The same trend was observed when diaryl diselenide (**2**) was employed as substrate, giving the products **12r–t** and **12k** efficiently, mainly when the reaction was performed under MW irradiation. Finally, expanding the methodology to electron-rich and electron-deficient diaryl disulfides (**4**), the products **13i–k** were accessed in excellent yields, keeping the superiority of MW irradiation as the reaction energy source.

Based on the obtained results and literature reports, authors have proposed a plausible reaction mechanism involved in this transformation. Initially, Zn reacts with diphenyl ditelluride (**3a**), giving the specie $Zn(TePh)_2$ (**I**), which, in the presence of the aryl diazonium tetrafluoroborate salts (**14a**), reacts to give the intermediate (**II**). Finally, N_2 is released to afford the respective product **15c** (Scheme 2.7).

In 2013, Ranu and coworkers[56] also reported a solvent-, ligand-, and metal-free synthesis of unsymmetrical diaryl chalcogenides (**3**), from aryl diazonium tetrafluoroborates salts (**14**) and diaryl dichalcogenides, under ball-milling reaction conditions. Through an optimization reaction study, the best reaction condition was reached by using 3.0 g of neutral Al_2O_3 as a grinding auxiliary, into a 25 mL stainless steel beaker with six balls (d = 10 mm) of the same material, accessing 21 unsymmetrical derivatives in moderate-to-excellent yields (Scheme 2.7).

In general, electron-rich and electron-deficient aryl diazonium tetrafluoroborates salts (**14**; R = 4-MeO, 4-NO_2, and 3-COMe) reacted smoothly to give the products **13l–o**. In addition, pyridyl and octadecanyl disulfides (**4**) reacted satisfactorily to give the products **13p** and **13q** in 82% and 72% yield, respectively. Expanding the methodology to unsymmetrical diaryl selenides (**3**), several electron-rich and electron-deficient aryl diazonium tetrafluoroborates salts (**14**) were employed with diphenyl diselenides, affording the products **12k, 12u, 12s,** and **12v** in good yields. Additionally, diphenyl ditelluride was also submitted to

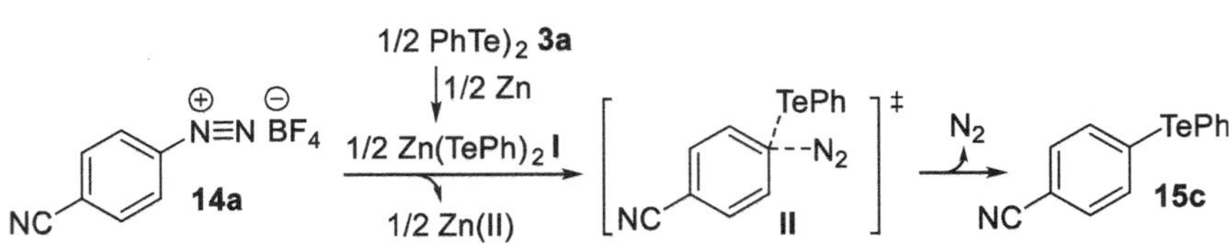

SCHEME 2.7 Plausible reaction mechanism for the Zn-promoted synthesis of chalcogenides.

the optimized reaction condition with the same diazonium tetrafluoroborates salt derivatives **14**, giving the respective products **15c, 15e, 15b,** and **15f** in moderate-to-good yields, confirming the extensive applicability of the developed protocol (Scheme 2.8).

Additionally, authors employed the protocol in the synthesis of *S*-aryl dithiocarbamates (**18**) through a one-pot protocol. Initially, carbon disulfide (**16**) reacts with secondary amines (**17**), under the optimized reaction condition at 0°C to 5°C, to give the intermediate **I**. Subsequently, aryl diazonium tetrafluoroborates salts **14** are poured out into the reaction media, and reacted for the tabled time, affording the desired products **18a–d** in good-to-excellent yields (Scheme 2.9).

In 2011, Alves and coworkers[57] presented a simple and efficient catalyst-free synthesis of unsymmetrical diaryl selenides, through the coupling between arylboron reagents (**19**) and arylselenium halides

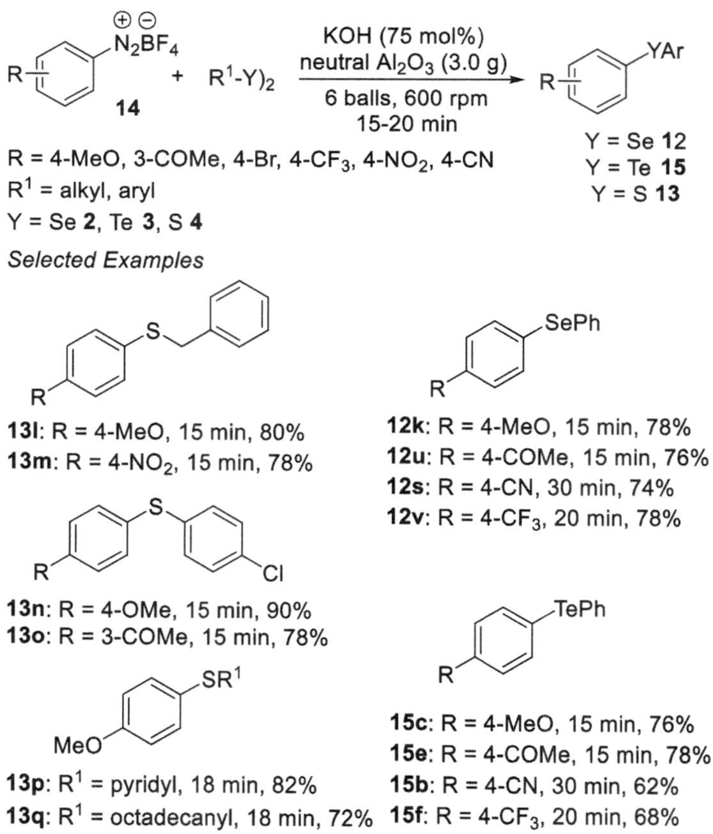

SCHEME 2.8 Synthesis of unsymmetrical diaryl chalcogenide under ball-milling conditions.

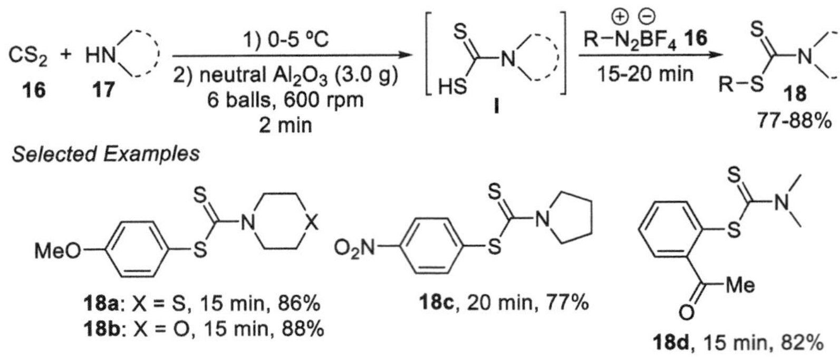

SCHEME 2.9 Synthesis of S-aryl dithiocarbamates (**18**) under ball-milling conditions.

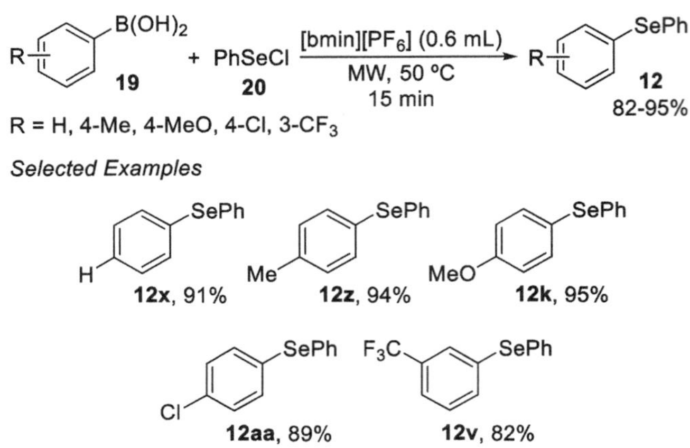

SCHEME 2.10 MW-mediated synthesis of diaryl selenides (**12**) using ionic liquid as a solvent.

(**20**; Cl and Br), in the presence of the ionic liquid [bmin][PF_6] as a reusable solvent. Despite the protocol involving reactions promoted at room temperature, in reaction times between 2 and 6 hours, the authors employed some substrates at MW irradiation conditions, in order to investigate the efficiency of this alternative energy source in the present transformation. Satisfactorily, the reaction time was drastically reduced from several hours to only a few minutes, while the yields were kept in a similar range. In addition, a wide electronic tolerance was achieved, since electron-rich and electron-deficient arylboron derivatives (**18**) were adequately employed. Finally, authors expanded the methodology efficiently to potassium aryltrifluorborates, instead of arylboronic acids, replacing the solvent by [bmim][BF_4], accessing the respective products (**12**) in very good yield. However, in this study, MW irradiation was not employed as energy source (Scheme 2.10).

In 2017, Braga and coworkers[58] presented an MW-assisted copper-catalyzed synthesis of unsymmetrical diorganyl chalcogenides from boronic acids (**19**) and diorganyl dichalcogenides through a ligand- and solvent-free protocol. The best reaction condition was set by using the system CuI (3 mol%) and DMSO (3 equivalent) as catalyst and oxidant species, respectively, being irradiated by MW (100 W), at 100°C for 15 minutes (Scheme 2.11).

Under the optimized reaction condition, the authors expanded the reaction scope to the synthesis of several unsymmetrical diaryl chalcogenides (**3**) in moderate-to-excellent yields. In general, electron-rich and electron-deficient diaryl diselenides and ditellurides reacted smoothly to give the products **12k**, **12ab–ac**, and **15j–k** in very good reaction yields. Satisfactorily, dialkyl diselenides and tellurides ($R^1 = Bu$) were also suitable to the protocol, affording the products **12ad** and **15l** in 77% and 87% yield, respectively (Scheme 2.11). In addition, the reaction scope was enlarged to diaryl disulfides, which have demonstrated a slight lower reactivity, affording the products **13r** and **13s** in 71% and 68% yield, respectively. It is worth mentioning that the compound **15c** was prepared in a 10 mmol scale, providing the product in 85% yield (Scheme 2.11).

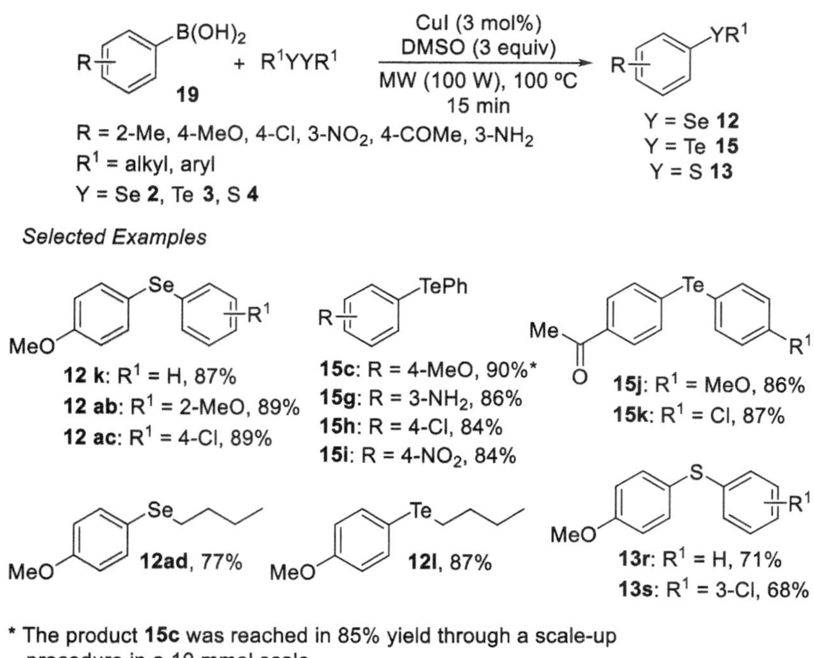

SCHEME 2.11 MW-assisted Cu-catalyzed synthesis of unsymmetrical diorganyl chalcogenides.

Based on the literature and some control experiments, the authors have proposed a plausible reaction mechanism. Initially, the intermediate **I** is generated from the addition reaction between CuI and arylboronic acid (**19**). After that, the specie **I** reacts with diorganyl diselenide (**2**) to afford the desired product **12** and the intermediate **II**. The specie **II** is oxidized to the $Cu^{(II)}$ intermediate **III**, which reacts with another equivalent of arylboronic acid (**19**), affording the intermediate **IV**. Finally, through an elimination reaction step, another equivalent of the desired product **12** is obtained and the catalyst is released to start a new reaction cycle (Scheme 2.12).

In 2012, the MW-assisted synthesis of symmetrical dichalcogenides **2–4** from organyl iodides (**20**) and elemental chalcogen was published.[59] Twenty-four diorganyl dichalcogenides were prepared by combining several organyl iodides (**20**) and elemental selenium, tellurium, or sulfur in the presence of CuO nanoparticles as catalyst and KOH at 80°C under MW irradiation. In this work, a wide range of substituted aryl iodides containing either donating or withdrawing groups giving diselenides, ditellurides, and disulfides in good-to-excellent yields in a very short reaction time. It is worth noting that the reaction between selenium and aryl iodides containing electron-donating groups, such as methyl, methoxyl, and amino groups, required more reaction time than withdrawing ones (from 7 to 15 minutes). Furthermore, the dialkyl dichalcogenides were synthesized in good yields using iodobutane as a starting material (Scheme 2.13).

In 2013, Jin and coworkers[60] presented an MW-assisted CuO_2-catalyzed synthesis of symmetrical diaryl selenides (**12**), starting from aryl iodine (**21**) and elemental selenium. The optimized reaction condition was set by using selenium powder (2 equivalent), CuO_2 (40 mol%) as catalyst, ethylenediamine (40 mol%) as ligand, and KOH (4 equivalent),

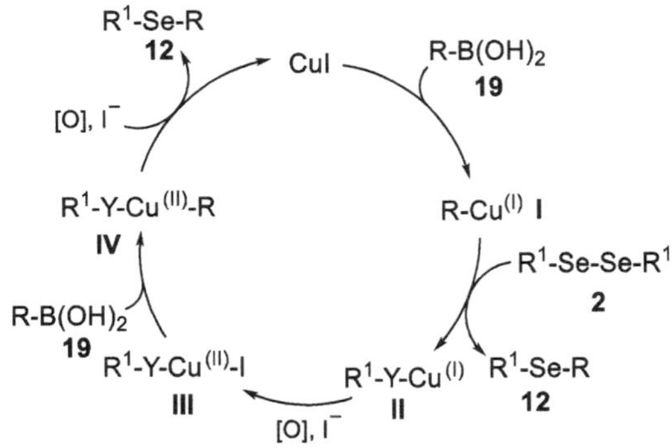

SCHEME 2.12 Reaction mechanism for the Cu-catalyzed synthesis of diorganyl chalcogenides.

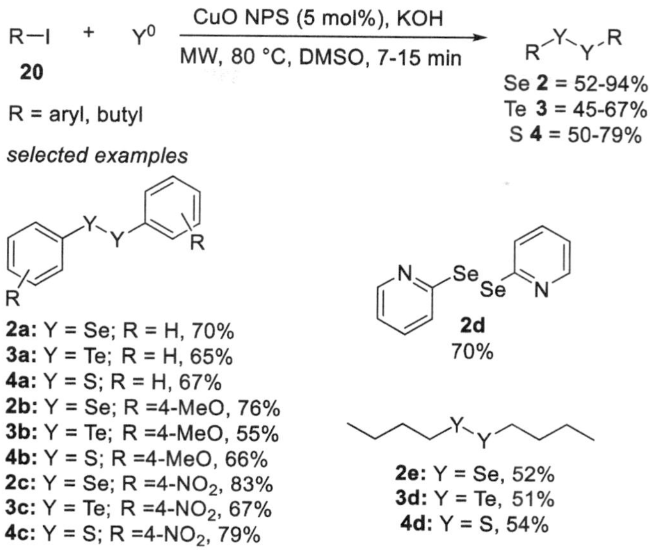

SCHEME 2.13 Synthesis of symmetrical dichalcogenides using CuO nanoparticles as catalyst.

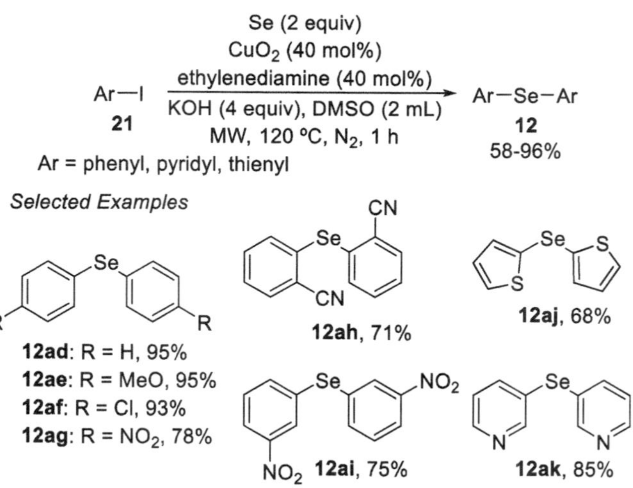

SCHEME 2.14 Synthesis of symmetrical diaryl selenide from elemental selenium.

in the presence of DMSO (2 mL), being irradiated at 120°C for 1 hour, under N_2 atmosphere (Scheme 2.14).

Under the optimized reaction condition, the authors synthesized a diversity of 14 symmetrical diaryl selenides derivatives (**12**) in moderate-to-excellent yields. Among them, neutral, electron-rich and electron-deficient aryl iodine (**21**) (R = H, MeO, and Cl) reacted smoothly to give the products **12ad–af** in high yields. Satisfactorily, aryl halides (**23**) bearing strong electron-withdrawing groups (R = NO_2 and CN) in *para-*, *meta-*, and *ortho*-positions demonstrated to be suitable substrates to give the products **12ag–ai** in good yields. Additionally, heteroaryl derivatives (**21**; Ar = 2-thienyl, 3-pyridyl) were efficient to give the products **12aj** and **12ak** in 68% and 85% yield, respectively (Scheme 2.14).

Based on the literature and obtained results, the authors have proposed a plausible reaction mechanism describing the present transformation. In the presence of a superbasic environment (DMSO–KOH), the elemental selenium undergoes a disproportionation reaction to give the selenolate anion (**I**). Parallelly, the aryl iodine (**21**) undergoes an oxidative addition to the CuO_2 catalyst, affording the intermediate **II**, which, in the presence of the selenolate **I**, is converted to the complex **III**. Thus, a reductive elimination gives the aryl selenolate anion (**IV**) and releases the CuO_2 to restart a new catalytic cycle. Following up, another unit of the intermediate **II**, in the presence of the aryl selenolate specie (**IV**), undergoes a ligand exchange reaction to give the intermediate (**V**). Finally, a reductive elimination releases the desired product **12** and restores the catalytic specie to start a new catalytic cycle (Scheme 2.15).

In 2006, Engman and coworkers[61] disclosed the synthesis of diaryl chalcogenides through a CuI/2,2-bipyridyl-catalyzed Mg-assisted coupling

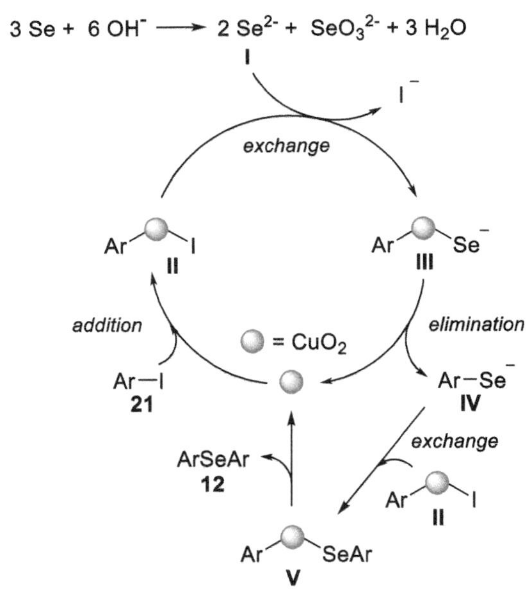

SCHEME 2.15 Reaction mechanism for the synthesis of diaryl selenide from Se^0.

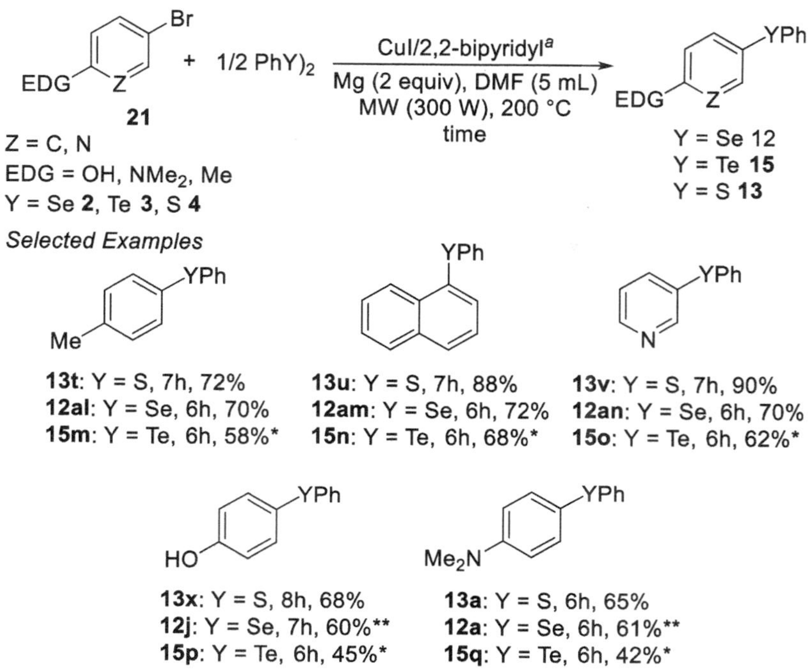

[a] 10, 15 and 20 mol%, respectively, for telluride, selenide and sulfide synthesis.

* reaction performed at 160 °C, under MW irradiation.

** by employing 4-chlorophenol as substrate, the product **12j** was reached in 45% yield.

SCHEME 2.16 MW-assisted synthesis of diaryl chalcogenides from activated aryl halides (**21**).

between aryl bromides (**21**) and diaryl dichalcogenides. The reaction protocol consists in an improvement of Taniguchi's method,[62] employing MW irradiation as an alternative energy source, reducing the reaction time from days to some hours, and avoiding high reaction temperatures (110°C). In addition, authors expanded the protocol to tellurides, as Taniguchi's method was limited to only sulfides and selenides (Scheme 2.16). In general, the protocol was suitable to several aryl bromides (**21**), including 4-bromotoluene, 1-bromonaphtalene, 3-bromopyridine, 4-bromophenol, and *N,N*-dimethyl-4-bromoanilines. On the other hand, diphenyl ditellurides demonstrated to be less reactive than disulfides and diselenides (Scheme 2.16).

2.3.2 Substitution reactions of heteroarenes

In 2015, Lenardão and coworkers[63] reported a US irradiation–mediated Cu(I)-catalyzed electrophilic selenylation of indoles (**22**), by using diorganyl diselenides (**2**), accessing the bioactive class of

3-selanylindoles (**23**). Through a detailed study of reaction optimization, authors set the best reaction condition by using N-*H*-indole (0.5 mmol), diphenyl diselenide (0.3 mmol), CuI (20 mol%), and DMSO (0.5 mL), with the mixture irradiated for 1.5 hours (20 kHz, 60% of amplitude), to give the desired product **3** in 74% yield. In order to compare the efficiency of US irradiation US with different energy sources, the authors employed a study of the reaction scope by employing US and MW irradiation, as well as conventional heating (oil bath) (Scheme 2.17).

In general, the results demonstrated that US irradiation and conventional heating were efficient to convert the starting materials **22** and **2**, to the respective products **23**, while MW irradiation was not appropriate as an alternative energy source. Among US and oil bath, the former was more efficient, giving the desired products **3** in shorter reaction time. Regarding the scope of indole (**22**) partner, 1-methyl-1*H*-indole, and 5-bromo-1*H*-indole were suitable for the standard reaction condition to give the products **23b** and **23c**, respectively. However, other *N*-protected indoles ($R^1 = $ Ts and Boc) were not appropriate substrates to be employed in this protocol. Concerning the scope of diorganyl diselenide reaction partner, aryl derivatives demonstrated good tolerance under the reaction condition. Among them, electron-rich derivatives ($R^2 = 4$-$MeOC_6H_4$) showed a lower efficiency in comparison to electron-deficient

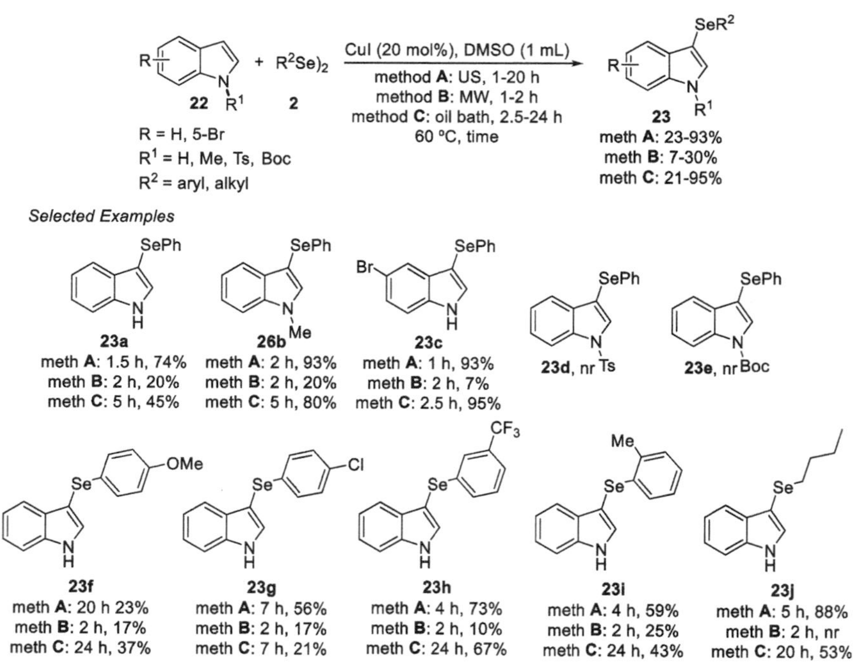

SCHEME 2.17 Cu(I)-catalyzed electrophilic selenylation of indoles.

derivatives ($R^2 = 4\text{-}ClC_6H_4$), giving the expected product in a higher yield range. This result was confirmed employing a strong electron-withdrawing group ($R^2 = 3\text{-}CF_3C_6H_4$), which gave the respective product **23h** in higher yields. Satisfactorily, the reaction was not drastically affected by steric hindrance effect, and the product **23i** was acceptably accessed. Finally, dialkyl diselenide was fit to the protocol, reaction smoothly to give the desired product **23j** (Scheme 2.17).

Two years later, Lenardão and coworkers[64] reported an updated study, affording a more efficient US-mediated Cu(I)-catalyzed protocol to access 3-selanylindoles (**23**), by employing SeO_2 (40 mol%) as additive. The addition of SeO_2 as cooxidant specie led to a remarkable improvement, greatly reducing the reaction times from hours to minutes, and enhancing the product yield (Scheme 2.18).

In addition, authors expanded the reaction scope, employing the imidazo[1,2-*a*]pyridines (**24**) as substrate, accessing the 3-(organylselanyl) imidazo[1,2-*a*]pyridines (**25**) in moderate-to-good yields. Regarding the indole scope, substituted indole (R = 5-Br) and *N*-protected indole (R^2 = Me) were satisfactorily employed as substrate, giving the desired product **23c** and **23b** in 87% and 96% yield, respectively. Similarly, several diphenyl diselenide derivatives (**2**) were successfully employed, including electron-rich and electron-deficient substrates, affording the respective products **23f–i**. Among them, it is worth to highlight that the

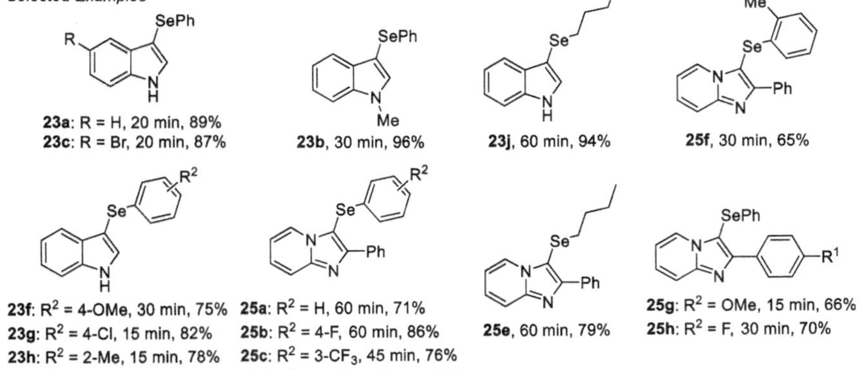

SCHEME 2.18 Cu(I)-catalyzed electrophilic selenylation of indoles using SeO_2 as additive.

ortho-substituted diphenyl diselenide ($R^2 = 2$-Me) reacted smoothly to give the desired product **23h** in 78% yield, after 15 minutes. Additionally, dialkyl diselenide ($R^2 =$ Bu) also demonstrated a good efficiency to the developed protocol, giving the desired product **23j** in 94% yield, after 60 minutes (Scheme 2.18).

Thus, regarding the imidazo[1,2-*a*]pyridines (**24**) reactivity compared to diorganyl diselenides (**2**) scope, several deficient derivatives (**2**) were satisfactorily employed to give the products **25a–c** in good-to-excellent yields, with the reaction less efficient when an electron-rich substrate ($R^2 = 4$-MeO) was employed to give the respective product **25d** in 57% yield. It is worth to mention that dialkyl ($R^2 =$ Bu) and *meta*-substituted diphenyl diselenide (**2**) reacted efficiently to give the products **25c** and **25e** in 76% and 79% yield. Finally, activated ($R^1 = 4$-MeO) and deactivated ($R^1 = 4$-F) imidazo[1,2-*a*]pyridines (**24**) were smoothly employed as substrate in front diphenyl diselenide (**2**) to reach the respective products **25g** and **25h** in 66% and 70% yield, respectively (Scheme 2.18).

Based on the obtained results and literature, the authors have proposed a plausible reaction mechanism to describe the main reaction pathways involved in both protocols, with the crucial role of SeO_2 described in the oxidation pathway (Scheme 2.19).

Initially, the CuI catalyst reacts with diphenyl diselenide (**2a**) to give the intermediate **I**, which undergoes a nucleophilic attack by the indole **22a**, affording the intermediates **II** and **III**. Thus, the desired product **23a** is formed through a deprotonation of the intermediate **II**, while the intermediate **III** is decomposed to phenylselenol, regenerating CuI. Finally, phenylselenol is oxidized to diphenyl diselenide (**2a**) in the presence of DMSO and air, as well as in the presence of DMSO and SeO_2, restarting a new reaction cycle (Scheme 2.19).

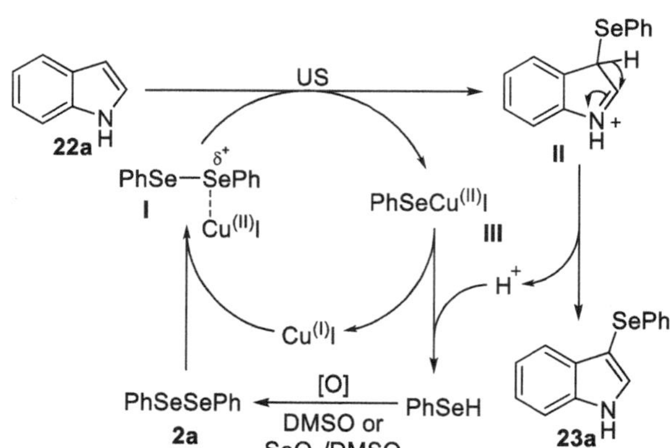

SCHEME 2.19 Reaction mechanism for the synthesis of 3-arylselanylindoles (**23**).

In 2014, Braga, Silveira, and coworkers[65] described an MW-promoted solvent- and metal-free I_2-catalyzed electrophilic chalcogenation of indoles **23**, by using DMSO as oxidant specie. Through a reaction optimization study, the best reaction condition was set by using indole **22** (0.5 mmol), diorganyl dichalcogenide **2** and **4** (0.25 mmol), iodine (5 mol%), and DMSO (3 equivalent), under MW irradiation (100 W) at 80°C, for 5 minutes (Scheme 2.20).

Thus, by employing the optimized reaction condition, the authors investigated the reaction scope, accessing the desired products in

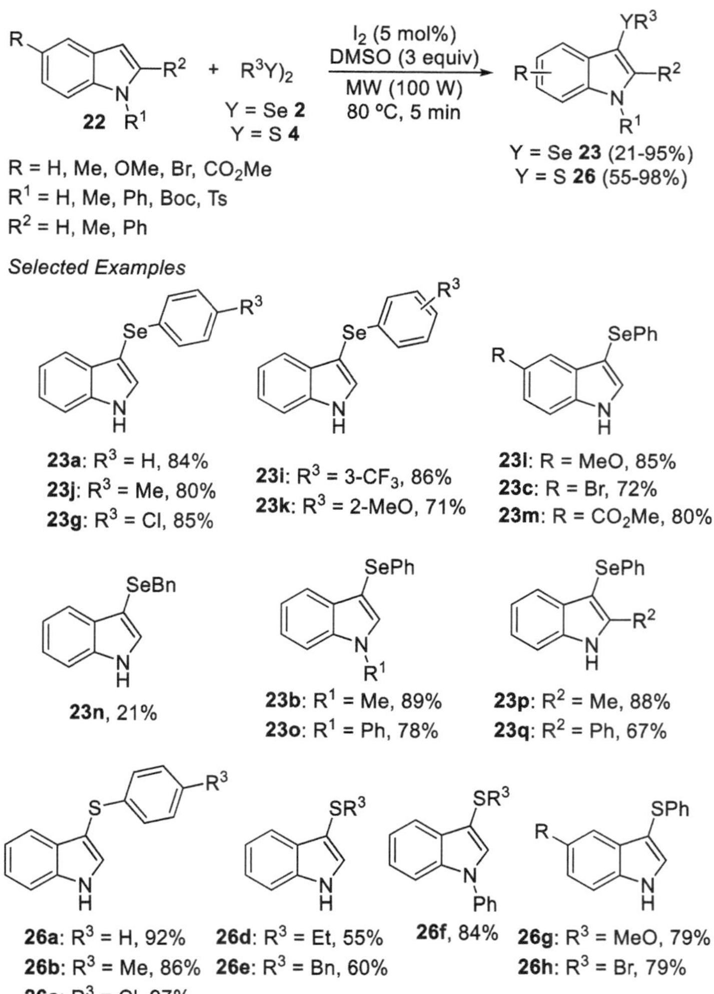

SCHEME 2.20 MW-mediated synthesis of 3-chalcogenylindoles **26** and **29**.

poor-to-excellent yields. Regarding the diphenyl diselenide (**2**) scope, *para-*, *meta-*, and *ortho*-substituted electron-rich ($R^3 = 4$-Me and 2-MeO) and electron-deficient ($R^3 = 4$-Cl and 3-CF_3) derivatives reacted smoothly to give the products **23a**, **23j**, **23g**, **23i**, and **23k** in moderate-to-very-good-yield. However, dibenzyl diselenide was not suitable for the developed approach, affording the product **23n** in only 21% yield. Hence, 5-substituted (R = MeO, Br and CO_2Me) and 2-substituted ($R^2 = $ Me and Ph) indoles **22**, as well as *N*-protected derivatives ($R^1 = $ Me and Ph) demonstrated to be highly useful under the optimized condition, giving the products in moderated-to-excellent yields. In addition, the protocol was extended to diorganyl disulfides (**4**), proving the applicability of the protocol. In general, electron-rich ($R^3 = $ Me) and electron-deficient ($R^3 = $ Cl) diaryl disulfides (**4**) were suitable, along with dialkyl derivatives ($R^3 = $ Et and Bn), which accessed the respective products **26a–e** in moderate-to-excellent yields. Finally, *N*-protected ($R^1 = $ Ph) and 5-substituted ($R^3 = $ MeO and Br) indoles (**22**) were apposite to give the products **26f–h** satisfactorily (Scheme 2.20).

Finally, based on literature reports, authors have proposed a plausible reaction mechanism, describing the reaction pathway. Initially, diphenyl diselenide (**2**) reacts with iodine to give the PhSeI as an electrophilic selenium specie, which undergoes an electrophilic attack, releasing the product (**23a**) and affording HI. Thus, a molecule of DMSO is protonated, giving the intermediate **I**, which is again protonated and attacked by an iodine anion, affording the intermediate **II**. Lastly, the intermediate **II** undergoes a nucleophilic substitution by iodine, releasing water and dimethyl sulfide as byproduct and recovering the I_2 to restart the catalytic cycle (Scheme 2.21).

In 2015, Wu, Zhang, and coworkers[66] reported the preparation of several 3-(3,4,5-trimethoxyphenylselenyl)-1*H*-indoles derivatives (**23**), in order to be applied as starting materials toward selenoxide derivatives for anticancer biological assays. The reaction optimization study set the best reaction condition by using the indole derivatives **22**, bis-(3,4,5-

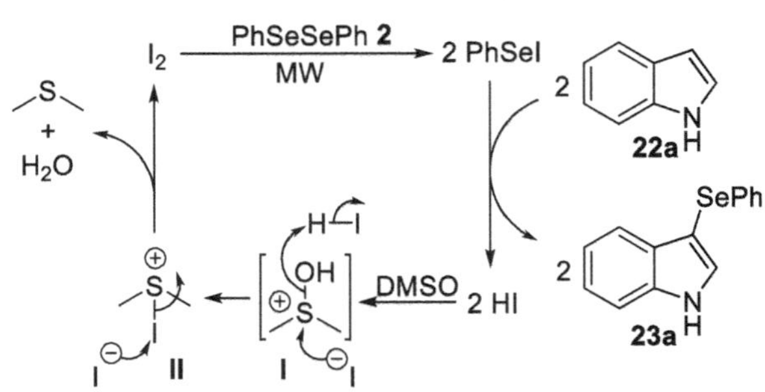

SCHEME 2.21 Reaction mechanism for the synthesis of **23a**.

trimethoxyphenyl) diselenide **2b** in the presence of $FeCl_3$ (20 mol%), I_2 (1 mol%) and MeCN (1 mL), once the obtained mixture was irradiated by MW irradiation (150 W) at 80°C, for 30 minutes. The methodology has shown to be widely applicable, since several electron-rich and electron-deficient indoles (**22**) were employed satisfactorily, with the bearing electron-donating groups (**23s** and **23t**) being more reactive. In addition, *N*-protected and 2-substituted indoles also demonstrated to be suitable for the developed protocol, giving the products **23x–aa** in moderate yields (Scheme 2.22).

In 2017, Savegnago, Lenardão, and coworkers[67] reported a US-promoted Cu(I)-catalyzed reaction between chrysin derivatives **27** and diaryl diselenides **2**, affording the semisynthetic bis-arylselanyl chrysin derivatives **28** and demonstrating the biological applicability of this class of compounds. By using 5 mol% of CuI as catalyst, in the presence of DMSO as solvent and oxidant specie, the authors synthesized six different analogs (**28**), in moderate-to-very-good yields (Scheme 2.23).

In this way, electron-rich diphenyl diselenides **2** (R = MeO) were efficient to give the product **28b** in 60%, while electron-deficient derivatives (R = F) were more suitable to the optimized reaction condition, affording the product **28c** in 80% yield after 3.5 hours of reaction. Satisfactorily,

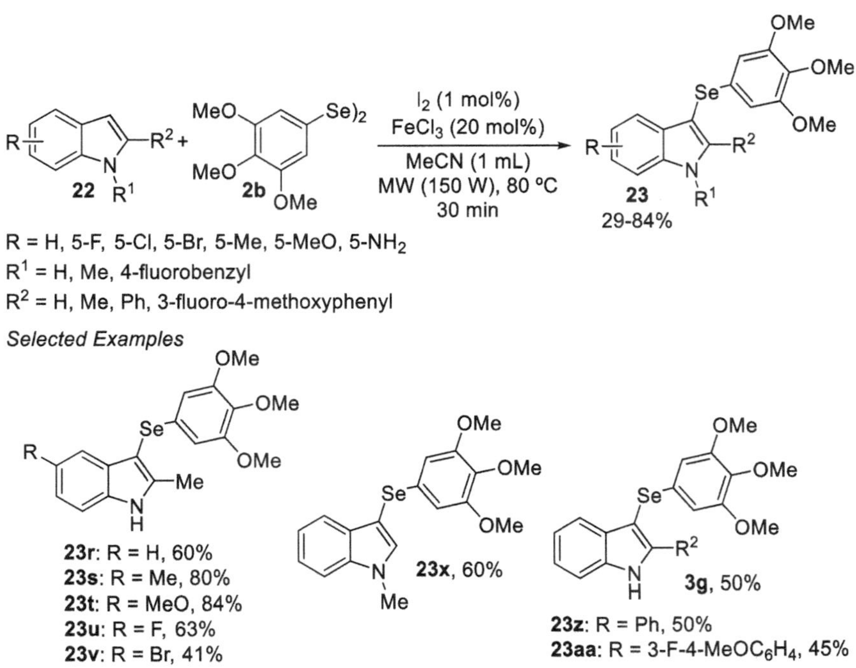

SCHEME 2.22 MW-assisted synthesis of 3-(3,4,5-trimethoxyphenylselenyl)-1H-indoles derivatives **23**.

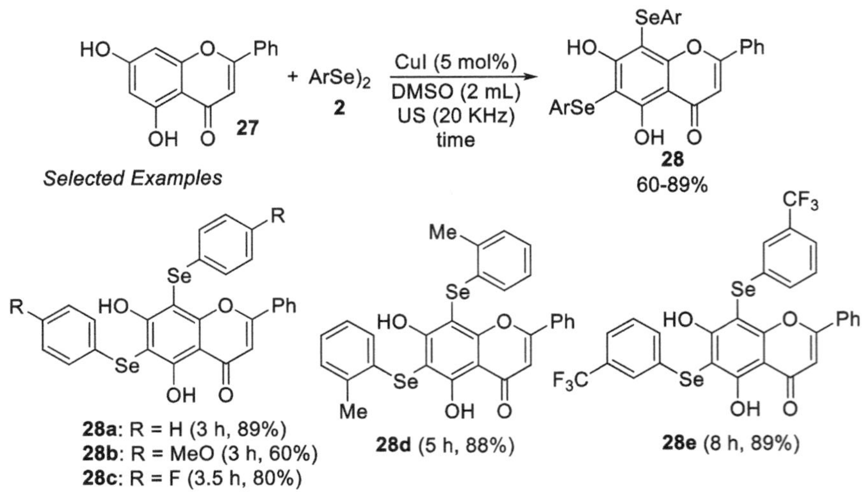

SCHEME 2.23 US-assisted synthesis of semisynthetic selenium derivative compounds.

ortho-substituted diphenyl diselenide **2** (R = 2-Me) was efficiently applied to reach the product **28d** in 88% yield after 5 hours, while the *meta*-CF_3 derivative reacted sleekly to afford the product **28e** in 89% yield, after 8 hours. It is worth mentioning that the reaction mechanism involved in this transformation, described by the authors, follows the mechanism involved in the US-mediated synthesis of 3-selanylindoles (Scheme 2.19)

2.3.3 Miscellaneous

In 2012, Braga and coworkers[68] reported a simple and eco-friendly MW-mediated Zn-promoted synthesis of selenol esters **30**, through an addition/elimination reaction between acyl chloride (**29**) and diorganyl diselenide (**2**). Thus, based on a reaction optimization study, authors set the best reaction condition by using acyl chloride (**29**) (0.5 mmol), diorganyl diselenide (**2**) (0.25 mmol) and Zn (0.25 mmol), with the resultant mixture irradiated by MW energy source (100 W) for 2 minutes at 80°C. The optimized reaction condition was applied to a reaction scope study, and 13 different selenol esters derivatives (**30**) were accessed in moderate-to-excellent yield (Scheme 2.24).

Thus, throughout the reaction scope study, *para*- and *ortho*-substituted electron-deficient diaryl diselenides **2** (R^1 = Cl) were found more suitable in comparison to electron-rich derivatives **2** (R^1 = Me and MeO) to access the products **30b–e**. Additionally, benzyl and alkyl (R^1 = Me) diselenides (**2**) were satisfactorily employed as substrate to give the products **30f** and **30g** in 71% and 70% yield, respectively. Regarding the

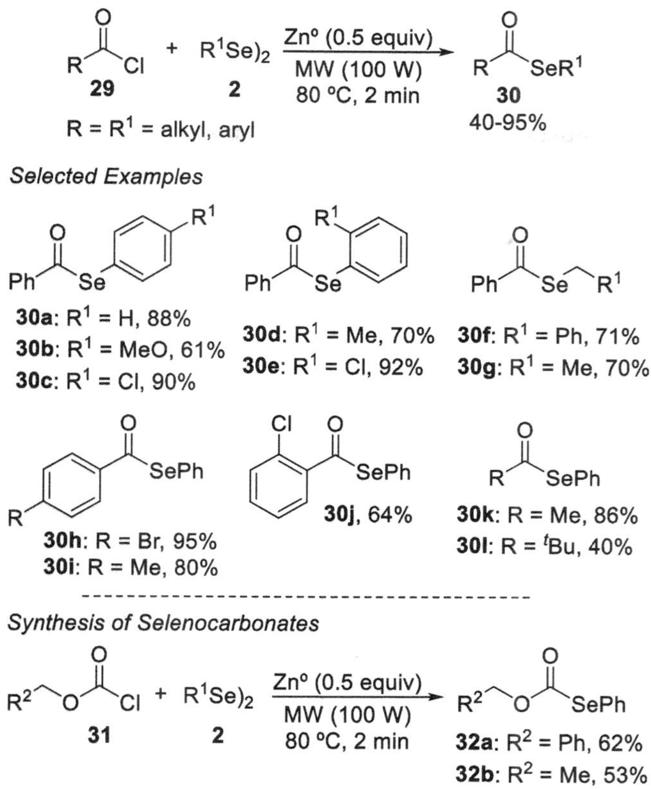

SCHEME 2.24 MW-mediated synthesis of selenol esters (**30**) and selenocarbonates (**32**).

acyl chloride (**29**) scope, electron-deficient derivative (R = 4-Br) was lightly more efficient than the electron-rich one (R = 4-Me), giving the products **30h** and **30i** in 95% and 80% yield, respectively. *Ortho*-substituted acyl chloride (**32**) (R = 2-Cl) was also suitable to the protocol to give the compound **30j** in 64% yield, similar to alkyl derivatives (R = Me and *t*-Bu), which reacted to afford the product **30k** and **30l** in 86% and 40% yield, respectively. Additionally, the authors expanded the reaction scope to the synthesis of selenocarbonates **32a** and **32b**, which were accessed in 62% and 53% yield, respectively, proving the wide applicability of the developed method (Scheme 2.24).

Based on the literature and obtained results, the authors proposed a plausible reaction mechanism to describe the synthesis of selenol esters **30a**. Initially, there is the formation of the reactive specie PhSeZnSePh **I**, through a well-established insertion of Zn_0 along the Se–Se bond. Thus, the zinc complex **I** reacts with the acyl chloride (**29a**) to give a molecule of product **30a** and PhSeZnCl complex **II**, which is also nucleophilic and attacks a new molecule of acyl chloride **29a** to give a second molecule of selenol esters **30a** (Scheme 2.25).

In 2000, Wang and coworkers[69] expanded the developed methodology to the synthesis of dibenzoyl diselenides (**33**), starting from elemental selenium and benzoyl chlorides (**29**). The strategy is based on the formation of a reactive sodium diselenide (**I**) specie under MW irradiation condition (750 W), by the treatment of elemental selenium with NaOH, and subsequent addition of the acyl electrophilic specie **29** under room temperature condition. Under the set reaction condition, nine dibenzoyl diselenides **33** were accessed satisfactorily, employing several electron-rich and electron-deficient benzoyl chloride (**29**), as well as alkenyl derivatives (Scheme 2.26).

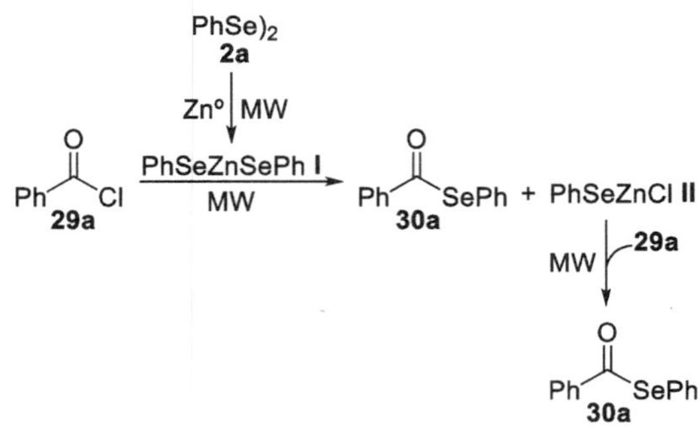

SCHEME 2.25 Reaction mechanism for the synthesis of **33a**.

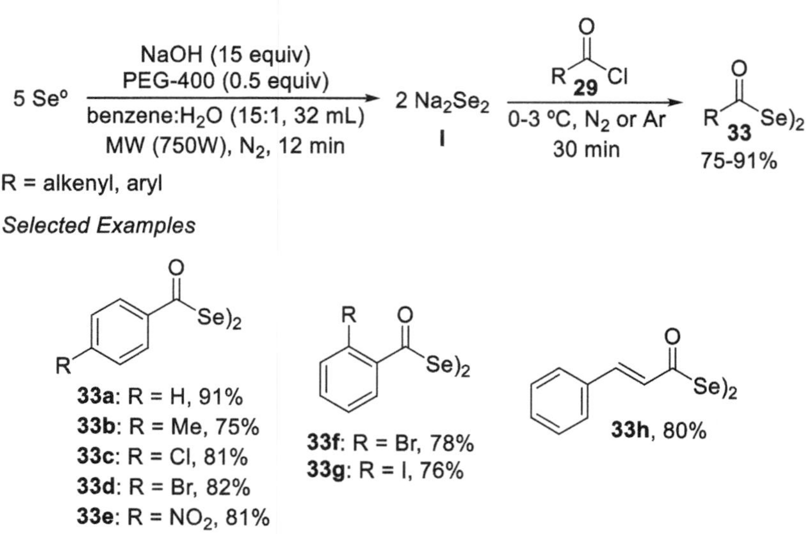

SCHEME 2.26 MW-assisted generation of Na_2Se_2 as reactive specie.

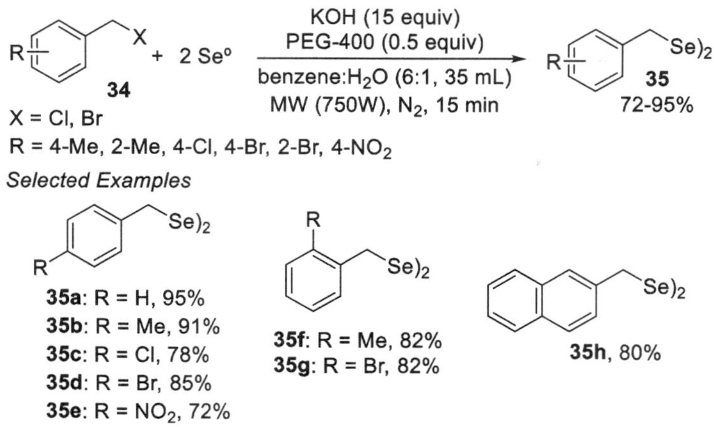

SCHEME 2.27 MW-assisted synthesis of dibenzyl diselenides (**35**) from elemental selenium.

In 2000, Wang and coworkers[70] reported a simple *one-pot* synthesis of dibenzyl diselenides (**38**) under MW irradiation, starting from benzyl halides (**34**) and elemental selenium. The optimization study set the best reaction condition by using KOH (15 equivalent) and PEG-400 (0.5 equivalent), in the presence of the solvent mixture benzene:H_2O (6:1, 35 mL), being irradiated (MW, 750 W) for 15 minutes under N_2 atmosphere. Under the optimized reaction condition, nine substituted dibenzyl diselenides (**35**) were accessed in good-to-excellent yields, employing several electron-rich and electron-deficient benzyl halides (**34**). The results confirm the robustness of the developed methodology (Scheme 2.27).

In 1986, Ley and coworkers[71] reported a US-assisted generation of sodium phenylselenolate (**36**), through the reaction between diphenyl diselenide (**2**) with 2.0 equivalents of metallic sodium, in the presence of tetrahydrofuran as solvent. The obtained specie **36** was employed as nucleophile to different electrophilic species, including alkyl and alkenyl halides, tosylates, and epoxides. Thus, a wide diversity of selenium compounds (**37**) were synthesized in good-to-excellent yields (Scheme 2.28).

2.4 Ring-opening reaction of epoxides

In 1998, Wang and coworkers[72] described the synthesis of 1,3-bis(arylseleno)-2-propanol (**38**) under MW irradiation conditions (Scheme 2.29). In this protocol, the diorganyl diselenides **2** were rapidly reduced to organylselenolate ion by using $NaBH_4$/NaOH, and then, the organylselenolate ion **2′** reacted with epichlorohydrin **39** to give 10

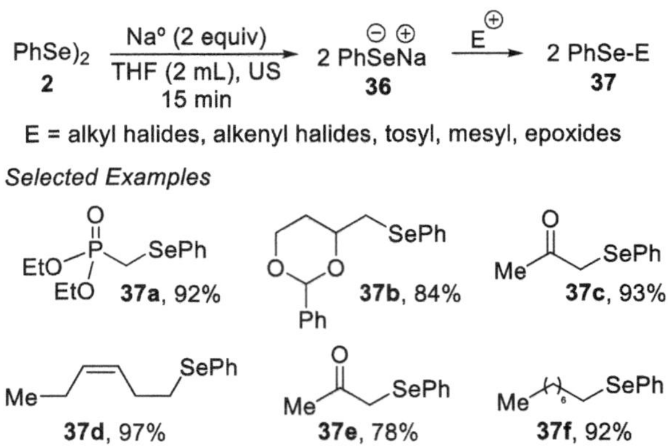

SCHEME 2.28 US-assisted generation of sodium phenylselenolate (**36**) as a selenium nucleophilic reagent.

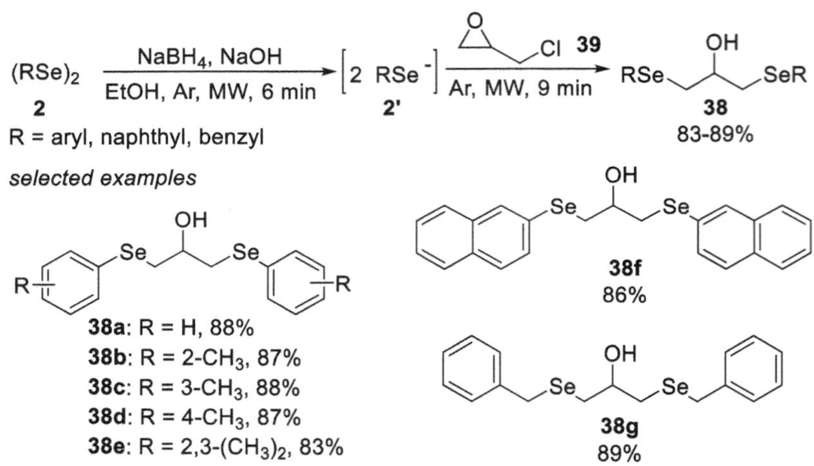

SCHEME 2.29 Synthesis of 1,3-bis(arylseleno)-2-propanol (**38**) under MW irradiation conditions.

compounds in good yields (83%–89%). In this study, the MW irradiation was carried out with an improved reflux microwave oven using an inert gases protector, under 750 W of power in two steps reaction of 6 and 9 minutes.

Later, the same research group[73] related the conversion of epoxypropoxyalkoxyls (**40**) to the corresponding 1-arylseleno-3-alkoxy-2-propanol derivatives **41** under MW irradiation at 750 W (Scheme 2.30). First, the arylselenolate ions were generated from reaction among diaryl diselenides (**2**), sodium hydroxide, and sodium borohydride; second, the epoxypropoxyalkoxyls **40** with a long carbon chain in ethanol solution

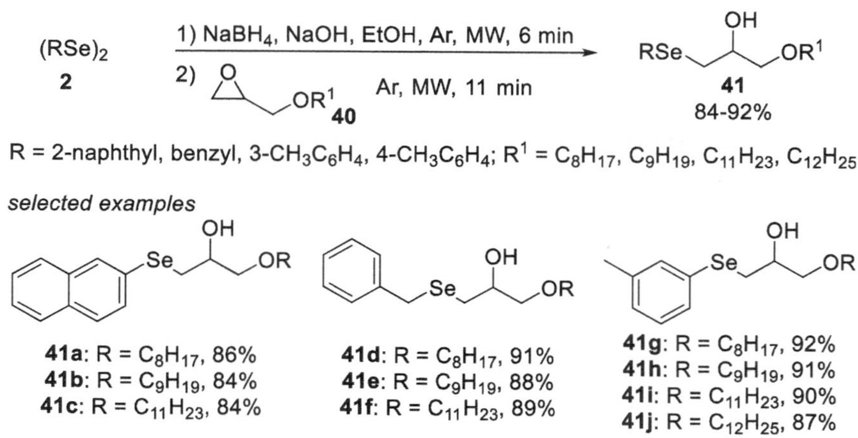

SCHEME 2.30 Synthesis of 1-arylseleno-3-alkoxy-2-propanols (**41**) under MW irradiation conditions.

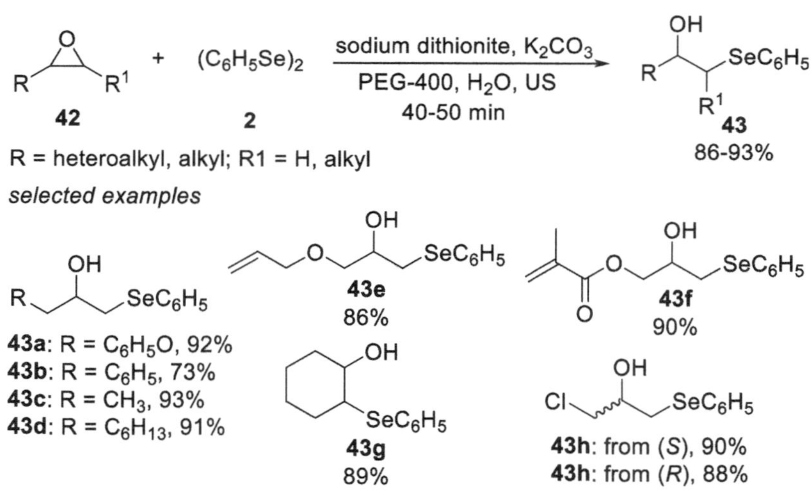

SCHEME 2.31 US-assisted ring-opening reaction of epoxides with 1,2-diphenyldiselenide.

was dropwise added to the above mixture under the same irradiation conditions. By this very simple and rapid method, the 11 products (**41**) were afforded in good yield (84%–92%) with high regioselectivity, without affecting the ether function. In this work, the authors combined aryl and benzyl diselenides with four kinds of epoxypropoxyalkoxyls.

In 2011, the US-assisted ring-opening of epoxides **42** with diphenyl diselenide **2** using PEG-400/H_2O as solvent in presence of sodium dithionite and K_2CO_3 was published (Scheme 2.31).[74] The authors have described the importance of US irradiation, once the optimal condition reaction without this source energy needs more time and a lower yield was obtained. Eleven β-hydroxy selenides (**43**) were isolated from good-to-excellent yield (86%–93%), and in all cases, highly regioselective nucleophilic attack of selenide anions occurred on the less hindered β-carbon of the oxirane ring. The efficiency of this protocol was demonstrated by the compatibility with epoxides containing an unsaturated bond, an ether, and an ester group. When a symmetrical epoxide was used, such as cyclohexene oxide, the *trans*-2-(phenylselanyl)cyclohexanol **43g** was afforded with high stereoselectivity in good yield. Furthermore, the (*R*)-1-chloro-3-(phenylselanyl)propan-2-ol **43h** and (*S*)-1-chloro-3-(phenylselanyl)propan-2-ol **43h** were prepared from optically pure epoxides without any racemization.

The procedure to prepare β-hydroxy sulfides (**44**) under US irradiation were proposed by reaction between epoxides **42** and diaryl disulfides **4** in the presence of 3.0 equivalent of Rongalite (Scheme 2.32).[75] In this case, the Rongalite (sodium hydroxymethanesulfinate) promoted cleavage of diaryl disulfides **4** generated the corresponding thiolate species in situ. Seventeen *β*-hydroxy sulfides (**44**) were isolated in good-to-excellent yields (70%–99%) with high regioselectivity, attributable to the exclusive attack of the sulfide anions on the less hindered carbon of the epoxide. Alkyl-substituted unsymmetrical epoxides and either electron-donating or electron-withdrawing groups bonded in the epoxides and disulfides were efficiently used. It is also necessary to highlight that the procedure proved to be successful in the presence of unprotected reactive functional groups, such as $-NH_2$ and with

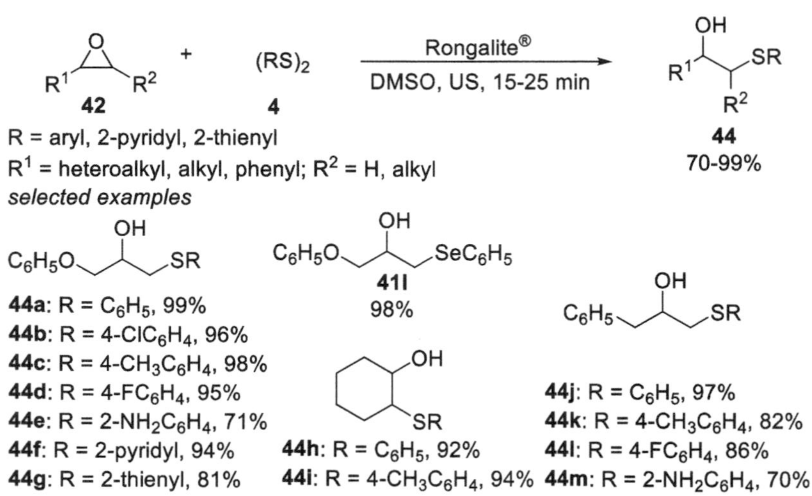

SCHEME 2.32 Synthesis of β-hydroxy chalcogenides under US irradiation.

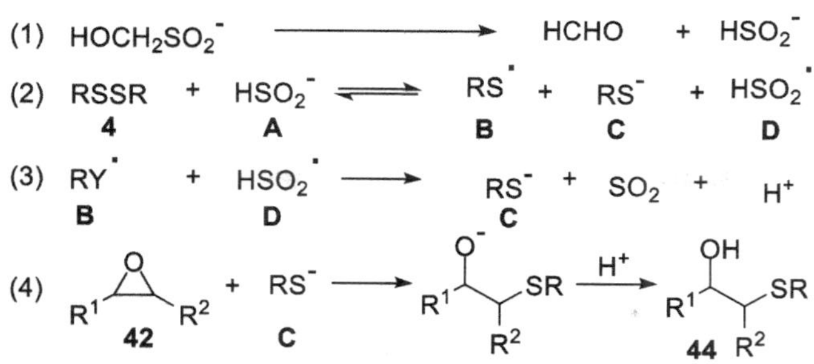

SCHEME 2.33 Mechanism for tandem base-free synthesis of β-hydroxy sulfides **44**.

heterocyclic disulfides such as 1,2-di(thiophen-2-yl)disulfide and 1,2-di(pyridin-2-yl)disulfide. Furthermore, the ring-opening reaction of 2-(phenoxymethyl)oxirane with diphenyl diselenide afforded 1-phenoxy-3-(phenylselanyl)propan-2-o (**41l**) in 98% yields after 25 minutes under the optimal reactional conditions.

Also, the authors proposed a mechanism for Tandem base-free synthesis of β-hydroxy sulfides (**44**) under US irradiation (Scheme 2.33). First, Rongalite can be decomposed into HCHO and HSO_2^- anion **A**. Then, the HSO_2^- anion **A** reacts with disulfide to generate two radicals (**B** and **D**) and an anion **C** as intermediates. The radical **B** can also be converted into the anion **C** by reacting with intermediate **D**. After that, the nucleophilic attack of the anion **C** occurs at the less hindered position of the epoxide to give the desired products.

2.5 Cyclization reactions

2.5.1 Chalcogen-based heterocycle synthesis

In 1996, Šibor and Pazdera[76] described the synthesis of some five-membered heterocycles containing selenium and tellurium under MW irradiation (Scheme 2.34). In this study, chalcogenazoles **45** and chalcogenophenes **46** and **47** were prepared from elementary selenium and tellurium, R-isothiocyanate (R = phenyl, allyl, benzoyl, and pivaloyl), and malonic acid functional derivatives **48**, either cyclohexanone **49** or camphor **50**. The authors reported the comparison among the classic condition, US, and MW irradiation, and US accelerated the described reaction significantly than conventional heating, and the MW irradiation manifested itself even more significantly. Through this optimal reaction condition, 28 chalcogenazoles (**45**) derivatives with phenyl and acyl substituents (pivaloyl, benzoyl) were isolated in high yields and short times, in addition to 12 chalcogenophenes **46** and **47** derivatives with a

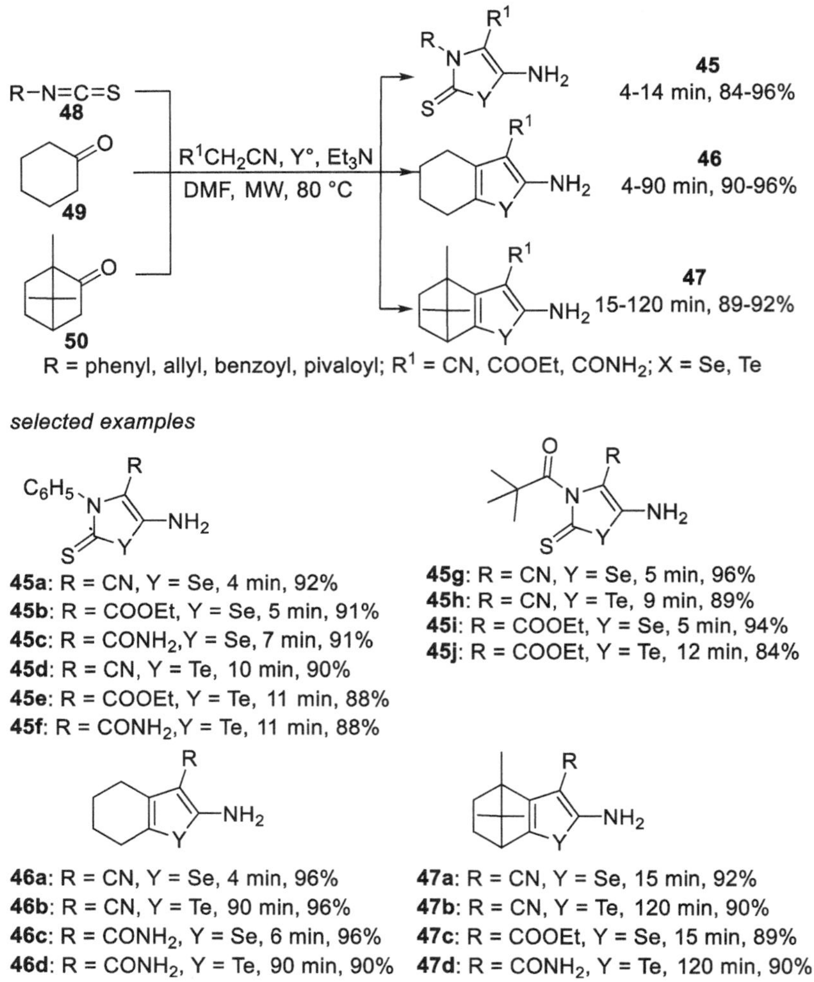

SCHEME 2.34 Five-membered heterocycles containing selenium and tellurium.

fused cyclohexane or bornane skeleton. Unfortunately, it was not possible to expand the MW irradiation procedure for the synthesis of sulfur analogs.

Wang and Zhao[77] described the use of US irradiation to prepare *cis*- and *trans*-2,6-diphenyl-1,4-diselena-fulvene (**51**) from phenylacetylene (**1**) and elemental selenium (Scheme 2.35). This process involves two steps: in the first one, the reaction of selenium powder, PEG-400, sodium hydroxide, water, and ethanol takes place and this mixture stays at 45°C–50°C for 25 minutes with ultrasonic irradiation. After that, the phenylacetylene is added in the reaction system, for more than

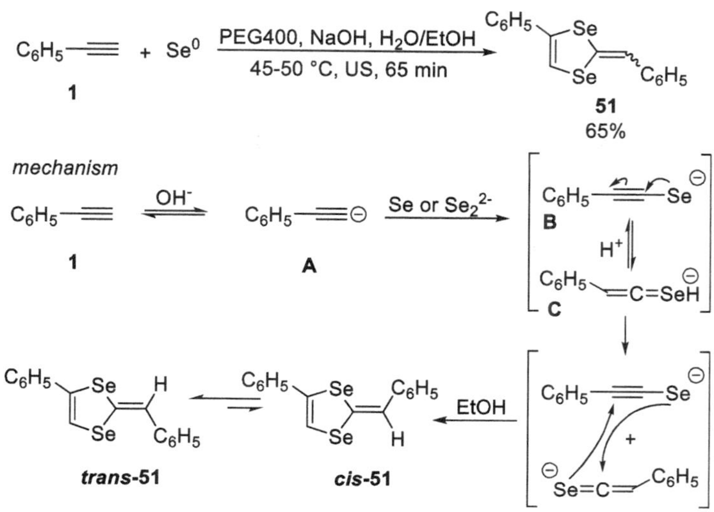

SCHEME 2.35 Synthesis and mechanism of *cis-* and *trans*-2,6-diphentyl-1,4-diselenafulvenes.

30 minutes. Scheme 2.35 describes the proposed mechanism to afford these isomers. Initially, the interaction of phenylacetylene **1** with OH^- gives intermediated **A**, which subsequently reacts with selenium to give the intermediate ethynylselenolate ion **B** that can form the selanylketene **C** by isomerization. *Cis*-2,6-diphenyl-1,4-diselenafulvenes (**51**) could be afforded by 1,3-dipolar addition of **C** to the ethynylselenolate ion **B**. The *cis*-isomer **51** can be transformed into the *trans*-isomer. The authors reported that the phase-transfer catalyst is not only involved in production of the ethynylselenolate ion in solution but also catalyzes the substitution reaction on the ethynylselenolate ion with selenium.

Raghavendra and coworkers[78] reported the use of MW to synthetize selenopheno[2,3-*b*]quinoline derivatives **52** or **53** via the reaction between 2-seleno-3-formyl-quinolines **54** and phenacylbromide **55** or 2-chloroacetamide **56** under basic and solvent-free condition (Scheme 2.36). Due the great interest from the chemical and biological point of view, this paper described the preparation of eight heterocyclic compounds fused to quinoline ring in 85%–90% yield and in only 7–8 minutes. Further, the easy isolation of selenopheno[2,3-*b*]quinoline derivatives **52** or **53** is highlighted since it occurred just by adding water to the reaction mixture.

Hesse and coworkers[79] published the synthesis of 4-chloroselenolo [3,2-d]pyrimidines (**57**) in two steps with high yields under MW irradiation (Scheme 2.37). To access these compounds, firstly the preparation of selenolo[3,2-d]pyrimidin-4-ones (**58**) was proposed by intermolecular

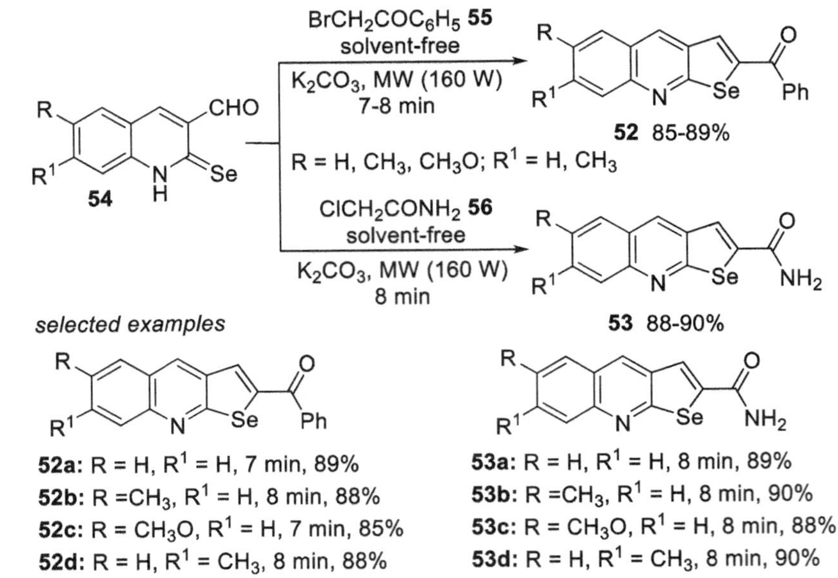

SCHEME 2.36 MW-induced synthesis of some novel selenolo[2,3-b]quinoline derivatives.

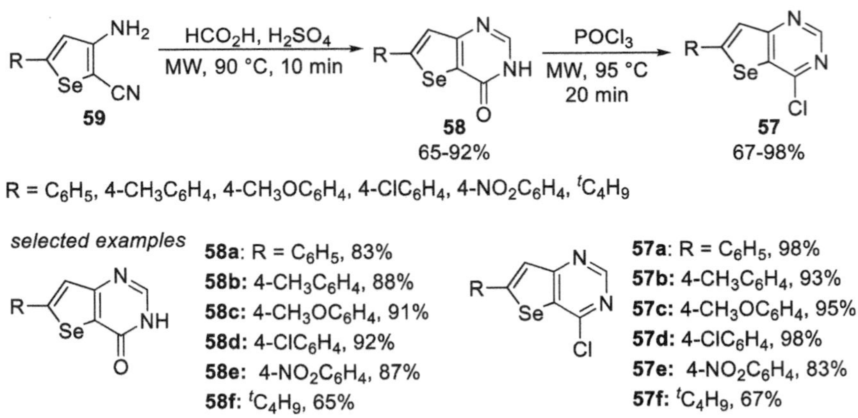

SCHEME 2.37 MW-assisted synthesis and functionalization of selenolopyrimidines.

cyclization using 3-aminoselenophene-2-carbonitriles (**59**) and formic acid in the presence of catalytic amount of sulfuric acid. After only 10 minutes of irradiation, the compounds **58** were obtained in good yields. Compounds **58** were converted into 4-chloroselenolo[3,2-*d*]pyrimidines (**57**) by reaction with phosphoryl chloride under MW irradiation for only 20 minutes at 95°C. The authors described that the use of MW irradiation allowed shorter reaction times and higher yields as compared with the classical heating, and it was compatible with aliphatic and aromatic groups bonded the selenophene core.

Sashida and coworkers[80] reported the practical one-pot preparation of (Z)-methylene-3-selenaquinolines (**60**) by the solvent- and catalyst-free reaction of 2-ethynylanilines (**61**) and isoselenocyanates (**62**; Scheme 2.38). The MW irradiation promoted acceleration of the intramolecular cyclization reaction of a selenol intermediate, which is generated in situ from the reaction between the isoselenocyanate (**62**) and 2-ethynylanilines (**61**), showing that the isoselenocyanates (**62**) can be a useful reagent into cyclization reaction. The authors synthesized many examples (**60**) in good yields under conventional heating (oil bath); however, when the reaction was carried out under MW irradiation, the compounds **60** were selectively obtained in less time and in some cases in a better yield.

The intermolecular cyclo-condensation of *N*-urethane protected amino acid-derived bromomethyl ketones (**63**) with selenourea (**64**) were proposed by Lalithamba and coworkers in 2010 (Scheme 2.39).[81] In this study, 2-amino-1,3-selenazole analogs of protected (Fmoc/Boc/Z) amino acids (**65**) were prepared under the influence of US operated at 35 kHz using acetone as reaction solvent. After 10 minutes of reaction, the solvent was evaporated under reduced pressure and isolated crude were purified by column chromatography. All examples synthetized **65** by this protocol were isolated in good yields (89%–90%). The current protocol showed that the reaction sequence was free from racemization through nuclear magnetic resonance analysis and compatible with different amino protecting group.

In 2013, Grozav et al. carried out MW-assisted Hantzsch type condensation reactions of selenosemicarbazones (**66**) or benzoylselenosemicarbazides (**67**) with carbonyl derivatives (**68**) to prepare aryl-hydrazinyl-1,3-selenazoles (**69**) and aryl-hydrazonyl-1,3-selenazoles (**70**)

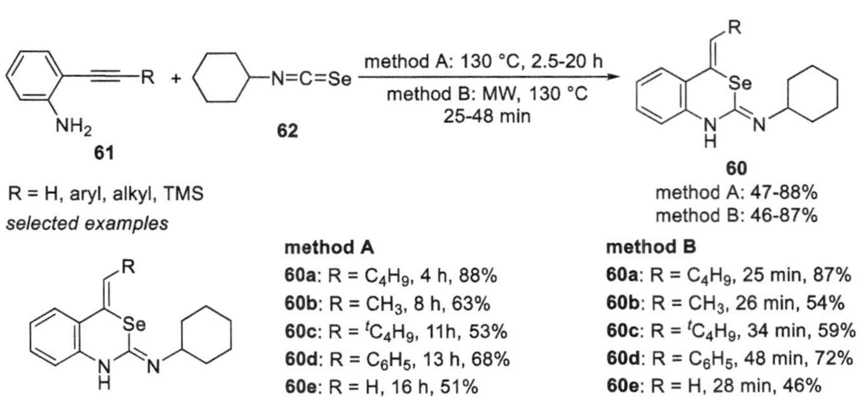

SCHEME 2.38 Synthesis of (Z)-4-methylene-3-selenaquinoline derivatives.

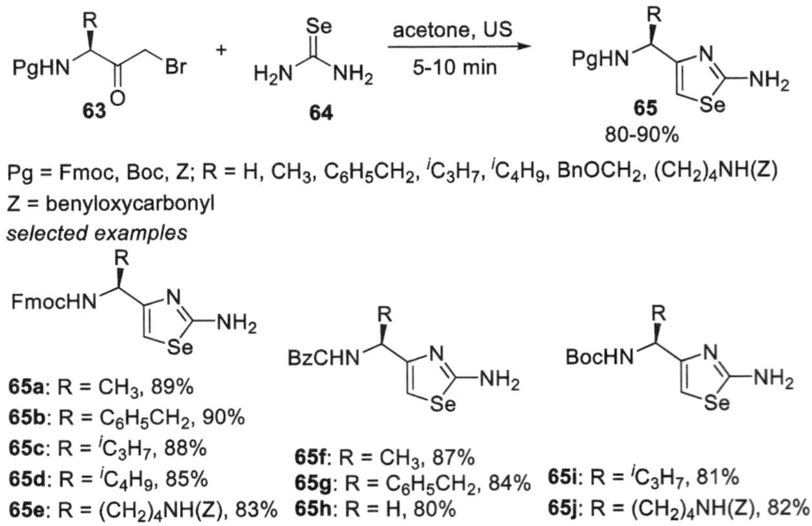

SCHEME 2.39 US-mediated synthesis of 2-amino-1,3-selenazoles.

(Scheme 2.40).[82] The authors presented the comparison under classical versus MW heating conditions. In both cases, eight compounds were formed; however the MW accelerated the reaction from 24 to 1 hour, and the yields were higher is this condition. Furthermore, the new 1,3-selenazole derivatives showed moderate antiproliferative in vitro effects against two leukemia cell lines (CCRF-CEM and HL60) and three carcinoma cell lines (MDA-MB231, HCT116 and U87MG).

The first synthetic pathway to prepare benzoselenazole (**71**) series under MW irradiation was described by Redon and coworkers[83] in 2014 (Scheme 2.41). Their proposed method involves the reaction between *N*-(acetyl)benzoyl-2-iodoanilines (**72**) and Woollins' reagent under basic conditions for 3 hours at 160°C. The 2-(alkyl)arylbenzoselenazoles (**71**) were synthesized in moderate-to-good yield, containing different alkyl, aryl, and heteroaryl groups, including functional groups as electron-withdrawing groups (chlorine and trifluoromethyl) and electron-donating groups (methoxy, furan, thiophene, and pyrrole) on the benzoyl moiety. In addition, the selenation/cyclization process was also applied to the substrate with bromine or methoxyl groups in *para* position of iodine. The corresponding products **71d** and **71e** were isolated in 44% and 21%, respectively.

In 2017, Balaguez and coworkers[84] reported the domino oxidative cyclization of methyl ketones (**68**) with bis(2-aminophenyl) diselenide (**2c**) to prepare 2-acyl-benzo[1,3-*d*]selenazoles (**73**) under MW irradiation (Scheme 2.42). Using this one-pot procedure, 12 2-acyl-benzo[1,3-*d*]selenazoles (**73**) were isolated in good-to-excellent yields, through in situ generation of 2-arylethane-1,2-diones (**74′**) from the nontoxic reducing agent $Na_2S_2O_5$ and

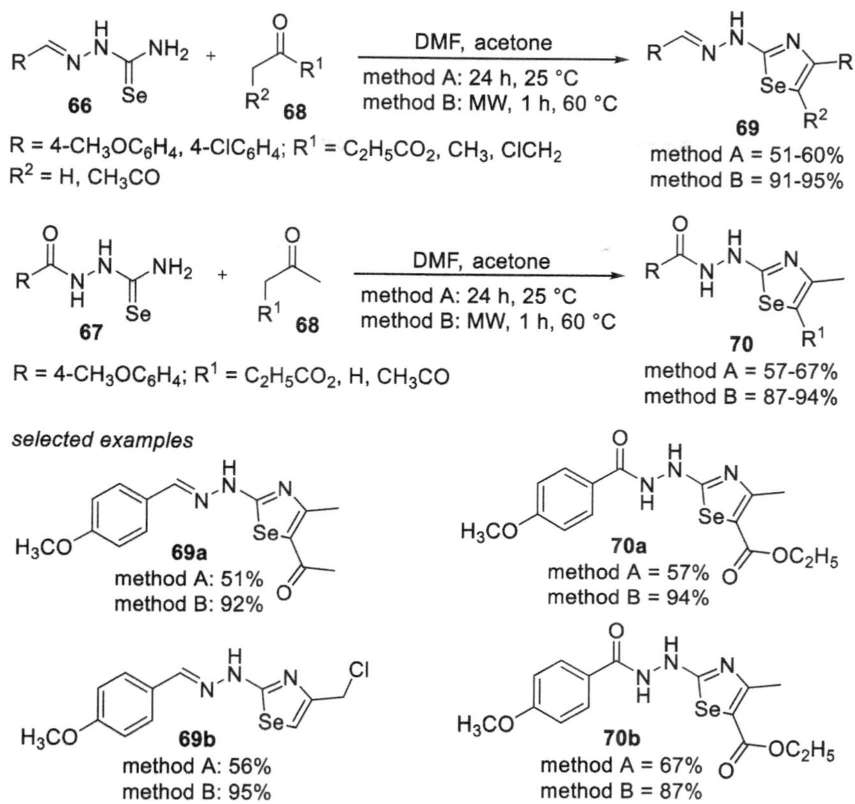

SCHEME 2.40 MW-assisted synthesis of new selenazole derivatives.

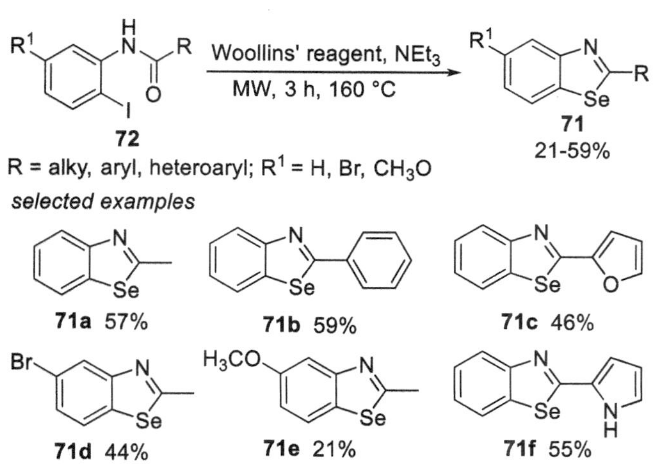

SCHEME 2.41 Preparation of 2-(alkyl)arylbenzoselenazoles from the corresponding *N*-(acetyl)benzoyl-2-iodoanilines.

commercially available arylmethyl ketones (**68**) using DMSO as solvent. For comparison, the authors described the use of MW irradiation and conventional heating method to promoted this intramolecular cyclo-condensation. In terms of time and energy economy, the MW irradiation shows more efficiency, giving the products **73** in good-to-excellent yields (60%–94%) after 2.3 hours, while in moderate-to-excellent yields (33%–94%) after 48 hours under conventional heating. Still, a possible mechanism was proposed to prepare 2-acyl-benzo[1,3-*d*]selenazoles (Scheme 2.43). The whole process requires three steps, in the first one occurs the formation of the ethane-1,2-dione derivative (**74′**) through Kornblum oxidation. Next, the amino group

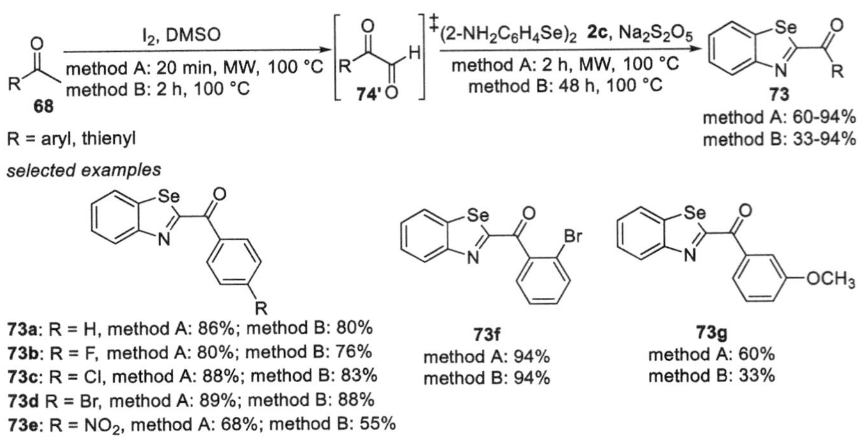

SCHEME 2.42 Synthesis of 2-acyl-benzo[1,3-*d*]selenazoles.

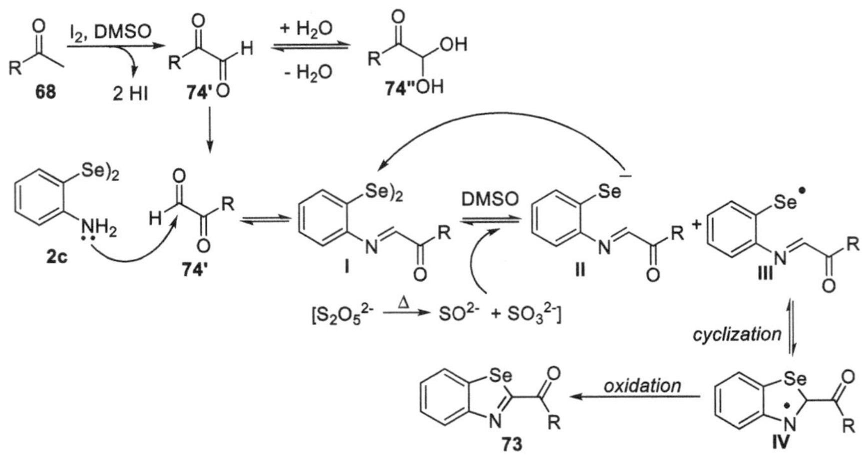

SCHEME 2.43 Domino oxidative cyclization mechanism of methyl ketones with bis(2-aminophenyl) diselenide.

of bis(2-aminophenyl) diselenide (**2c**) reacts with ethane-1,2-dione (**74′**), to form the imine diselenide intermediate **I**, followed by Se–Se bond cleavage leading to the intermediates **II** and **III**. The radical **III** undergoes intramolecular cyclo-condensation affording to the aminyl radical **III**. In the last step, the oxidation of intermediate **IV** provides the desired selenazoles (**73**).

2.5.2 Chalcogen-containing cyclic compounds

The use of US-assisting multicomponent reactions of aldehydes **75**, malononitrile **76**, and benzeneselenol **77** was developed to prepare highly functionalized selenopyridines **78** using PEG-400 as solvent (Scheme 2.44).[85] In this work, several 2-amino-4-aryl-6-(phenylselanyl) pyridine-3,5-dicarbonitriles (**78**) were synthetized using different aromatic aldehydes tethered groups in the *meta* and *para* position with either electron-withdrawing or electron-donating. After some hours under sonication, the corresponding 2-amino selenopyridines (**78**) were isolated pure enough for the characterization without any column chromatography in moderate-to-good yields (67%–87%). This procedure was efficiently extended to thiophene-2-carbaldehyde and phenylacetaldehyde, yielding the corresponding compounds **78g** and **78h** in 87% and 67%, respectively. However, this procedure shows as limitation the use of sterically hindered 2,2′-disubstituted arylaldehydes, which gave the final product **79** as nonaromatized 1,4-dihydroselenopyridines.

In this methodology, a total of four new bonds (two C–C, one C–N, and one C–Se) form in a simple and benign one-pot protocol, as proposed in Scheme 2.45. The plausible reaction mechanism starts with the Knoevenagel condensation step in the presence of PEG-400 to form

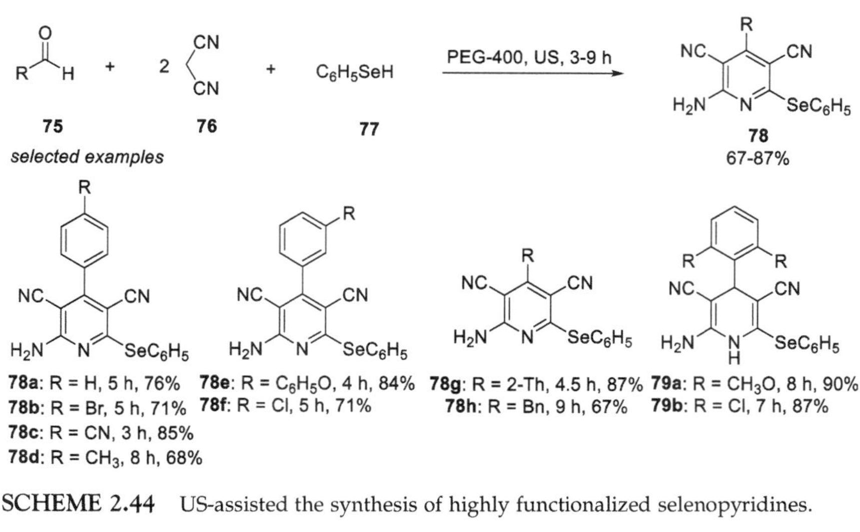

SCHEME 2.44 US-assisted the synthesis of highly functionalized selenopyridines.

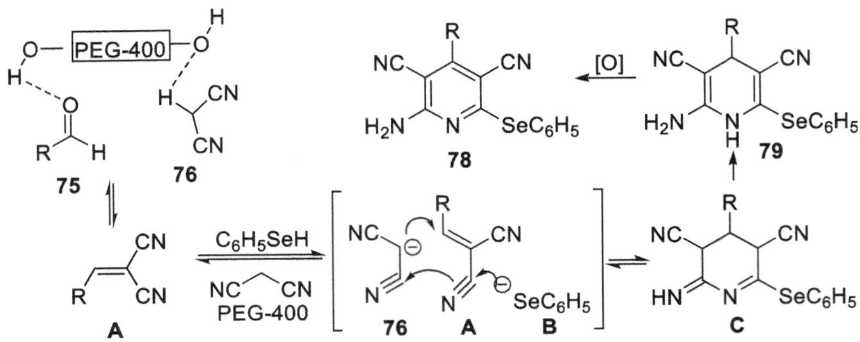

SCHEME 2.45 Mechanism for the synthesis of selenopyridines.

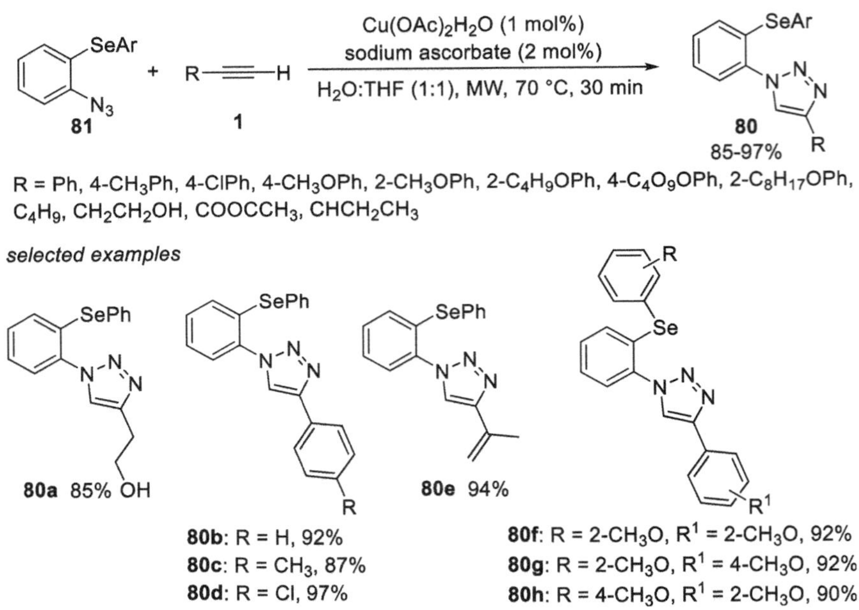

SCHEME 2.46 Synthesis of (arylselanyl)phenyl-1*H*-1,2,3-triazoles by copper-catalyzed 1,3-dipolar cycloaddition.

yilidine intermediate **A**. After that, the simultaneous nucleophilic attack of the benzeneselenol **B** and another equivalent of malononitrile **76** take place for an intramolecular cyclization and tautomerization to form **C**. This step leads to the formation of intermediate 1,4-dihydroselenopyridine (**79**), which undergoes an oxidation providing the selenopyridines (**78**).

In 2016, Alves and coworkers[86] described the use of MW irradiation as a nonclassic energy source to the preparation of (arylselanyl)phenyl-1*H*-1,2,3-triazoles (**80**) (Scheme 2.46). In this study, 16 products (**80**) were obtained in good-to-excellent yields (85%–97%) after short reaction time (30 minutes) through the copper-catalyzed azide–alkyne cycloaddition (CuAAC), in the presence of $Cu(OAc)_2.H_2O$ (1 mol%)/sodium ascorbate (2 mol%) system. A variety of terminal alkynes **1**, with different substituents, alkyl, aryl, vinyl, alcohols, and ester were used in the reaction with azidophenyl arylselenides (**81**). By this click chemistry protocol, new selenium compounds, which possess liquid–crystalline properties, were successfully prepared.

Later, the same research group[87] described the use of US to promote organocatalytic enamine-azide [3 + 2] cycloaddition reactions for the synthesis of [(arylselanyl)phenyl-1*H*-1,2,3-triazol-4-yl]ketones (**84**) (Scheme 2.47). In this work, eight novel selenium-containing 1,2,3-triazole compounds (**82**) were prepared in good-to-excellent yields by the reaction between aryl azidophenyl selenides (**81**) and 1,3-diketones (**83**), using 1.0 mol % Et_2NH as organocatalyst in DMSO as solvent under US irradiation (40% amplitude) at room temperature for 5 minutes. Furthermore, six selanyltriazoyl carboxylates, carboxamides, and carbonitriles **84** were synthetized in high yields using β-keto esters, β-keto amides, and α-cyano ketones **85**, as starting material instead of 1,3-diketones under the same reactional conditions (Scheme 2.48).

In 2018, Perin and coworkers[88] described the use of US irradiation to promote the cyclization reaction to obtain 2-organoselanyl-naphthalenes (**86**) (Scheme 2.49). Nine selenoethers **86** were prepared by the reaction between diorganyl diselenides **2** and alkynols **87** in the presence of Oxone as oxidant agent and water as the solvent. Specifically, Oxone

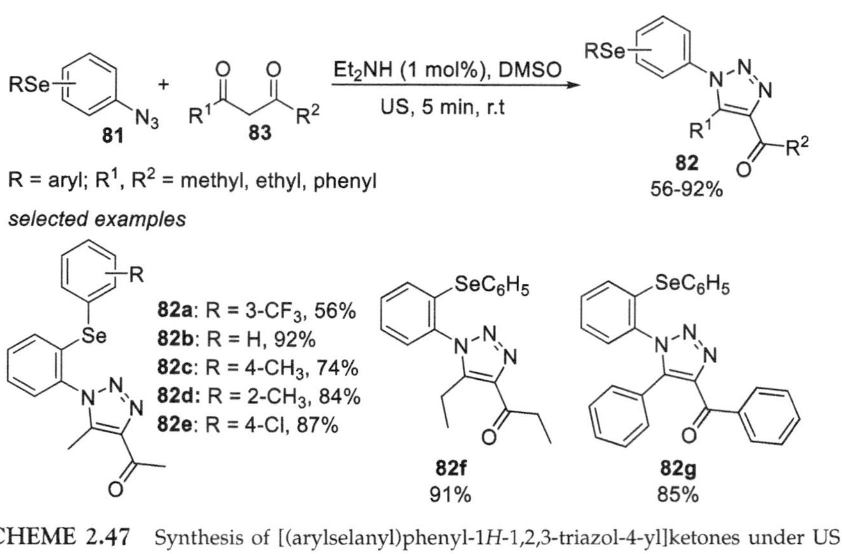

SCHEME 2.47 Synthesis of [(arylselanyl)phenyl-1*H*-1,2,3-triazol-4-yl]ketones under US conditions.

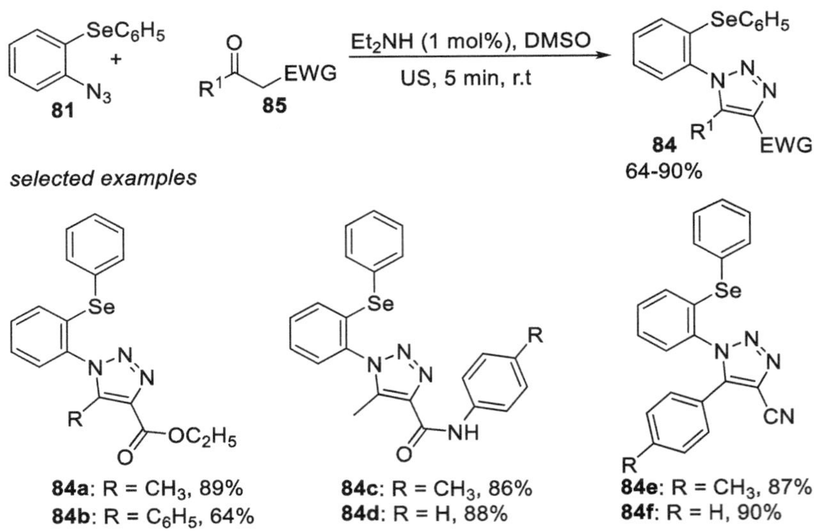

SCHEME 2.48 Synthesis of selanyltriazoyl carboxylates, carboxamides and carbonitriles under US conditions.

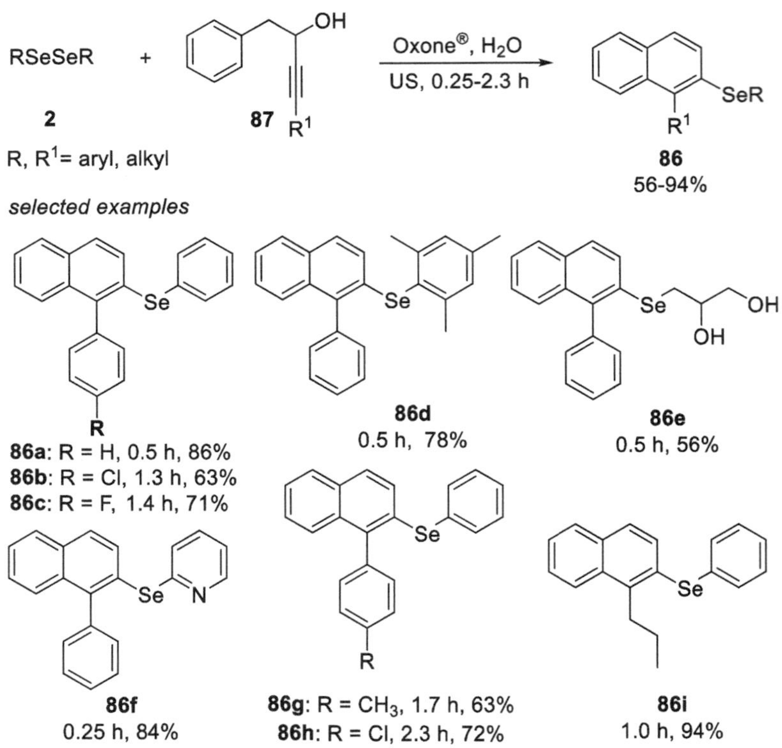

SCHEME 2.49 US-promoted synthesis of 2-organoselanyl-naphthalenes.

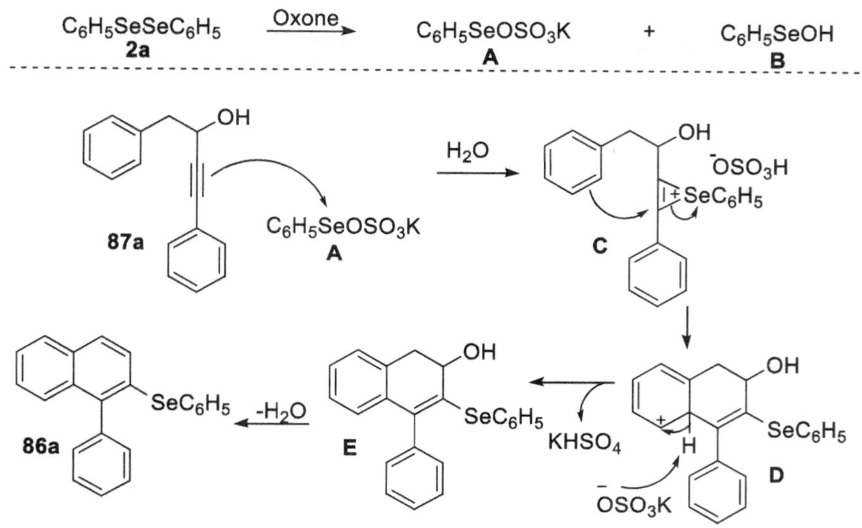

SCHEME 2.50 Mechanism for the reaction between alkynol, $(C_6H_5Se)_2$, Oxone in aqueous medium.

promoted the oxidative cleavage of the Se–Se bond of diselenides to form the electrophilic species of selenium in situ. After some studies, the best reaction condition was extended to different substrates, using electron-withdrawing and electron-donating groups attached to the aromatic ring of the alkynols and the diselenides. Additionally, aliphatic alkynols (**87**) and diselenides (**2**) also showed good starting material for this carbo-cyclization, yielding the respective 2-organoselanyl-naphthalenes (**86**) in moderate-to-excellent yields (56%–94%).

Scheme 2.50 describes a plausible mechanism for the reaction between alkynol (**87**), $(C_6H_5Se)_2$ (**2a**), and Oxone in aqueous medium. First, there occurs the oxidative cleavage of the Se–Se bond of diphenyl diselenide **2a** by Oxone, forming intermediates **A** and **B**. After that, the electrophilic selenium species **A** reacts with the carbon–carbon triple bond of the alkynol **87** to produce the seleniranium intermediate **C**. In the next step, an intramolecular 6-*endo*-dig cyclization gives the intermediate **D**, which undergoes deprotonation to restoring the aromaticity of the system, forming the dihydronaphthalene **E**. The water is eliminated in the last step to give the desired product **86a**.

2.6 Miscellaneous synthetic methodologies

In a study described by Martins and coworkers,[89] the authors related the synthesis and biological activity of 6-selenocaffeine (**88**; Scheme 2.51).

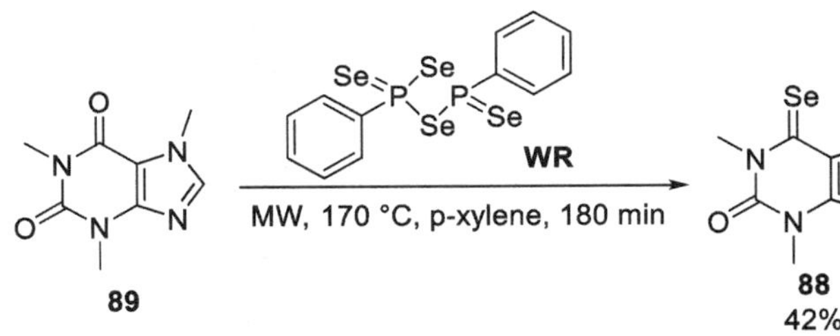

SCHEME 2.51 Synthesis of 6-selenocaffeine under MW radiation.

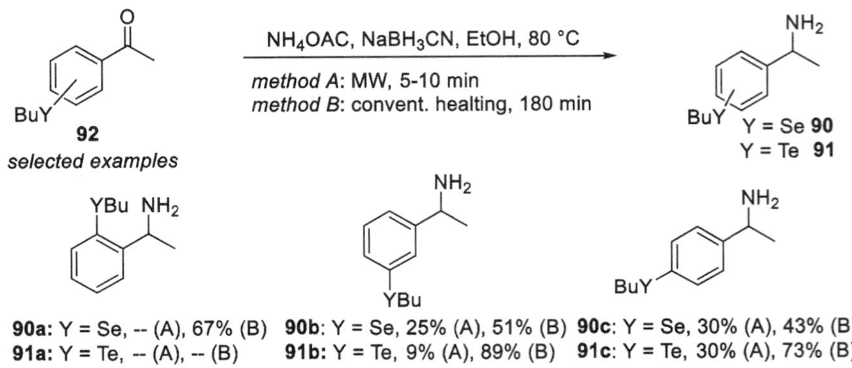

SCHEME 2.52 MW-assisted interconversion of ketones to primary amines.

The compound of interest was prepared by interconversion of one carbonyl group of the caffeine (**89**) into selenocarbonyl moieties using the Woollins' reagent (2,4-diphenyl-1,3-diselenadiphosphetane-2,4-diselenide, WR). The procedure consisted in the reaction between these both reagents solubilized in *p*-xylene under MW irradiation at 300 W (170°C) for 180 minutes in a sealed Pyrex MW vial. After, the solvent was evaporated under reduced pressure, the residue was purified by PTLC on silica, yielding selectively the compound **88** in 42% yield. Still, the authors reported a mild antioxidant activity for 6-selenocaffeine (**88**), contrasting with caffeine, along with low cytotoxicity potential toward normal mammary cells and potential applications in cancer therapy.

Considering synthetic and biological applications of organoselenium and organotellurium amines, Dalmolin and coworkers[90] described the MW-assisted reductive amination reaction by no need protective groups. The chalcogen amines **90** and **91** were efficiently and quickly prepared by reaction between primary, secondary, and tertiary amines **17** and arylchalcogen ketones **92**. For the interconversion of ketones to primary amines, the authors used arylchalcogen ketones, $AcONH_4$, and $NaBH_3CN$ and ethanol as a green solvent (Scheme 2.52). By comparation of MW irradiation versus conventional heating, almost all tests showed that the MW irradiation was faster and capable of preparing the products in good yields. Straightforward reductive amination of secondary and tertiary amines assisted by MW were explored in the same reactional condition (Scheme 2.53). Several chalcogen amines **90** and **91** were afforded in acceptable yields, and in these cases, the tertiary amines were the most sterically hindered than the secondary ones employed, hindering the progress of the reaction.

An environmentally benign and safe procedure using aqueous sodium hypochloride on solid supports under MW irradiation was proposed to promote the oxidation of sulfides **13**, selenides **12**, and tellurides **15** to the corresponding monoxides **93**, **94**, and **95** or the corresponding dioxides **96**, **97**, and **98** (Scheme 2.54).[91] When neutral alumina was used as a solid support, sulfoxides **93** were isolated after 5–7 minutes of MW irradiation (77%–91% yields), while selenoxides **94** and telluroxides **95** were obtained in 84%–89% yield after 7–10 minutes. By this solvent-free procedure, the authors prepared sulfones **96**, selenones **97**, and tellurones **98** in 76%–93% yield with the aid of neutral alumina. The authors synthesized 18 dioxides and 17 monoxides using a range of substituted aromatic and aliphatic sulfides **13**, selenides **12**, or tellurides **15** with very good results. For the efficiency of the oxidation of chacogenides **13**, **12**, or **15** to monoxides **93**, **94**, or **95**, the

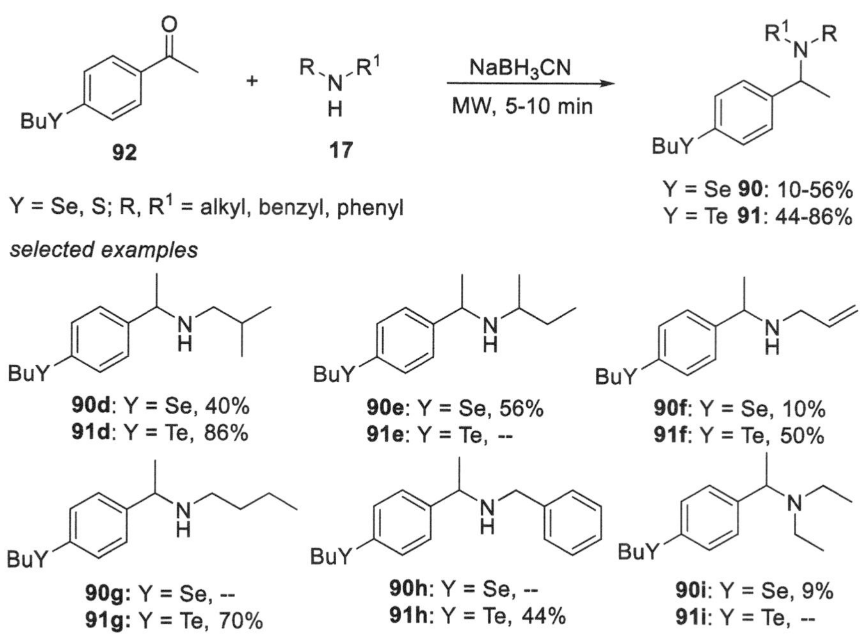

SCHEME 2.53 MW-assisted synthesis of organochalcogen amines by reductive amination.

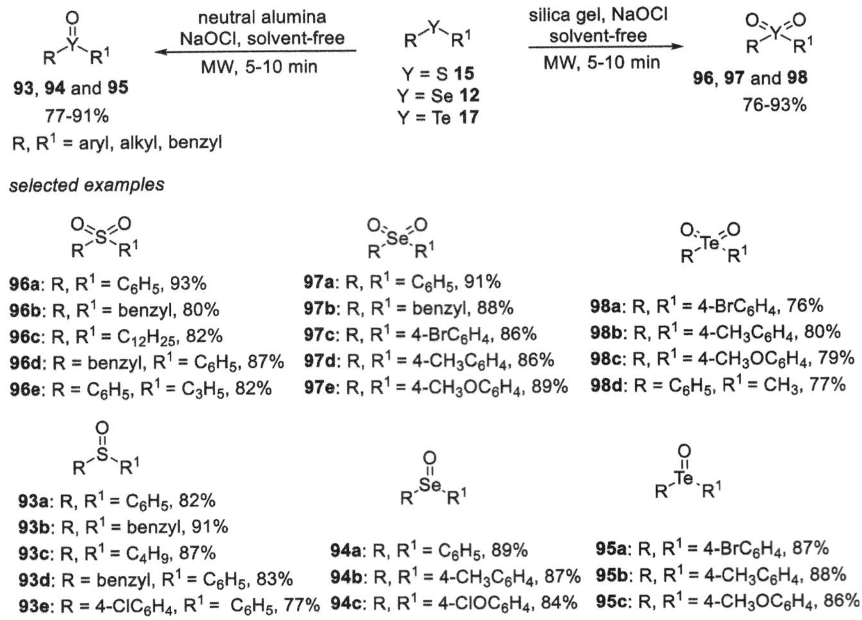

SCHEME 2.54 Chemoselective oxidation of chalcogenides to monoxides and dioxides.

addition of sodium hypochlorite was done in two portions with a break of 1 minutes for the addition of the reagent and cooling the reaction system.

2.7 Conclusions and perspectives

In conclusion, we have observed the advances and results of alternative energy sources on organoselenium chemistry, specially related to reactions containing diselenides as starting materials. US, MW, and mechanochemical mixing are effective partners in green and sustainable chemistry, because of their economical and safer alternatives in practical and industrial processes. The prominent advantages of these distinct energy sources include wide substrate scope, short reaction times, good yields, mild reaction conditions, and the capability of modulating the reactivity of some kind of reactions according to the desirable products. These outcomes provide new possibilities of pharmacological studies when combined with highly effective and sustainable reactions with structural versatility. Additionally, experimental details about selectivity and reactivity related to mechanochemical energy source are still absent, giving new opportunities to develop synthetic methodologies to prepare organoselenium compounds.

Acknowledgments

This study was financed in part by the Coordenação de Aperfeiçoamento de Pessoal de Nível Superior – Brasil (CAPES) – Finance Code 001. Fundação de Amparo à Pesquisa do Estado do Rio Grande do Sul (FAPERGS), Conselho Nacional de Desenvolvimento Científico e Tecnológico (CNPq) and Financiadora de Estudos e Projetos (FINEP) are acknowledged for financial support. CNPq and FINEP are acknowledged for financial support. CNPq is also acknowledged for Fellowships to G. P. and D. A.

Conflict of interest

There are no conflicts to declare.

Credit author statement

Filipe Penteado and Angelita Barcellos: Writing- Original draft preparation; D. Alves and G. Perin: Writing- Reviewing and Editing; M. S. Silva: Supervision. Writing-Reviewing and Editing.

References

1. Anastas, P. T.; Warner, J. C., Eds. *Green Chemistry: Theory and Practice;* Oxford University Press: Oxford, 1998.
2. Anastas, P. T.; Heine, L. G.; Williamson, T. C., Eds. *Green Chemical Syntheses and Processes;* American Chemical Society: Washington, DC, 2000.
3. Clark, J. H.; Macquarrie, D. J. *Handbook of Green Chemistry and Technology;* Black-well: Abingdon, 2002.
4. Matlack, A. S. *Introduction to Green Chemistry;* Marcel Dekker: New York, 2001.
5. Lancaster, M. *Green Chemistry: An Introductory Text;* Royal Society of Chemistry: Cambridge, 2002.
6. Eguaogie, O.; Vyle, J. S.; Conlon, P. F.; Gílea, M. A.; Liang, Y. Mechanochemistry of Nucleosides, Nucleotides and Related Materials. *Beilstein J. Org. Chem.* **2018,** *14,* 955–970.
7. Leonardi, M.; Estévez, V.; Willacampa, M.; Menéndez, J. C. The Hantzsch Pyrrole Synthesis: Non-conventional Variations and Applications of a Neglected Classical Reaction. *Synthesis* **2019,** *51,* 816–818.
8. Koike, T.; Akita, M. Visible-Light-Induced Photoredox Catalysis: An Easy Access to Green Radical Chemistry. *Synlett* **2013,** *24,* 2492–2505.
9. Kendall, J. L.; Canelas, D. A.; Young, J. L.; DeSimone, J. M. Polymerization in Supercritical Carbon Dioxide. *Chem. Rev.* **1999,** *99,* 543–564.
10. Gallezot, P. Process Options for Converting Renewable Feedstocks to Bioproducts. *Green. Chem.* **2007,** *9,* 295–302.
11. Sheldon, R. A. In Industrial Environmental Chemistry; Sawyer, D. T., Martell, A. E., Eds.; Plenum: New York, 1992.
12. Sheldon, R. A. Atom Efficiency and Catalysis in Organic Synthesis. *Pure Appl. Chem.* **2000,** *72,* 1233–1246.
13. Sheldon, R. A.; Arends, I.; Hanefeld, U. *Green Chemistry and Catalysis;* Wiley-VCH Verlag GmbH & Co: Weinheim, Germany, 2007.

14. Tundo, P.; Perosa, A.; Zecchini, F., Eds. *Methods and Reagents for Green Chemistry: An Introduction;* Wiley-Interscience, John Wiley & Sons: New Jersey, 2007.
15. Lenardão, E. J.; Santi, C.; Sancineto, L. New Frontiers in Organoselenium Compounds, *Chapter 2:* Bioactive Organoselenium Compounds and Therapeutic Perspectives; Springer: Gewerbestrasse, Switzerland, 2018.
16. Reilly, C. *Selenium in Food and Health;* Springer Science + Business Media, LLC: , New York, 2006.
17. Shao, S.; Zheng, B. The Biogeochemistry of Selenium in Sunan Grassland, Gansu Northwest China, Casts Doubt on the Belief that Marco Polo Reported Selenosis for the First Time in History. *Env. Geochem. Health* **2008,** *30,* 307–314.
18. Nogueira, C. W.; Zeni, G.; Rocha, J. B. T. Organoselenium and Organotellurium Compounds: Toxicology and Pharmacology. *Chem. Rev.* **2004,** *104,* 6255–6286.
19. Sancineto, L.; Mariotti, A.; Bagnoli, L.; Marini, F.; Desantis, J.; Iraci, N.; Santi, C.; Pannecouque, C.; Tabarrini, O. Design and Synthesis of DiselenoBisBenzamides (DISeBAs) as Nucleocapsid Protein 7 (NCp7) Inhibitors with Anti-HIV Activity. *J. Med. Chem.* **2015,** *58,* 9601–9614.
20. Sartori, G.; Jardim, N. S.; Sari, M. H. M.; Dobrachinski, F.; Pesarico, A. P.; Rodrigues, L. C.; Cargenlutti, J.; Flores, E. F.; Prigol, M.; Nogueira, C. W. Antiviral Action of Diphenyl Diselenide on Herpes Simplex Virus 2 Infection in Female BALB/c Mice. *J. Cell. Biochem.* **2016,** *117,* 1638–1648.
21. Wójtowicz, H.; Kloc, K.; Maliszewska, I.; Mlochowski, J.; Pietka-Ottlik, M.; Piasecki, E. Azaanalogues of Ebselen as Antimicrobial and Antiviral Agents: Synthesis and Properties. *Farmaco* **2004,** *59,* 863–868.
22. Harris, M. T.; Walker, D. M.; Drew, M. E.; Mitchell, W. G.; Dao, K.; Schroeder, C. E.; Flaherty, D. P.; Weiner, W. S.; Golden, J. E.; Morris, J. C. Interrogating a Hexokinase-Selected Small-Molecule Library for Inhibitors of *Plasmodium falciparum* Hexokinase. *Antimicrob. Agents Chemother.* **2013,** *57,* 3731–3737.
23. Alves, D.; Goldani, B.; Lenardão, E. J.; Perin, G.; Schumacker, R. F.; Paixão, M. W. Cooper Catalysis Showing the Way: Synthesis of Selenium-Containing Highly Functionalized 1,2,3-Triazoles. *Chem. Rec.* **2018,** *18,* 527–542.
24. Beletskaya, I. P.; Ananikov, V. P. Transition-Metal-Catalyzed C-S, C-Se, and C-Te Bond Formation via Cross-Coupling and Atom-Economic Addition Reactions. *Chem. Rev.* **2011,** *111,* 1596–1636.
25. Xavier, M. C. D. F.; Goldani, B.; Schumacher, R. F.; Perin, G.; Schneider, P. H.; Alves, D. Silver-catalyzed Direct Selenylation of Terminal Alkynes Trough C-H Bond Functionalization. *Mol. Catal.* **2017,** *427,* 73–79.
26. Recchi, A. M. S.; Back, D. F.; Zeni, G. Sequential Carbon-Carbon/Carbon-Selenium Bond Formation Mediated by Iron (III) Chloride and Diorganyl Diselenides: Synthesis and Reactivity of 2-Organoselenyl-Naphthalenes. *J. Org. Chem.* **2017,** *82,* 2713–2723.
27. Perin, G.; Lenardão, E. J.; Jacob, R. G.; Panatieri, R. B. Synthesis of Vinyl Selenides. *Chem. Rev.* **2009,** *109,* 1277–1301.
28. Santi, C., Ed. Organoselenium Chemistry: Between Synthesis and Biochemistry, Bentham Science: Sharjah, 2014. Available from: http://doi.org/10.2174/97816080583891140101.
29. Santi, C. Tidei in *Electrophilic Se/Te Reagents: Reactivity and their Contribuition to "Green Chemistry.* In *The Chemistry of Organic Selenium and Tellurium Compounds;* Rappoport, Z., Ed.; Vol. 4; John Wiley & Sons: Chichester, 2013; pp 569–655.
30. Perin, G.; Alves, D.; Jacob, R. G.; Barcellos, A. M.; Soares, L. K.; Lenardão, E. J. Synthesis of Organochalcogen Compounds using Non-Conventional Reaction Media. *ChemistrySelect* **2016,** *2,* 205–258.
31. Santi, C.; Jacob, R. G.; Monti, B.; Bagnoli, L.; Sancineto, L.; Lenardão, E. J. Water and Aqueous Mixtures as Convenient Alternative Media for Organoselenium Chemistry. *Molecules* **2016,** *21,* 1482–1498.

32. Bruckmann, A.; Krebs, A.; Bolm, C. Organocatalytic Reactions: Effects of Ball Milling, Microwave and Ultrasound Irradiation. *Green. Chem.* **2008,** *10,* 1131–1141.
33. Varma, R. S. "Greener" Chemical Syntheses using Mechanochemical Mixing or Microwave and Ultrasound Irradiation. *Green. Chem. Lett. Rev.* **2007,** *1,* 37–45.
34. Baig, R. B. N.; Varma, R. S. Alternative Energy Input: Mechanochemical, Microwave and Ultrasound-Assisted Organic Synthesis. *Chem. Soc. Rev.* **2012,** *41,* 1559–1584.
35. Lenardão, E. J.; Santi, C.; Sancineto, L. *New Frontiers in Organoselenium Compounds, Chapter 1. :* Organoselenium Compounds as Reagents and Catalysts to Develop New Green Protocols; Springer: Gewerbestrasse, Switzerland, 2018.
36. Penteado, F.; Monti, B.; Sancineto, L.; Perin, G.; Jacob, R. G.; Santi, C.; Lenardão, E. J. Ultrasound-Assisted Multicomponent Reactions, Organometallic and Organochalcogen Chemistry. *Asian J. Org. Chem.* **2018,** *7,* 2368–2385.
37. Santoro, S.; Azeredo, J. B.; Nascimento, V.; Sancineto, L.; Braga, A. L.; Santi, C. The Green Side of the Moom: Ecofriendly Aspects of Organoselenium Chemistry. *RSC Adv.* **2014,** *4,* 31521–31535.
38. Kobel, A.; Schäfer, C.; Török, B. Application of Microwave-Assisted Heterogeneous Catalysis in Sustainable Synthesis Design. *Green. Chem.* **2017,** *19,* 3729–3751.
39. Chattopadhyay, S. K. Recent Advancement in the Synthesis of 1,2- and 2,1-Benzothiazines. *Synt. Commun.* **2018,** *48,* 3033–3078.
40. Sharma, A.; Appukkuttan, P.; Eycken, V. D. Microwave-Assisted Synthesis of Medium-Sized Heterocycles. *Chem. Comm.* **2012,** *48,* 1623–1637.
41. Gising, J.; Odell, L. R.; Larhed, M. Microwave-Assisted Synthesis of Small Molecules Targeting the Infectious Diseases Tuberculosis, HIV/AIDS, Malaria And Hepatitis C. *Org. Biomol. Chem.* **2012,** *10,* 2713–2729.
42. Strukil, V. Mechanochemical Synthesis of Thioureas, Ureas and Guanidines. *Beilstein J. Org. Chem.* **2017,** *13,* 1828–1849.
43. Belenguer, A. M.; Friscié, T.; Day, G. M.; Sanders, J. K. M. Solid-State Dynamic Combinatorial Chemistry: Reversibility and Thermodynamic Product Selection in Covalent Mechanosynthesis. *Chem. Sci.* **2011,** *2,* 696–700.
44. Howard, J. L.; Brand, M. C.; Browne, D. L. Switching Chemoselectivity: Using Mechanochemistry to Alter Reaction Kinetics. *Angew. Chem. Int. (Ed.)* **2018,** *57,* 16104–16198.
45. Ruano, J. L. G.; Parra, A.; Alemán, J. Efficient Synthesis of Disulfides by Air Oxidation of Thiols under Sonication. *Green. Chem.* **2008,** *10,* 706–711.
46. Fritze, U. F.; Delius, M. V. Dynamic Disulfide Metathesis Induced by Ultrasound. *Chem. Commun.* **2016,** *52,* 6363–6366.
47. Menezes, P. H.; Zeni, G. *Vinyl Selenides in Patai's Chemistry of Functional Groups;* John Wiley & Sons: New York, 2011.
48. Kuligowski, C.; Bezzenine-Lafolle, S.; Chaume, G.; Mahuteau, J.; Barrire, J.-C.; Bacqu, E.; Pancrazi, A.; Ardisson, J. Approach Toward the Total Synthesis of Griseoviridin: Formation of Thioethynyl and Thiovinyl Ether-Containing Nine-Membered Lactones through a Thioalkynylation – Macrolactonization – Hydrostannylation Sequence. *J. Org. Chem.* **2002,** *67,* 4565–4568.
49. Marcantoni, E.; Massaccesi, M.; Petrini, M. A Novel Route to the Vinyl Sulfide Nine-Membered Macrocycle Moiety of Griseoviridin. *J. Org. Chem.* **2000,** *65,* 4553–4559.
50. Lam, H. W.; Cooke, P. A.; Pattenden, G.; Bandaranayake, W. M.; Wickramasinghe, W. A. Structure and Total Synthesis of Benzylthiocrellidone, a Novel Dimedone-Based Vinyl Sulfide from the Sponge *Crella spinulata. J. Chem. Soc., Perkin Trans.* **1999,** *1,* 847–848.
51. Lenardão, E. J.; Silva, M. S.; Mendes, S. R.; Azambuja, F.; Jacob, R. G.; Santos, P. C. S.; Perin, G. Synthesis of β-phenylchalcogeno-α,β-Unsaturated Esters, Ketones and Nitriles Using Microwave and Solvent-Free Conditions. *J. Braz. Chem. Soc.* **2007,** *18,* 943–950.

52. Lara, R. G.; Soares, L. K.; Jacob, R. G.; Schumacher, R. F.; Perin, G. Selective Synthesis of (Z)-Chalcogenoenynes and (Z,Z)-1,4-bis-Chalcogenbuta-1,3-Dienes Using PEG-400. *J. Braz. Chem. Soc.* **2016,** *27*, 2046–2054.
53. Vieira, A. A.; Azeredo, J. B.; Godoi, M.; Santi, C.; Júnior, E. N. S.; Braga, A. L. Catalytic Chalcogenylation Under Greener Conditions: A Solvent-Free Sulfur- and Seleno-Functionalization of Olefins via I_2/DMSO Oxidant System. *J. Org. Chem.* **2015,** *80*, 2120–2127.
54. Saba, S.; Rafique, J.; Braga, A. L. DMSO/Iodine-Catalyzed Oxidative C–Se/C–S Bond Formation: A Regioselective Synthesis of Unsymmetrical Chalcogenides with Nitrogen- or Oxygen-Containing Arenes. *Catal. Sci. Technol.* **2016,** *6*, 3087–3098.
55. Kundu, D.; Ahammed, S.; Ranu, B. C. Microwave-Assisted Reaction of Aryl Diazonium Fluoroborate and Diaryl Dichalcogenides in Dimethyl Carbonate: A General Procedure for the Synthesis of Unsymmetrical Diaryl Chalcogenides. *Green. Chem.* **2012,** *14*, 2024–2030.
56. Mukherjee, N.; Chatterjee, T.; Ranu, B. C. Reaction Under Ball-Milling: Solvent-, Ligand-, and Metal-Free Synthesis of Unsymmetrical Diaryl Chalcogenides. *J. Org. Chem.* **2013,** *78*, 11110–11114.
57. Freitas, C. S.; Barcellos, A. M.; Ricordi, V. G.; Pena, J. M.; Perin, G.; Jacob, R. G.; Lenardão, E. J.; Alves, D. Synthesis of Diaryl Selenides Using Electrophilic Selenium Species and Nucleophilic Boron Reagents in Ionic Liquids. *Green. Chem.* **2011,** *13*, 2931–2938.
58. Saba, S.; Botteselle, G. V.; Godoi, M.; Frizon, T. E. A.; Galleto, F. Z.; Rafique, J.; Braga, A. L. Copper-catalyzed Synthesis of Unsymmetrical Diorganyl Chalcogenides (Te/Se/S) from Boronic Acids under Solvent-Free Conditions. *Molecules* **2017,** *22*, 1367.
59. Botteselle, G. V.; Godoi, M.; Galetto, F. Z.; Bettanin, L.; Singh, D.; Rodrigues, O. E. D.; Braga, A. L. Microwave-Assisted One-Pot Synthesis of Symmetrical Diselenides, Ditellurides and Disulfides from Organoyl Iodides and Elemental Chalcogen Catalyzed by CuO Nanoparticles. *J. Mol. Catal. A Chem.* **2012,** *365*, 186–193.
60. Zhang, S.; Karra, K.; Heintz, C.; Kleckler, E.; Jin, J. Microwave-Assisted Cu_2O-Catalyzed One-Pot Synthesis of Symmetrical Diaryl Selenides from Elemental Selenium. *Tetrahedron Lett.* **2013,** *54*, 4753–4755.
61. Kumar, S.; Engman, L. Microwave-Assisted Copper-Catalyzed Preparation of Diaryl Chalcogenides. *J. Org. Chem.* **2006,** *71*, 5400–5403.
62. Taniguchi, N.; Onami, T. Magnesium-Induced Copper-Catalyzed Synthesis of Unsymmetrical Diaryl Chalcogenide Compounds from Aryl Iodide via Cleavage of the Se – Se or S – S Bond. *J. Org. Chem.* **2004,** *69*, 915–920.
63. Vieira, B. M.; Thurow, S.; Brito, J. S.; Perin, G.; Alves, D.; Jacob, R. G.; Santi, C.; Lenardão, E. J. Sonochemistry: An Efficient Alternative to the Synthesis of 3-Selanylindoles using CuI as Catalyst. *Ultrason. Sonochem.* **2015,** *27*, 192–199.
64. Vieira, B. M.; Thurow, S.; Costa, M.; Casaril, A. M.; Domingues, M.; Schumacher, R. F.; Perin, G.; Alves, D.; Savegnago, L.; Lenardão, E. J. Ultrasound-Assisted Synthesis and Antioxidant Activity of 3-Selanyl-*1H*-Indole and 3-Selanylimidazo[1,2-a] Pyridine Derivatives. *Asian J. Org. Chem.* **2017,** *6*, 1635–1646.
65. Azeredo, J. B.; Godoi, M.; Martins, G. M.; Silveira, C. C.; Braga, A. L. A Solvent- and Metal-Free Synthesis of 3-Chacogenyl-Indoles Employing DMSO/I_2 as an Eco-Friendly Catalytic Oxidation System. *J. Org. Chem.* **2014,** *79*, 4125–4130.
66. Wen, Z.; Xu, J.; Wang, Z.; Qi, H.; Xu, Q.; Bai, Z.; Zhang, Q.; Bao, K.; Wu, Y.; Zhang, W. *Eur. J. Med. Chem.* **2015,** *90*, 184–194.
67. Fonseca, S. F.; Padilha, N. B.; Thurow, S.; Roehrs, J. A.; Savegnago, L.; Souza, M. N.; Fronza, M. G.; Collares, T.; Buss, J.; Seixas, F. K. Ultrasound-Promoted Copper-Catalyzed Synthesis of Bis-Arylselanyl Chrysin Derivatives with Boosted Antioxidant and Anticancer Activities. *Ultrason. Sonochem.* **2017,** *39*, 827–836.

68. Godoi, M.; Ricardo, E. W.; Botteselle, G. V.; Galetto, F. Z.; Azeredo, J. B.; Braga, A. L. Synthesis of Selenol Esters From Diorganyl Diselenides and Acyl Chlorides under Solvent-Free Conditions and Microwave Irradiation. *Green. Chem.* **2012,** *14,* 456–460.
69. Wang, J.-X.; Bai, L.; Liu, Z. Synthesis of Sodium Diselenides and its Reaction with Benzoyl Chloride Under Phase Transfer Catalysis and Microwave Irradiation Conditions. *Synth. Commun.* **2000,** *30,* 971–977.
70. Wang, J.-X.; Bai, L.; Li, W.; Hu, Y. A Convenient One-Pot Synthesis of Dibenzyl Diselenides under Microwave Irradiation Conditions. *Synth. Commun.* **2000,** *30,* 325–332.
71. Ley, S. V.; O'Neil, I. A.; Low, C. M. R. Ultrasonic Formation and Reactions of Sodium Phenylselenide. *Tetrahedron* **1986,** *42,* 5363–5368.
72. Wang, J. X.; Xi, Y.; Wu, X.; Hu, Y.; Du, Z. The Synthesis of 1,3-Bis(Phenylseleno)-2-Propanol under Microwave Irradiation Conditions. *Synth. Commun.* **1998,** *28,* 4619–4627.
73. Wang, J. X.; Wu, X.; Huang, D.; Hu, Y. The Synthesis of 1-Arylseleno-3-Alkoxy-2-Propanols under Microwave Irradiation Conditions. *Synth. Commun.* **2002,** *32,* 1903–1908.
74. Cheng, T.; Zheng, X.; Ke, Q. Ultrasound Assisted Ring-Opening Reaction of Epoxides with 1,2-Diphenyldiselenide. *J. Chem. Res.* **2011,** *35,* 522–524.
75. Lv, G. S.; Duan, F.-J.; Ding, J.-C.; Cheng, T.-X.; Gao, W.-X.; Chen, J.-X.; Wu, H.-Y. Tandem Base-Free Synthesis of β-Hydroxy Sulphides under Ultrasound Irradiation. *J. Chem. Sci.* **2012,** *124,* 1057–1062.
76. Šibor, J.; Pazdera, P. Syntheses of Some New Five-Membered Heterocycles Containing Selenium and Tellurium. *Molecules* **1996,** *1,* 157–162.
77. Wang, J. X.; Zhao, K. Synthesis of Cis and Trans-2,6-Diphentyl-1,4-Diselena-Fulvenes from Phenyiacetylene with Selenium and Base under PTC-Ultrasound Conditions. *Synth. Commun.* **1996,** *26,* 1617–1622.
78. Raghavendra, M.; Bhojya-Naik, H. S.; Sherigara, B. S. A Facile One-Pot Microwave-Induced Synthesis of Some Novel Selenolo[2,3-b]Quinoline Derivatives under Solvent-Free Conditions, *Phosphorus. Sulfur. Silicon Relat. Elem.* **2008,** *183,* 1501–1509.
79. Hesse, S.; Chenet, C.; Thomae, D.; Kirsch, G. Microwave-Assisted Synthesis and Functionalization of Selenolopyrimidines. *Synthesis* **2009,** 1204–1208.
80. Sashida, H.; Pan, C.; Kaname, M.; Minoura, M. A Facile and Practical Solvent-Free One-Pot Synthesis of (Z)-4-Methylene-3-Selenaquinoline Derivatives from O-Ethynylanilines and Isoselenocyanates. *Synthesis* **2010,** 3091–3096.
81. Lalithamba, H. S.; Narendra, N.; Naik, S. A.; Sureshbabu, V. V. Ultrasound Mediated Synthesis of 2-Amino-1,3-Selenazoles Derived from Fmoc/Boc/Z-α-Amino Acids. *Arkivoc* **2010,** 77–90.
82. Grozav, A. I.; Gaina, L.; Kuete, V.; Silaghi-Dumitrescu, L.; Efferth, T.; Zaharia, V. Microwave-Assisted Synthesis of new Selenazole Derivatives with Antiproliferative Activity. *Molecules* **2013,** *18,* 4679–4688.
83. Redon, S.; Kabri, Y.; Crozet, M. D.; Vanelle, P. One-Pot Preparation of 2-(Alkyl) Arylbenzoselenazoles from the Corresponding N-(acetyl)Benzoyl-2-Iodoanilines via a Microwave-Assisted Methodology. *Tetrahedron Lett.* **2014,** *55,* 5052–5054.
84. Balaguez, R. A.; Betin, E. S.; Barcellos, T.; Lenardão, E. J.; Alves, D.; Schumacher, R. F. Synthesis of 2-Acyl-Benzo[1,3-d]Selenazoles via Domino Oxidative Cyclization of Methyl Ketones with bis(2-Aminophenyl) Diselenide. *N. J. Chem.* **2017,** *41,* 1483–1487.
85. Khan, M. N.; Karamthulla, S.; Choudhury, L. H.; Haque Faizi, M. S. Ultrasound Assisted Multicomponent Reactions: A Green Method for the Synthesis of Highly Functionalized Selenopyridines Using Reusable Polyethylene Glycol as Reaction Medium. *RSC Adv.* **2015,** *5,* 22168–22172.
86. Xavier, M.; Xavier, D.; Seus, N.; Lenardão, E.; Perin, G.; Alves, D. Microwave Assisted Rapid Synthesis of (Arylselanyl)phenyl-1H-1,2,3-triazoles by Copper Catalyzed 1,3-Dipolar Cycloaddition. *Curr. Microw. Chem.* **2015,** *3,* 14–23.

87. Costa, G. P.; Seus, N.; Roehrs, J. A.; Jacob, R. G.; Schumacher, R. F.; Barcellos, T.; Luque, R.; Alves, D. Ultrasound-Promoted Organocatalytic Enamine-Azide [3 + 2] Cycloaddition Reactions for the Synthesis of ((Arylselanyl)Phenyl-1H-1,2,3-Triazol-4-yl)Ketones. *Beilstein J. Org. Chem.* **2017,** *13,* 694–702.
88. Perin, G.; Araujo, D. R.; Nobre, P. C.; Lenardão, E. J.; Jacob, R. G.; Silva, M. S.; Roehrs, J. A. Ultrasound-Promoted Synthesis of 2-Organoselanyl-Naphthalenes Using Oxone ® in Aqueous Medium as an Oxidizing Agent. *PeerJ* **2018,** *6,* e4706.
89. Martins, I.; Miranda, J.; Oliveira, N.; Fernandes, A.; Gonçalves, S.; Antunes, A. Synthesis and Biological Activity of 6-Selenocaffeine: Potential Modulator of Chemotherapeutic Drugs in Breast Cancer Cells. *Molecules* **2013,** *18,* 5251–5264.
90. Dalmolin, M. C.; Bandeira, P. T.; Ferri, M. S.; de Oliveira, A. R. M.; Piovan, L. Straightforward Microwave-Assisted Synthesis of Organochalcogen Amines by Reductive Amination. *J. Organomet. Chem.* **2018,** *874,* 32–39.
91. Khurana, J.; Nand, B. Aqueous Sodium Hypochlorite Mediated Chemoselective Oxidation of Chalcogenides to Monoxides and Dioxides by Microwave Exposure. *Can. J. Chem.* **2010,** *88,* 906–909.

CHAPTER

3

Flow chemistry in the synthesis of organochalcogen compounds

Ornelio Rosati and Claudio Santi

Group of Catalysys, Synthesis and Organic Green Chemistry, Department of Phamaceutical Sciences, University of Perugia, Perugia, Italy

3.1 Introduction

In honor of IUPAC's 100th anniversary, in 2018, the Chemistry International Editorial Board initiated a special activity to initiate the search for the Top 10 Emerging Technologies in Chemistry.[1]

The ultimate intent was to show the value of chemistry in its broadest sense (including material science, nanotechnology, and biochemistry) and to inform the community about how chemical sciences can contribute to the well-being of society and the sustainability of the planet Earth.[2]

An international group of experts reviewed and discussed the pool of nominations and ultimately selected the final top emerging technologies in chemistry. These are (in alphabetical order): 3D-bioprinting, directed evolution of selective enzymes, enantio-selective organocatalysis, **flow chemistry**, metal organic frameworks for water harvesting, nanopesticides, plastics to monomers, reactive extrusion, reversible-deactivation of radical polymerization, and solid-state batteries.[3]

Interestingly, flow chemistry, one of the emerging technologies in chemistry, has attracted massive attention from chemists. This technology finds several synthetic applications both in academia and industrial production. In particular, many traditional synthetic processes can be considerably improved in terms of sustainability.[4] One of the sectors mainly affected by this technology is the pharmaceutical production, where the use of traditional reagents is still necessary to obtain specific final compounds.[5]

Organochalcogen Compounds
DOI: https://doi.org/10.1016/B978-0-12-819449-2.00007-0

Wirth asserted that flow chemistry and the use of microreactors, an efficient and extremely powerful tool, could be further integrated in medicinal chemistry and be applied in the routine production of pharmaceuticals in the future.[6]

3.2 Flow chemistry: a general introduction

The term "flow chemistry" refers to a chemical process that is performed in a special pipe system in a continuous flow.[7] This term has only been recently coined for its application on a laboratory scale.[8] Contrary to batch production, in a flow process, the fluids (generally reactants, catalysts, or their solutions) move through one or more of the pipe system by means of a pressure generated by a pump. The pipe system joins one another at a specific point where the fluids are mixed and the reaction takes place under precise reaction conditions. In such a system, the reaction stoichiometry is determined by the concentration of reagents and the ratio of corresponding flow rate, while the duration of the reaction is determined by volumetric residence time that is set by the ratio of volume of the reactor and the overall flow rate.

Thus, the residence time represents a key parameter of continuous reaction technology.[9] The possibility of fine tuning of this parameter is of a fundamental importance in determining the reaction rate, higher conversion, and product selectivity of a chemical reaction. It is important to remark that such a fine reaction control cannot be performed in conventional batch reaction setups, where instead, the reaction stoichiometry is determined by the concentration and ratio of reactants, and the residence time is determined by how long a reactor is held at a given temperature.

Generally, a flow process can offer many benefits by providing easier scale up, better and precise control of reaction conditions, better mixing, easier handling of unstable intermediates, and in-system purification. Another important point is the safety of chemical processes, which can be improved by flow systems.[4,10] Moreover, flow systems can be further improved by immobilization of reagents or catalysts, or by using fixed-bed reactors in parallel process.[7]

Obviously, running a chemical reaction under flow conditions can offer both a variety of advantages and disadvantages (Table 3.1).

A schematic example of a flow device system is shown in Fig. 3.1.

3.3 Flow chemistry and chalcogens

Since great progress has been made in terms of flexibility and robustness for large-scale processes,[16] as well as in bioorganic catalysis,[17]

TABLE 3.1 Advantages and disadvantages of flow processes.

Advantages	
Higher yields and/or selectivities	
Precise control of reaction temperature	The small volume of the laboratory device allows to increase the reaction temperature above the boiling point of the solvent.
Small-scale mixing	Mixing can be achieved within seconds at the smaller scales used in flow chemistry.[11]
Heat-transfer intensification	Endothermic and exothermic reactions can be thermostated easily and continuously due to the large area to volume ratio. The temperature gradient can be steep, allowing efficient control over reaction time.
Safety increase[12]	Thermal mass of the system is dominated by the apparatus making thermal runaways unlikely. Smaller reaction volume is also considered a safety benefit.[4] The reactor operates under steady-state conditions.
Automated processes	Flow reactions can be automated with far less effort than batch reactions.[13] An automated system can be realized by the coupling of a detector system with the output of the reactor. Thus, several reaction parameters, such as stoichiometry, residence time, and temperature, can be sequentially investigated.
Multi-step reactions continuous sequence	Particularly important for unstable, toxic, or sensitive to air intermediate compounds, since they will exist only momentarily and in very small quantities
Adaptability of the flow apparatus	The position along the flowing stream and reaction time point are directly related to one another. This means that it is possible to arrange the system such that further reagents can be introduced into the flowing reaction stream at a precise time point that is desired.
Possible coupling with purification devices	Solid phase scavenging Chromatographic separation Liquid/liquid extraction
Handling of reagents containing dissolved gases	
Multiphase liquid reactions	
Possible scale-up of validated reaction	It can be achieved rapidly with little or no process development work[14] by either changing the reactor volume or by running several reactors in parallel, provided that flows are recalculated to achieve the same residence times.

(*Continued*)

TABLE 3.1 (Continued)

Advantages	
Disadvantages[15]	
Dedicated equipment	They are needed for precise continuous dosing (e.g., pumps), connections, etc.
Start-up and shut-down procedures	They have to be established.
Scale-up of micro effects	For example, the high area to volume ratio is not possible and economy of scale may not apply. Typically, a scale-up leads to a dedicated plant.
Safety issues	The problem of the storage of reactive materials remains.

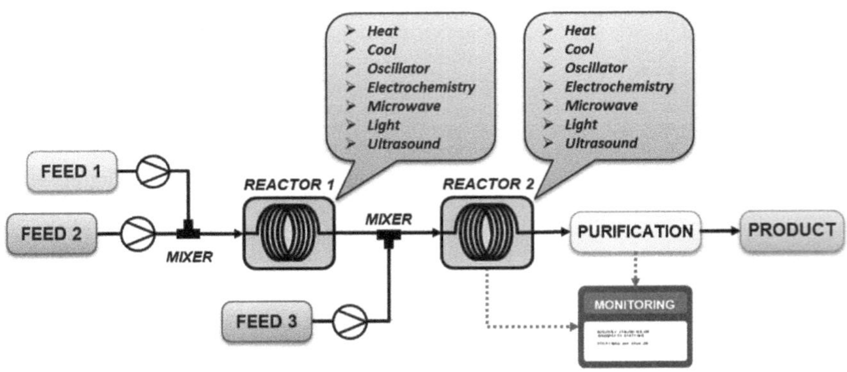

FIGURE 3.1 Schematic example of a flow device system.

photochemistry,[18] and electrochemistry,[19] combined with on-line analysis,[20] the discussion in this chapter is a sampling of developments regarding the features and typical benefits of continuous flow chemistry technology applied to the synthesis of organochalcogen derivatives.

Sulfoxides and sulfones are chemical compounds characterized by the presence of a sulfur as a sulfinyl (SO) or a sulfonyl (SO_2) functional group, respectively, attached to two carbon atoms (Fig. 3.2).[21] Sulfoxide bears a lone pair of electrons on the sulfur atom, and thus, tetrahedral electron-pair geometry and trigonal pyramidal shape exists. Accordingly, when two different organic substituents are bond to the sulfinyl group, sulfur becomes a chiral center (Fig. 3.2). For example, (R)-(+)-methylphenylsulfoxide has $[\alpha]_D^{20} = +147$ degrees (acetone).[22] Chiral sulfoxides are generally optically stable at room temperature with a slow racemization rate. The extent of the enthalpy of activation for racemization is around 35–42 kcal/mol.[23] Usually, sulfones are colorless and odorless in the pure state and are highly stable both thermally and chemically.

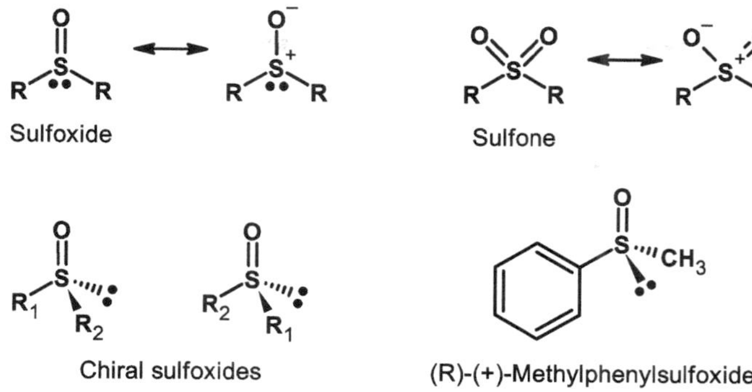

FIGURE 3.2 General representation of sulfoxides and sulfones.

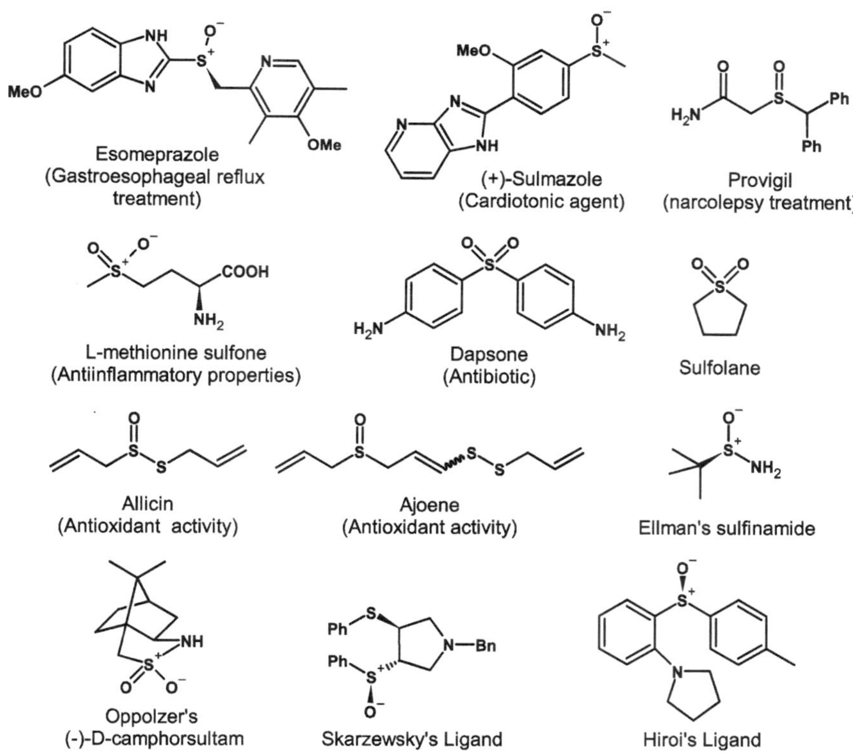

FIGURE 3.3 Examples of interesting sulfoxides and sulfones, ranging from pharmaceuticals to chiral ligands.

Sulfoxides and sulfones are important chemical compounds that can be found in a wide range of functional organic molecules.[24] Fig. 3.3 shows several examples of these kind of compounds such as pharmaceutical compounds (e.g., Esomeprazole, Provigil, Sulmazole, Methionine sulfone, and Dapsone),[25] biologically active components of garlic extracts (e.g., Allicin and Ajoene),[26] solvents (e.g., dimethylsulfoxide and sulfolane: an industrial solvent used in the extraction of aromatic hydrocarbons from hydrocarbon mixtures and in purifying natural gas) and chiral sulfoxides employed as chiral auxiliaries (e.g., Ellman's sulfinamide, Oppoltzer camphorsultam)[27] and chiral ligands in asymmetric transition metal–catalyzed transformations (e.g., Skarzewzki's and Hiroi's ligands).[28]

However, given the importance of sulfoxides and sulfones, chemists have paid particular attention to the synthetic aspect of these kinds of derivatives. Examples on the preparation of sulfoxide derivatives include the synthesis of allyl sulfoxides from allylsilanes via silyl sulfonates,[29] the bora-ene reaction of sulfur dioxide and prop-2-ene-1-boronic esters,[30] the one-pot synthesis exploiting a sulfinyl-dication equivalent generated from a DABSO/trimethylsilyl chloride sequence,[31] the preparation of both, aryl sulfinyl chlorides and diaryl sulfoxides from arenes through reaction with thionyl chloride in the presence of Lewis acid catalysts such as $BiCl_3$, $Bi(OTf)_3$, $LiClO_4$, or $NaClO_4$.[32]

A convenient synthesis of sulfones can be achieved starting from sulfur dioxide in cycloaddition reactions with dienes,[33] from sulfonyl and sulfuryl halides (or sulfonic acid anhydrides) under Friedel–Crafts reaction conditions to generate the electrophile RSO_2^+ in the presence of Lewis acid catalysts such as $AlCl_3$ and $FeCl_3$ or bismuth(III) trifluoromethanesulfonate[34] and through nucleophilic displacement of halides by sulfonates.[35] The synthesis of sulfoxides and sulfones can also be performed by oxidation processes starting from the corresponding thioethers using a wide variety of oxidants.[21,36]

The more common and green oxidizing agent used in the oxidative processes is hydrogen peroxide in combination with a variety of catalysts such as metals,[37] *m*-CPBA,[38] $NaIO_4$,[39] CrO_3,[39] $KMnO_4$,[40] and dioxiranes.[41] Unfortunately, when the target is the synthesis of sulfoxides, a particular attention to avoid overoxidation to sulfone is required.[42]

To avoid this drawback and undesired side products, regioselective oxidation processes of sulfides to sulfoxides were accomplished.[43] These processes can implement several oxidizing agents such as hydrogen peroxide,[44] nitric acid, *m*-chloroperbenzoic acid,[45] urea/hydrogen peroxide,[46] sodium hypochlorite,[47] and sodium periodate.[48]

Since unsymmetrical thioethers are prochiral, their oxidation to sulfoxides lead to chiral derivatives. This process can be performed enantioselectively, as reported, for example, by Ye et al.[49]

A more sustainable selective oxidation procedure of sulfides implements Oxone ($KHSO_5 \times 0.5KHSO_4 \times 0.5K_2SO_4$)[50] as oxidant without using any catalyst/additive under mild conditions, as reported by Li et al.[51] In this case, the use of specific solvent determined the preferential oxidation to sulfoxide.

Besides classical synthetic approaches, the use of innovative technology, such as flow apparatus, has been proven to significantly improve several synthetic aspects, mainly in the field of sulfur compounds chemistry.

Chemoselective oxidation of sulfides with a silica-supported peracid in supercritical carbon dioxide under flow conditions was proposed by Núñez et al.[52] In this procedure, supercritical carbon dioxide ($scCO_2$) was used as reaction media for the selective oxidation of sulfides (**1**) to either sulfoxides (**2**) or sulfones (**3**) using a silica-supported peracid (2-[percarboxyethyl]-functionalized silica) as heterogeneous oxidant system, under flow conditions allowing two consecutive steps of oxygen transfer to the substrates (Scheme 3.1).

Although the first step of this process, that is, oxidation of sulfide (**1**) to sulfoxide (**2**), is kinetically favored over the second step, that is, oxidation of sulfoxide (**2**) to sulfone (**3**), it is possible to change the reaction chemoselectivity using different reaction conditions.

Even though particular conditions of temperature and pressure are required (Tc = 31.0°C; Pc = 73.8 bar), the use of $scCO_2$ as a reaction medium has considerable advantages over classical liquid solvents, owing to its low surface tension, low viscosity, high diffusivity, benign character, and low cost.[53] Furthermore, the solvating ability of $scCO_2$ can be tuned just varying the operative pressure. This characteristic is of particular importance in determining the course of the reaction, since in this condition, there are changes in the interaction of the active surface with catalysts, reactants, products, or transition states.[54]

In order to verify the reactivity of the supported peracid (**4**) toward sulfides (**1**) and sulfoxides (**2**), Núñez et al. performed initial oxidation experiments with a 1:2 molar ratios of sulfide (**1**)/peracid (**4**) by flowing $scCO_2$ (0.10–0.12 mL $scCO_2$ min^{-1}) at 250 bar and 40°C for 3 hours through a reservoir containing the substrate (**1**) and then through a column packed with 25% wt./wt. hydrated reagent (**4**) (Scheme 3.2; Table 3.2: method A).

As reported in Table 3.2, the results showed quite fast reactions of both the sulfides (**1**) and the sulfoxides (**2**) with the supported peracid (**4**), which allowed the formation of the sulfone (**3**) as the main final compound. A dependence of the reaction efficiency on the substrate structure was also observed. Thus, the presence of electron-withdrawing or sterically demanding substituents on sulfides (**1**) produced lower conversions or lower sulfone/sulfoxide ratios (Table 3.2, # 4–7). To obtain the selective

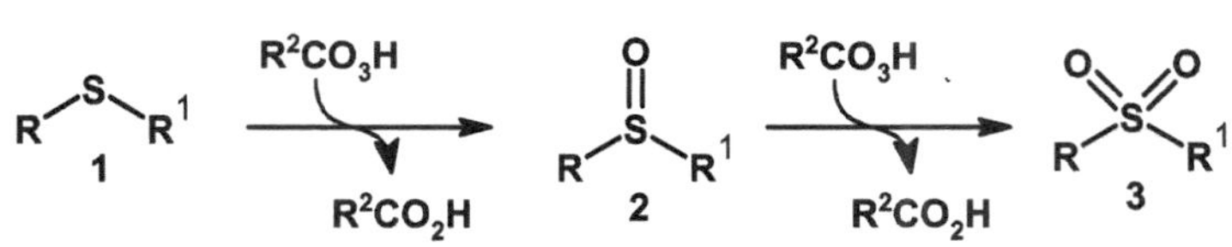

SCHEME 3.1 Consecutive steps of oxygen transfer in the oxidation of sulfides with a peracid.

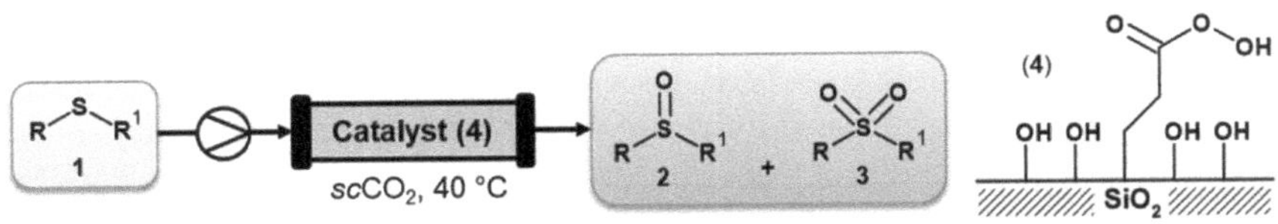

SCHEME 3.2 Chemoselective oxidation of sulfides with a silica-supported peracid in supercritical carbon dioxide under flow conditions.

TABLE 3.2 Oxidation of sulfides (**1**) with the silica-supported peracid (**4**) in $scCO_2$ under flow conditions.

			Method A[a]		Method B[b]	
#	Sulfide (1)		Conv. (%)	3/2	Conv. (%)	3/2
1	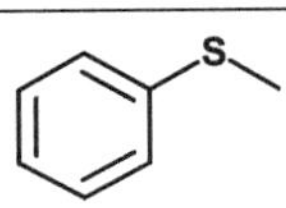	**1a**	95	29.8	73	0.27
2	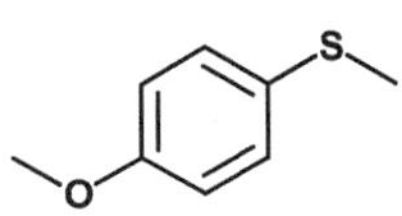	**1b**	>99	270	98	0.13
3	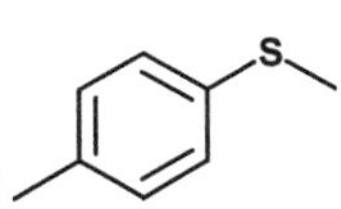	**1c**	99	13.6	98	0.46
4	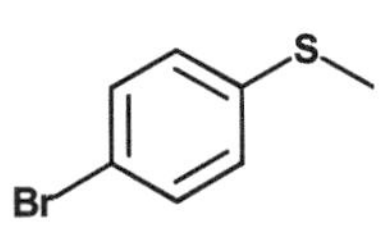	**1d**	87	11.1	89	0.17
5	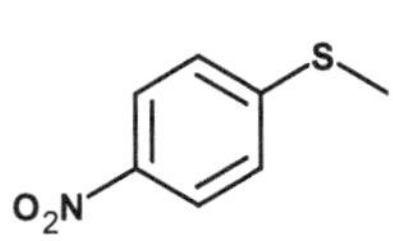	**1e**	95	4.2	90	0.02
6	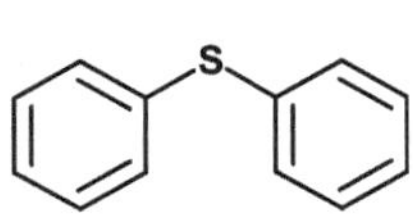	**1f**	85	4.4	99	0.11
7	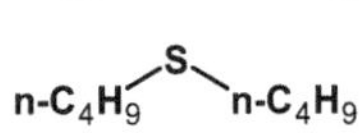	**1g**	93	5.8	85	0.15

[a] *Reactions at 250 bar, 40°C, flow rate 0.10–0.12 mL $scCO_2$ min^{-1}, load of peroxide: 1 mmol of peracid g^{-1}, hydration of **4**: 25% wt./wt., molar ratio of **1/4** was 1:2.*

[b] *Reactions in $scCO_2$ at 100 bar, with anhydrous **4**, 1:1 molar ratio of **1/4**; products were eluted with a solution of methanol in $scCO_2$.*

oxidation of sulfides (**1**) to the corresponding sulfoxides (**2**), Núñez et al. performed the oxidation reactions with a 1:1 molar ratio of sulfide (**1**)/ peracid (**4**) by flowing *sc*CO_2 (0.12–0.10 mL *sc*CO_2 min^{-1}) at 100 bar and at 40°C for 3 hours through a reservoir containing the substrate (**1**) and then through a column packed with anhydrous silica-supported peracid (**4**) (Table 3.2, method B).

The influence of operative pressure and the hydration grade of the silica surface on the reaction chemoselectivity under flow condition, was determined by Núñez et al. carrying out the oxidation of thioanisole (**1a**) flowing *sc*CO_2 (0.12 mL *sc*$CO_2$$min^{-1}$) at 40°C and at pressures ranging from 100 to 300 bar, for 3 hours using anhydrous, 25% and 50% wt./wt. silica-supported peracid (**4**) (initial molar ratio sulfide/peracid 1:1) (Scheme 3.3).

The results reported in Fig. 3.4 indicate that the reaction chemoselectivity is the result of the different diffusion rates of sulfide (**1**) and

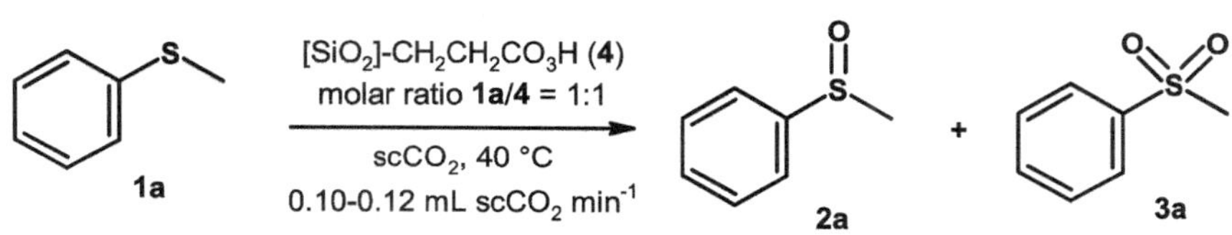

SCHEME 3.3 Oxidation of thioanisole (**1a**).

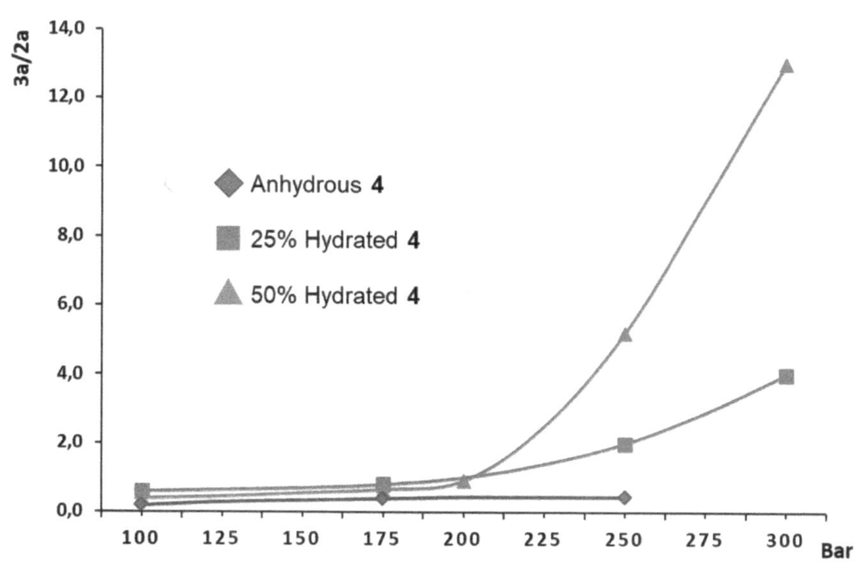

FIGURE 3.4 The influence of pressure and hydration grade of the silica surface on the reaction chemoselectivity of thioanisole (**1a**) under flow condition.

sulfoxide (**2**) over the reagent bed, and it can be controlled by changing the pressure and the hydration of the silica surface. Thus, anhydrous silica-supported peracid (**4**) is required to obtain sulfoxide (**2a**) as the main reaction product, regardless of the pressure used (low **3a/2a** ratio). To reverse the chemoselectivity, hydrated silica-supported peracid (**4**) and high pressure are required. It is possible to observe this trend at pressure higher than 200 bar. Thus, the use of 25 and 50% hydrated silica-supported peracid (**4**), flowing *sc*CO_2 at 300 bar, increased the ratio of **3a/2a** to 2.5 and 12.6, respectively.

With respect to other traditional procedures, the most important advantage of this approach is the possibility to recover the final compounds just by depressurization without the necessity of a work-up. Besides, the reactivation with hydrogen peroxide under acidic conditions allows the silica-supported peracid (**4**) to be recycled for further oxidative rounds.

In 2018, Wirth et al. implemented the use of flow-chemistry as an efficient system for a rapid, versatile and selective oxidation of sulfur compounds.[55] For this purpose, a packed-bed reactor with Oxone (Scheme 3.4) was used to obtain sulfoxides without the formation of sulfones or other side products. This methodology displayed high tolerance for several functional groups such as halogens, nitro, aldehyde/ketone, and C–C double bond. Besides, compounds bearing both sulfur and selenium atoms were selectively oxidized to the sulfoxides.

Preliminary experiments with thioanisole (**1**) allowed the authors to find suitable reaction conditions. Thus, 0.05–0.2 M dichloromethane solution of thioanisole (**1**) and trifluoroacetic acid (TFA; 15 vol.%) can be pumped through an Omnifit column packed with Oxone with a flow rate of 0.6 mL/min (4 minutes residence time) obtaining quantitative conversion. The use of an alternative reaction media and the increment of temperature led to an important decrease of conversion and a lower selectivity (sulfoxide versus sulfone), respectively. Under optimized reaction condition, the packed-bed reactor loaded with 5.6 g of Oxone (approximately 9.1 mmol Oxone) can oxidize up to 10 mmol of sulfide in a single reaction cycle or reused after washing with dichloromethane with a different substrate until the Oxone is completely consumed.

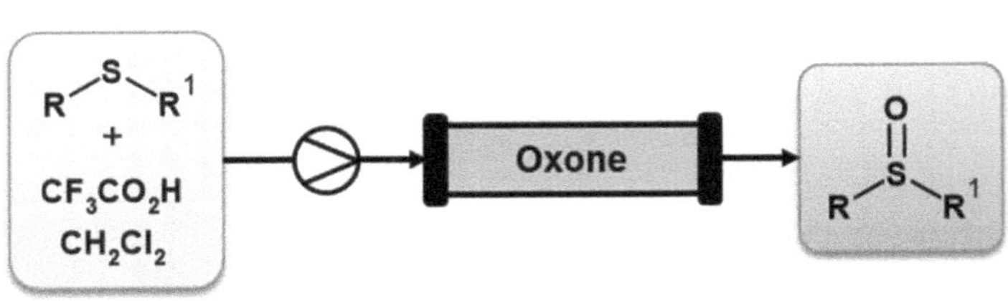

SCHEME 3.4 Selective oxidation of sulfur compounds by oxone.

The optimized reaction conditions were used with different sulfides affording the desired sulfoxides in good to excellent yields (Scheme 3.5). When this protocol was applied to the oxidation of allyl sulfides (**1n**, **1p**), no epoxide formation was observed.[56]

High chemoselectivity was observed when starting material contained both a phenyl sulfide and a phenyl selenide moiety such as (**1m**) led to the sulfoxide (**2m**) in 76% yield (Scheme 3.2), whereas the treatment of (**1m**) with one equivalent of *m*-CPBA (*meta*-chloroperbenzoic acid) under batch conditions produced sulfoxide (**2m**) only in 7% of yield.

With small changes to the previous optimal reaction conditions, this protocol was also found to be suitable for the oxidation of disulfides. Thus, the decrease of the concentration of (**5a**) to 0.025 M and the percentage of TFA in the mixture to 10% resulted in almost quantitative conversion of sulfide (**5a**) to thiosulfinate (**6a**). Under these reaction conditions, the oxidation of various symmetric disulfides afforded thiosulfinate derivatives with yields from fair to excellent (Scheme 3.6).

Among the various oxidants used for the oxidation of sulfides, hydrogen peroxide represents one of the most acceptable from a sustainability point of view.[44] Furthermore, when hydrogen peroxide is implemented in flow process, significant improvements compared to conventional batch processes can be obtained. Maggi et al. developed a continuous flow process in which the chemoselective oxidation of sulfides (**1**) to sulfoxides

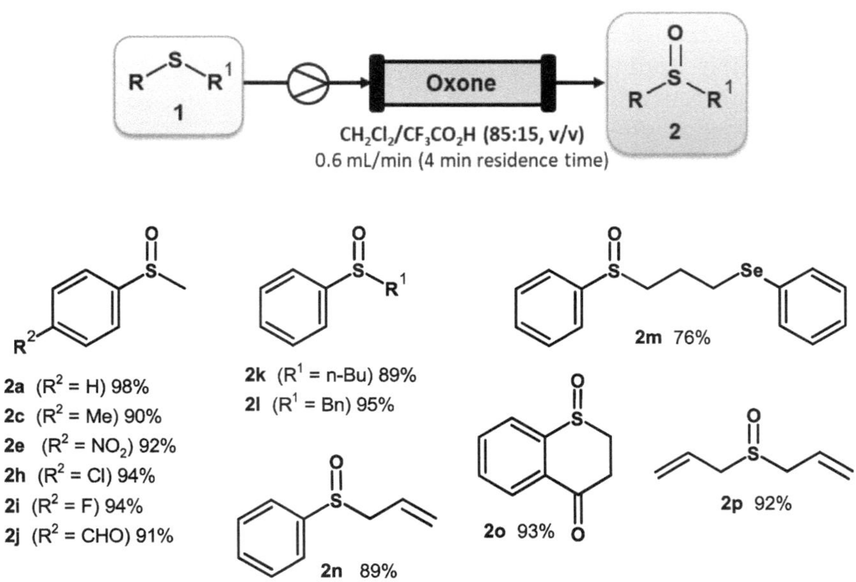

SCHEME 3.5 Sulfoxides obtained in flow condition.

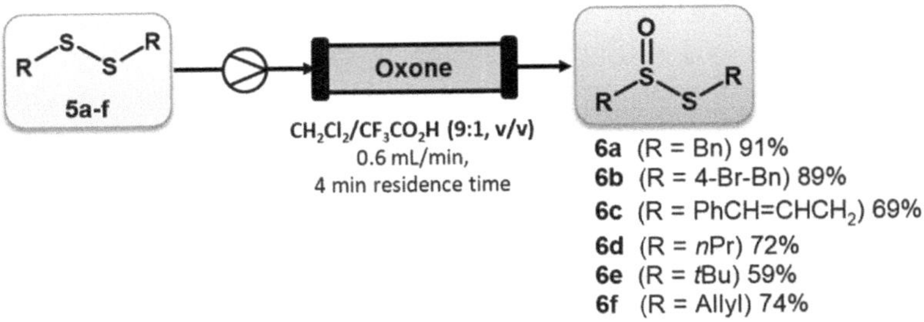

SCHEME 3.6 Examples of oxidation of disulfides (**5**) (0.025 M) to sulfinothioate (**6**) (0.025 M) in flow condition.

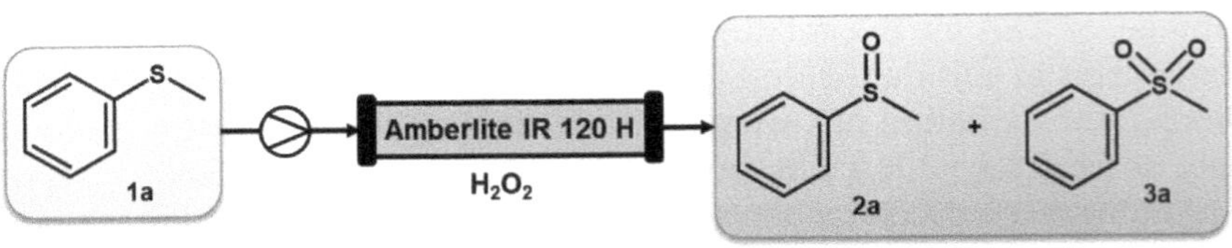

SCHEME 3.7 Chemoselective oxidation of sulfides (**1**) to sulfoxides (**2**) by hydrogen peroxide and Amberlite IR-120H.

(**2**) is obtained by the use of hydrogen peroxide as the oxidant and Amberlite IR-120H as the catalyst (Scheme 3.7).[57]

For this purpose, preliminary continuous flow experiments with thioanisole (**1a**, 0.89 mol L^{-1} in methanol) were carried out in order to establish the optimized reaction conditions, using a glass column packed with Amberlite IR 120 H (6.80 g) as a reactor. Thus, a set of parameters such as 30% aqueous H_2O_2/**1a** ratio (1.0–1.7), temperature (6°C–30°C) and residence time (17–50 minutes) were evaluated and compared with an analogous batch experiment. Both processes were carried out using methanol as solvent (Scheme 3.8).

Under optimized reaction conditions (22°C, stoichiometric amount of 30% aqueous H_2O_2, residence time of 25 minutes) sulfoxide (**2a**) was obtained in 94% of yield and 99% of selectivity. This result confirmed the higher performance of the reaction carried out under flow condition when compared to conventional batch processes that furnished sulfoxide (**2a**) only with 55% of yield. It is worth noting that under the best conditions, less concentrated hydrogen peroxide (3% aqueous solution) can also be used in the continuous flow process obtaining (**2a**) as the only product with a minimal loss of the yield (Table 3.3, #1). This new process condition was then implemented to other differently substituted aryl alkyl sulfides using an equimolecular amount of 3% hydrogen peroxide. As

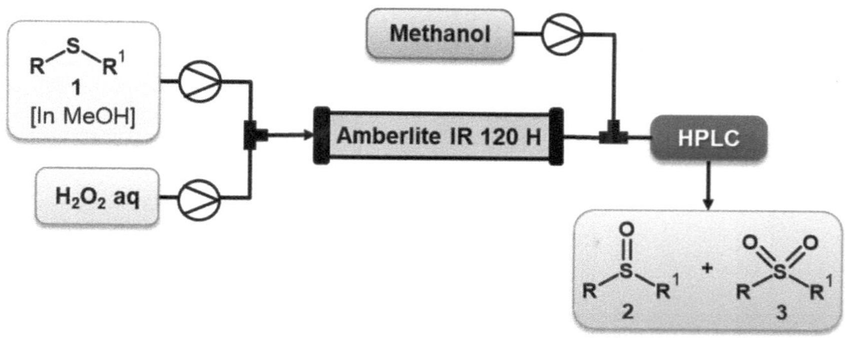

SCHEME 3.8 Representation of the reactor setup for the continuous flow oxidation of sulfides with hydrogen peroxide.

shown in Table 3.3, in all cases, sulfoxides were obtained with excellent yield and selectivity also in the presence of electron-withdrawing or electron-donating groups in the para position of aryl sulfides.

Another example of sulfide oxidation in the presence of hydrogen peroxide under flow condition was reported by Santi et al. in 2020.[58] In this process, a diluted aqueous solution of hydrogen peroxide was used for the in situ generation of the catalytic specie perseleninic acid from selenium (IV) oxide to selectively afford sulfoxides and sulfones in line with the principles of efficiency, safety, mild reaction conditions, and overall atom economy. This process was developed on the basis of a study about the applicability of selenium derivatives as catalysts in biomimetic oxidations using peroxides as stoichiometric oxidants, miming the ability of selenium atom of the glutathione peroxidase enzyme.[59] A previous report demonstrated that phenylselenozinc chloride (Santi's reagent)[60] can be used as a catalyst in the biomimetic oxidation of thiols mediated by hydrogen peroxide.[61] Further development in organoselenium chemistry revealed that aqueous media or biphasic systems can be implemented with beneficial effects on the reactivity and selectivity,[62] and that bioinspired selenium-catalyzed reactions can be easily translated into continuous mode.[63]

The protocol proposed by Santi et al. represents a bioinspired approach in which a continuous and chemoselective oxidation of sulfides (**1**) into their corresponding sulfoxides (**2**) or sulfones (**3**) is carried out using SeO_2 as the most atom-economical Se-based pre-catalyst, in a liquid–liquid biphasic system at room temperature with hydrogen peroxide as the green oxidant (Scheme 3.9). This continuous procedure overcomes some drawbacks normally associated with the stoichiometric use of selenium (IV) oxide (toxicity and tendency to generate volatile and malodorous side products)[64] and represents an improvement of the corresponding batch version reported by Drabowicz et al.[65]

TABLE 3.3 Continuous flow oxidation of various sulfides with 3% aqueous hydrogen peroxide catalyzed by Amberlite IR 120 H.

#	Sulfide (1)	Conversion (%)[a]	Sulfoxide (2) yield (%)[a]	Sulfone (3) yield (%)[a]
1	**1a**	90	90	–
2[b]	**1b**	99	96	3
3[b]	**1d**	94	93	1
4[b]	**1e**	97	93	4
5[b]	**1f**	87	84	3
6	**1h**	81	81	–

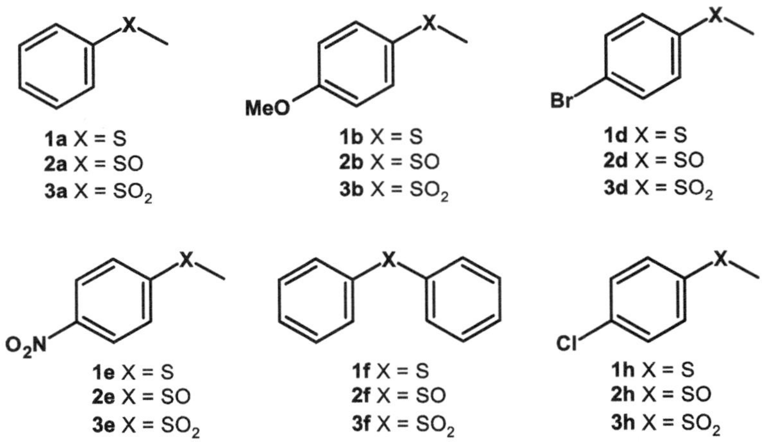

[a]*Determined by HPLC analysis.*
[b]*For the less soluble sulfides, hydrogen peroxide was diluted with methanol.*
Reaction parameters: residence time 25 min, T = 22°C, molar ratio $H_2O_2/1$ = 1:1.

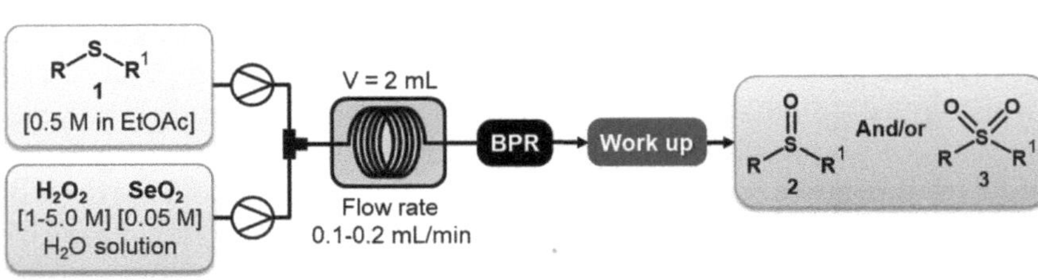

SCHEME 3.9 Bioinspired approach of a continuous and chemoselective oxidation of sulfides (**1**) to sulfoxides (**2**) or sulfones (**3**) using SeO_2 as Se-based pre-catalyst, in a liquid–liquid biphasic system. *BPR*, Back pressure regulator.

In order to establish the best reaction conditions, preliminary investigations were carried out fluxing the ethylacetate solution of phenylethylsulfide (**1q**; 0.5 M) and the aqueous solution of SeO_2 (0.05 M) with different amounts of H_2O_2, through a Y junction in a 2 mL tubular reactor, at different flow rates. It was found that the chemoselectivity is influenced predominantly by the amount of hydrogen peroxide, thus generating sulfoxide (**2q**) at low concentration of the oxidant (Table 3.4, # 1–5) and sulfone (**3q**) at the higher concentration of the oxidant (Table 3.4, # 6–8). Notably, under best reaction conditions, the reaction without SeO_2 did not occur, confirming the catalytic role of selenium.

The optimal reaction conditions for the preparation of sulfoxides (**2**) (Solutions fluxed at 0.05 mL/min, SeO_2 [0.05 M, 0.1 equivalent] and H_2O_2 [1.0 M, 2 equivalent] in H_2O) and sulfones (**3**) (Solutions fluxed at 0.1 mL/min), SeO_2 [0.05 M, 0.1 equivalent] and H_2O_2 [5.0 M, 10 equivalent] in H_2O were employed to evaluate general applicability to a set of sulfides (Fig. 3.5).

As reported in Table 3.5, the aryl sulfides afforded the corresponding sulfoxide in very high yields and with complete chemoselectivity, with the exception of naphtylethylsulfide (**1r**), which was converted to sulfoxide (**2r**) with a yield of only 20%, and tetrahydrothiophene (**1t**), which led to an inseparable mixture of (**1t**), (**2t**), and (**3t**). To obtain a quantitative conversion of naphtylethylsulfide (**1r**), it was necessary to halve its concentration and increase the amount of H_2O_2 to 5 equivalent. Under these conditions, the sulfoxide (**2r**) was isolated by chromatographic purification with a yield of 75% from an 80:20 mixture of (**2r/3r**). In the

TABLE 3.4 Optimization of the continuous flow oxidation conditions by fluxing ethylacetate solution of phenylethylsulfide 1q (0.5 M) and aqueous solution of SeO_2 (0.05 M) and H_2O_2 at room temperature, each at the half-total flow rate.

#	H_2O_2 equivalent [concentration]	Total flow rate mL/min	Residence time (min)	Conversion (%)[a]	Selectivity 2/3
1	1 [0.5 M]	0.3	6.5	59	95:5
2	1 [0.5 M]	0.02	100	61	100:0
3	2 [1.0 M]	0.3	6.5	60	100:0
4	2 [1.0 M]	0.2	10	78	100:0
5	2 [1.0 M]	0.1	20	85	100:0
6	5 [2.5 M]	0.1	20	>99	71:29
7	10 [5.0 M]	0.1	20	>99	0:100
8	10 [5.0 M]	0.2	10	>99	0:100

[a]*Calculated by ^{1}H-NMR.*

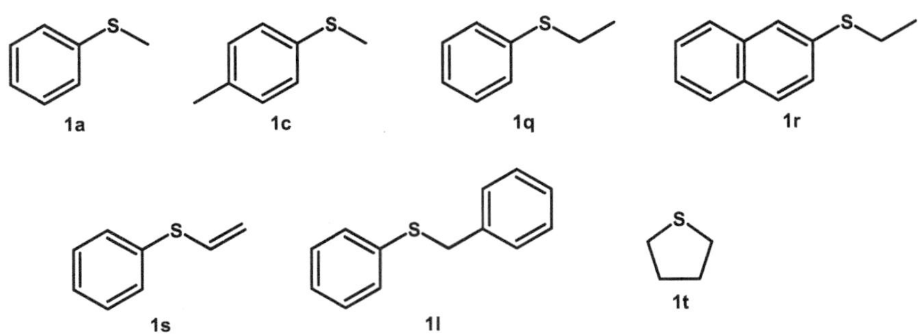

FIGURE 3.5 Set of sulfides used as starting materials in the chemoselective oxidation to sulfoxides (**2**) and sulfones (**3**) using SeO_2 as Se-based pre-catalyst, in a liquid–liquid biphasic system.

TABLE 3.5 Continuous flow oxidation of sulfides to sulfoxides or sulfones.

	Sulfide (1)	Sufoxide (2)[a]			Sufone (3)[b]	
#	(0.5 M in EtOAc)		Conversion (%)[c] [yield (%)]	Ratio 2/3		Yield (%)
1	**1q**	**2q**	85 [85]	100:0	**3q**	>99
2	**1a**	**2a**	91 [91]	100:0	**3a**	>99
3	**1c**	**2c**	93 [89]	100:0	**3c**	>99
4	**1r**	**2r**	20 99[d] [75]	100:0 80:20	**3r**	>99[e]
5	**1s**	**2s**	80 [74]	100:0	**3s**	>99
6	**1l**	**2l**	99 [99]	100:0	**3l**	>99
7	**1t**	**2t**	Nd[f]	nd[c]	**3t**	>99

[a]*Each solution was fluxed at 0.05 mL/min, (SeO_2 [0.05 M, 0.1 eq] and H_2O_2 [1.0 M, 2 eq] in H_2O).*
[b]*Each solution was fluxed at 0.1 mL/min, (SeO_2 [0.05 M, 0.1 eq] and H_2O_2 [5.0 M, 10 eq] in H_2O).*
[c]*Conversion was evaluated by ^1H-NMR from the crude product; isolated yields are shown in brackets.*
[d]*Solution of 1r in EtOAc = 0.25 M, (SeO_2 [0.025 M, 0.1 eq] and H_2O_2 [1.25 M, 5.0 eq] in H_2O).*
[e]*Solution of 1r in EtOAc = 0.25 M, (SeO_2 [0.025 M] and H_2O_2 [2.5 M] in H_2O).*
[f]*Not determined.*

presence of a 5.0 M solution of hydrogen peroxide (10 equivalent), all the tested sulfides were quantitatively converted to sulfones (Table 3.5) in pure form. Also in this case, a 0.25 M solution of the sulfide (**1r**) and a 0.025 M solution of the catalyst were required for the complete conversion to sulfone (**3r**). It is worth noting that the reaction of vinyl sulfide **1s**, in both conditions for the synthesis of the sulfoxide or the sulfone, was not affected by the oxidation of vinyl group, leading to the sulfoxide (**2s**) or to the sulfone (**3s**) as the only product.

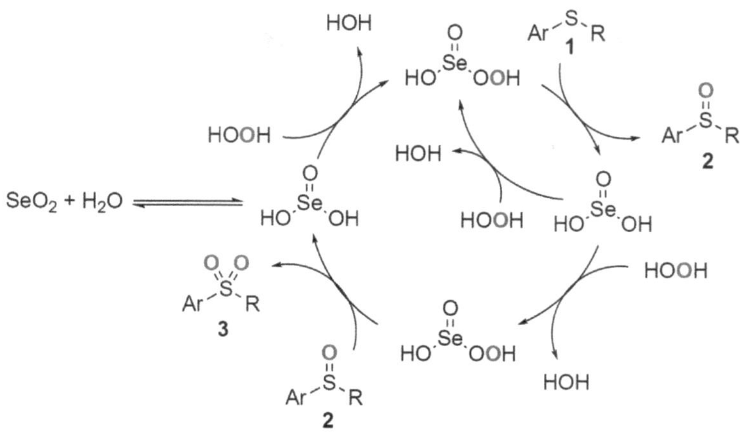

FIGURE 3.6 Proposed catalytic mechanism (Creative commons attribution (CC BY) license).[58]

The mechanism proposed by the Santi et al. (Fig. 3.6) shows that the selenium (IV) oxide in water is in equilibrium with the hydrated form, which is particularly prone to oxidation by hydrogen peroxide to peroxyselenic (IV) acid. The latter is the actual catalyst in the oxidation of sulfides (**1**) to the sulfoxide (**2**) and/or sulfone (**3**), depending on the amount of the hydrogen peroxide.

Organic electrosynthesis is an alternative environment-friendly method of performing redox processes by electrons, thus avoiding toxic or dangerous oxidizing/reducing reagents and stoichiometric redox reagents.[66] Despite electrochemical transformations requiring specific apparatus and suffering several drawbacks, an improvement by a combination with flow condition system can be obtained.[67] For example, supporting electrolyte, typically required for batch electrochemistry, can be used in low concentrations or avoided completely, owing to short distance between electrodes, which enables a more efficient mass transfer. Interestingly, significant advantages in downstream processing are provided by electrolyte-free solvent systems with a residual conductivity such as acetonitrile/water, as described by Waldvogel et al.[68]

To overcome the drawbacks of classical oxidation-mediated synthesis of sulfoxides and sulfones, Noël et al. adopted an environmentally benign and selective electrochemical oxidation of thioethers and thiols in a continuous-flow microreactor. This procedure presents a great flexibility as it allows to access both compounds, sulfoxides, and sulfones, simply by tuning the applied potential in the electrochemical flow cell whilst keeping the reaction mixture the same.[69]

Preliminary experiments with thioanisole (**1**) allowed Noël et al. to establish the optimal conditions for the direct electrochemical oxidation of thioethers, using a commercially available electrochemical flow module equipped with cheap stainless-steel electrodes (Scheme 3.10). The use of tetrabutylammonium perchlorate ($TBAClO_4$) as an electrolyte gave optimal results to maintain a stable current, and an aqueous acidic solvent was also necessary to lower the pH and increase the redox potential of the Fe electrodes (Table 3.6, entries 3 and 4).

Under these reaction conditions (Table 3.6), the oxidation of thioanisole (**1a**) to sulfoxide (**2a**) was obtained by an applied potential of 2.5 V (corresponding to a current of approximately 2.1 mA/cm^2) with a yield of 63% (Table 3.6, entry 1), which increased to 75% by rising the residence time while keeping the flow rate constant (Table 3.6, entry 2). The selectivity shift toward sulfone derivative (**3a**) was obtained by increasing the applied potential as shown in Table 3.6 (entries 7–9). Furthermore, it was found that O_2 likely serves as the [O]-source (Table 3.6, entry 6).

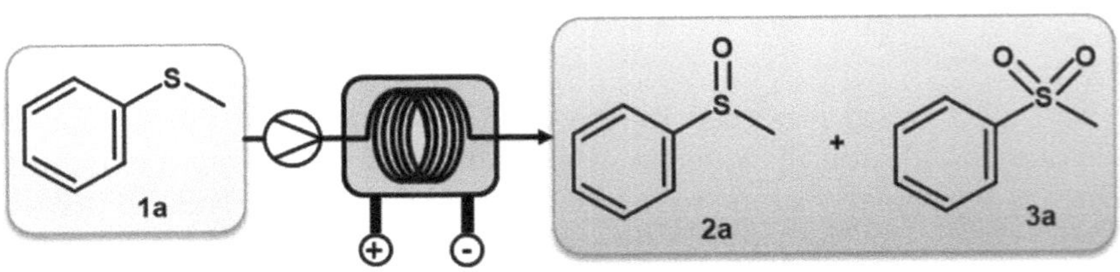

SCHEME 3.10 Electrochemical oxidation of thioanisole (**1a**) in a continuous-flow microreactor.

TABLE 3.6 Optimization of reaction condition for the electrochemical oxidation of thioanisole (**1a**) in continuous flow: Thioanisole (1 mmol, 0.1 M), $TBAClO_4$ (0.1 mmol), MeCN/HCl (10 mL, 3:1 vol./vol., with 0.1 M HCl in H_2O), applied potential: 2.5 V (2.1 mA/cm^2, Fe anode/cathode, Vint: 300 μL), residence time: 5 min (at a flowrate of 0.06 mL/min).

#	Deviation from above reaction condition	Conv.[a] (%)	% 2a[a]	% 3a[a]
1	None	63	62	0
2	*Residence time of 10 min*	75	72	2
3	No H_2O/HCl (0.1 M)	n.r.	–	–
4	No electrolyte	n.r.	–	–
5	0.2 M H_2O/HCl	37	34	–
6	No H_2O/HCl, MeCN saturated with O_2	39	39	0
7	Potential at 3.0 V	92	65	27
8	Potential at 3.5 V	98	59	39
9	Potential at 4.0 V	99	7	92

[a]*Conversion and selectivity evaluation was performed by GC-MS with an internal standard.*

Further investigation to confirm the importance of the applied potential on the oxidation selectivity was carried out with a single sweep voltammetry experiment at a constant residence time of 5 minutes. The polarogram shows two plateaus at 2.0–2.5 and 3.5–4.0 V, which correspond to the oxidation of thioanisole (**1a**) to sulfoxide (**2a**) and sulfone (**3a**), respectively.

The investigation of flow rate and residence time allowed to establish the optimal operating range and to avoid mass-transfer limitation. The optimal flow rate has been found to be in the range of 0.04–0.06 mL/min, which refers to residence times of 7.5–5 minutes. A standard residence time of 5 minutes was considered to be optimal for a single run. Under this condition, re-injection of reaction mixture is required to increase the overall residence time when the sulfides conversion is incomplete in a single run.

Nöel et al. applied the optimized reaction conditions for the selective conversion of several sulfides (**1**) (Fig. 3.7) to the corresponding sulfoxide (**2**) or sulfone (**3**) using methods A and B, respectively (Scheme 3.11). As shown in Table 3.7, sulfoxides (**2**) and sulfones (**3**) were obtained in moderate to excellent yield with a good functional group tolerance and high chemoselectivity. Interestingly, sulfides bearing aromatic heterocycle moiety, such as pyridine (**1ac**) or benzimidazole (**1ad**), were selectively

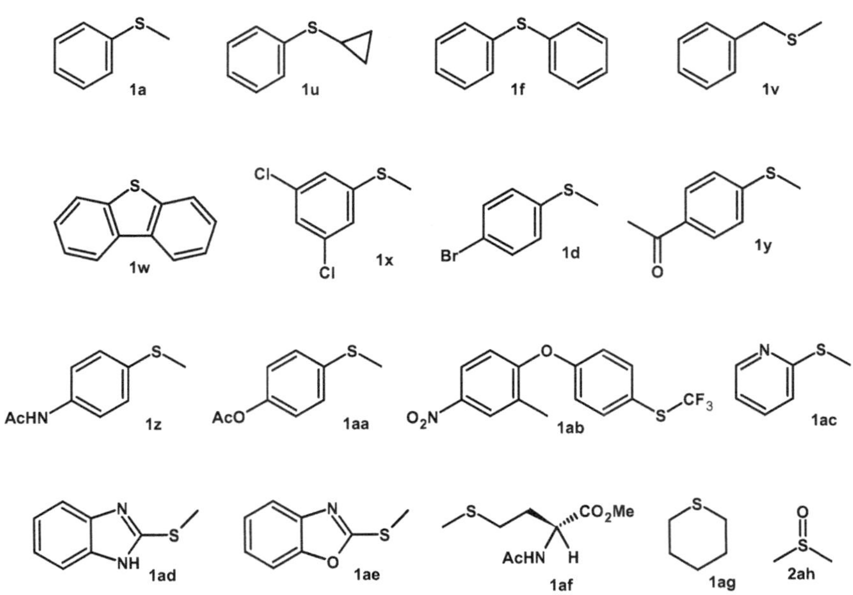

FIGURE 3.7 Set of sulfides (**1**) used as starting materials in the electrochemical oxidation to sulfoxides (**2**) and sulfones (**3**).

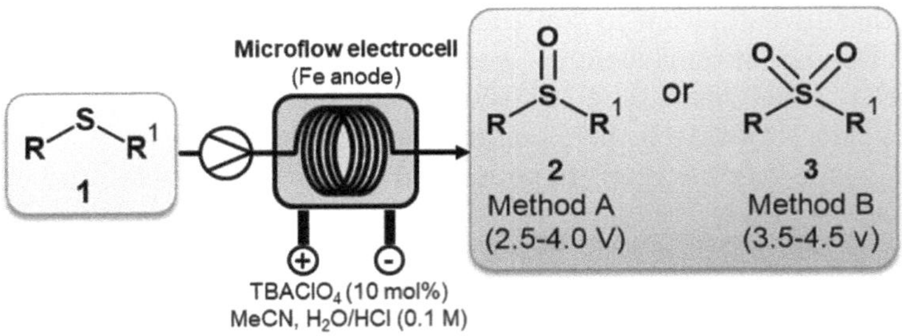

SCHEME 3.11 Selective electrochemical oxidation of sulfides (**1**) to sulfoxide (**2**) or sulfone (**3**).

TABLE 3.7 Oxidation of sulfides (**1**) to sulfoxides (**2**) and sulfones (**3**).

#	R-S-R′	R-SO-R′ (2.5–4.0 V)	Yield (%)	t_r (min)	R-SO_2-R′ (3.5–4.5 V)	Yield (%)[a]	t_r (min)
1	**1a**	**2a**	65	5	**3a**	92	10
2	**1u**	**2u**	61	10	**3u**	31	10
3	**1f**	**2f**	62	10	**3f**	81	10
4	**1v**	**2v**	59	10	**3v**	86	15
5	**1w**	**2w**	n.r.	–	**3w**	52	5
6	**1x**	**2x**	55	10	**3x**	64	10
7	**1d**	**2d**	83	20	**3d**	52	10
8	**1y**	**2y**	72	20	**3y**	68	10
9	**1z**	**2z**	32	10	**3z**	51	10
10	**1aa**	**2aa**	60	10	**3aa**	73	10
11	**1ab**	**2ab**	48	40	**3ab**	n.r.	–
12	**1ac**	**2ac**	51	10	**3ac**	58	15
13	**1ad**	**2ad**	40	15	**3ad**	54	15
14	**1ae**	**2ae**	46	5	**3ae**	n.r.	–
15	**1af**	**2af**	46	15	**3af**	36	10
16	**1ag**	**2ag**	46[b]	10	**3ag**	84	15
17	**2ah**	–	–	–	**3ah**	80	10

[a]*Isolated yield.*
[b]*The yield of* ***2ag*** *was calculated by* ^{1}*H-NMR. n.r. = no reaction. The electrodes used are identical (Fe).*
Sulfides = 0.15 M dissolved in an oxygen saturated 3:1 (vol./vol.) mixture of acetonitrile (DMF for 1w) and a 0.1 M HCl solution in water, with 10 mol.% tetrabutylammonium perchlorate as supporting electrolyte.

oxidized to the corresponding sulfoxides and sulfones without N-oxide formation, with the exception of compound **1ae**, in which the nitrogen oxidation of the benzoxazole moiety occurred before the formation of the sulfone. The presence of free amines, alcohols, and carboxylic acids groups or nitriles, aldehydes, and hydroxyls at the γ-position on starting sulfides, is not tolerated and, in this case, the reaction did not occur or lead to the formation of by-products. Furthermore, it is not always possible to access both the sulfoxide and sulfone, as occurred for the more reactive dibenzothiophene (**1w**), which is directly oxidized to the sulfone (**3w**), or for the less reactive sulfide (**1ab**), which was oxidized only to the corresponding sulfoxide (**2ab**), owing to the presence of the strong electron-withdrawing CF_3 on sulfur.

Nöel et al. applied the optimized reaction conditions also to the oxidation of thiols (**7**) (Fig. 3.8) to symmetric disulfides (**5**) using the proper voltage value (Scheme 3.12, method C: 2.5–3–0 V). As reported in Table 3.8, thiols were generally converted to the corresponding disulfide in good to excellent yield.

The interest in disulfides is principally due to chemical properties of S–S bond (also known as disulfide bridge) that is formed by the coupling between two thiol groups. In biology, the disulfide bridges are formed between two cysteine residues and play an important role in determining the secondary and tertiary structure of proteins.[70] For example, the high disulfide bridges content in keratins is the responsible of the stiffness of hair and feather. The presence of disulfide bridges is also important in determining tensile strength in rubbers and elastomers. Generally, disulfides find applications in several fields as anti-oxidants, pharmaceuticals, pesticides and rubber vulcanization reagents.[71] The importance of the disulfide compounds is also demonstrated by the numerous synthetic approaches reported in the literature,[72] among which the oxidative coupling of thiols is the most relevant.[73]

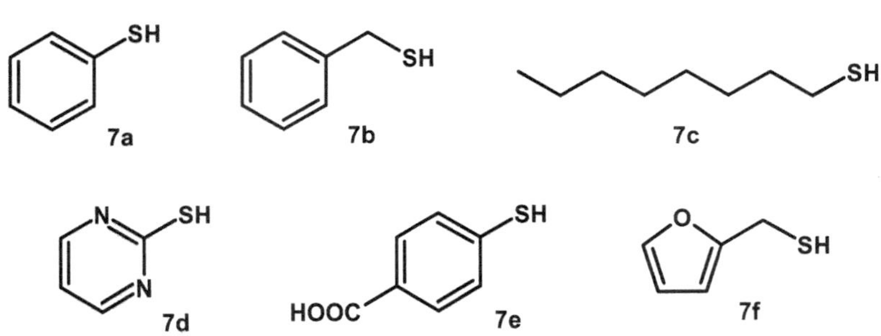

FIGURE 3.8 Set of thiols (**7**) used as starting materials in the oxidation to symmetric disulfides (**5**).

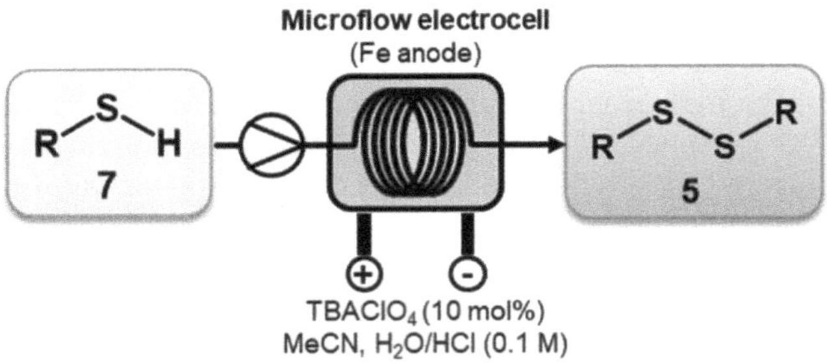

SCHEME 3.12 Oxidation of thiols (7) to symmetric disulfides (5).

TABLE 3.8 Oxidation of thiols (7) to disulfides (5).

#	R-S-H (7)	Method C (2.5–3.0 V) R-S-S-R (8)	Yield (%)[a]	t_r (min)
1	**7a**	**5g**	91	10
2	**7b**	**5a**	69	10
3	**7c**	**5h**	65	10
4	**7d**	**5i**	83	15
5	**7e**	**5j**	36	10
6	**7f**	**5k**	85	15

[a]*Isolated yield. The electrodes used are identical (Fe).*
Thiols = 0.15 M (0.075 M for **7e**) substrate dissolved in an oxygen saturated 3:1 (vol./vol.) mixture of acetonitrile (DMF for **7e**) and a 0.1 M HCl solution in water, with 10 mol.% tetrabutylammonium perchlorate as supporting electrolyte.

In a precedent paper,[69] Noël et al. reported a mild metal-free photocatalytic aerobic oxidation of thiols to disulfides using Eosin Y and visible-light irradiation in flow.[74] The continuous flow procedure allows to accelerate the photocatalytic process leading to disulfides with high purity with reaction time in the order of minutes. The authors previously investigated the possibility to perform photocatalytic aerobic oxidation of thiols to disulfides using *p*-methoxythiophenol as a substrate in batch process.

The reactions were initially carried out under air atmosphere and irradiated with a standard 24 W compact fluorescent light bulb at room temperature in the presence of ruthenium or iridium-based photocatalysts and metal-free photocatalytic systems, such as 9-mesityl-10-methylacridinium perchlorate or the organic dye Eosin Y, using ethanol as a green solvent. They found that disulfide is not formed in the absence of light and that only the Eosin Y catalyzed the conversion of *p*-methoxythiophenol (**7g**) to the corresponding disulfide (**5i**) with an excellent yield of 95% over 16 hours without over-oxidation. Furthermore, the addition of tetramethylethylenediamine, also in substoichiometric quantities, increased the reaction rate shortening the reaction time from 16 to 24 hours. Under these conditions, several thiophenols (**7**) were converted to disulfides (**5**) with high yields (86%–99%) within 2–16 hours, with a high tolerance toward several functional groups.

It was also observed that the batch reaction rate was greatly improved by the mixing efficiency. This phenomenon can be explained by the increased oxygen dissolution in the liquid phase at higher stirring rate (Fig. 3.9).

In this case, the implementation of continuous flow microreactor technology introduced several advantages over the batch process, such as:

- bypassing the problem of gas-liquid mass-transfer limitations, allowing enhancement of mixing and increase of mass transfer from the gas to the liquid phase,[75] due to segmented flow regime;
- improving of the reaction irradiation, due to the confined dimension;[76]
- obtaining a safe handling of oxygen.[77]

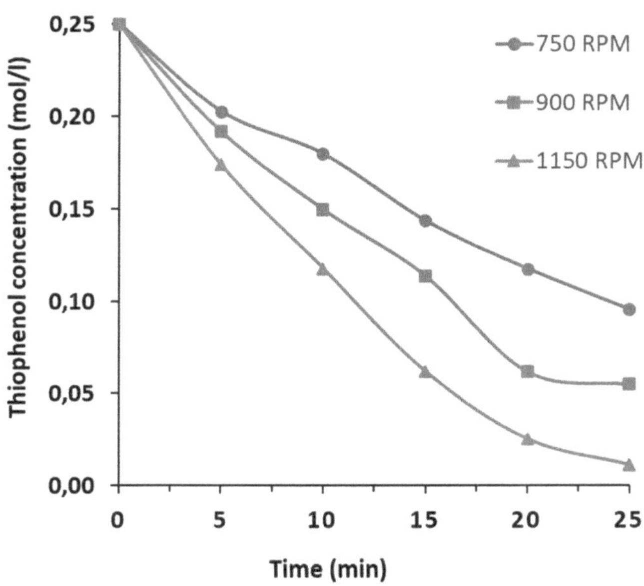

FIGURE 3.9 Effect of the stirring rate on the consumption of thiophenol in the photocatalytic aerobic oxidation of thiophenol using Eosin Y (1 mol.%) and tetramethylethylenediamine (1 equivalent) in ethanol under irradiation with a 24 W compact fluorescent light. *RPM*, Revolutions per minute.

For this purpose, a high purity perfluoroalkoxyalkane capillary microreactor (750 μm, 208 cm length, 950 μL volume) and a Tefzel mixing tee were used as microflow system. A representation of the photo-microreactor flow system is shown in Scheme 3.13. The oxidation of thiophenols (**7**) were carried out mixing pure oxygen, introduced through a mass-flow controller, and the solution of the reagents. Under these conditions, a gas-liquid segmented flow regime was obtained. A white LED light (3.12 W) was used for the photo activation of the reaction. A dilution of the emerging mixture with EtOH was at the end required to avoid obstruction.

As reported in Table 3.9, all the thiophenols (Fig. 3.10) tested under continuous flow protocol afforded the corresponding disulfides in excellent yield and high purity with a substantial reduction of the reaction time to 20 minutes respect to the batch process.

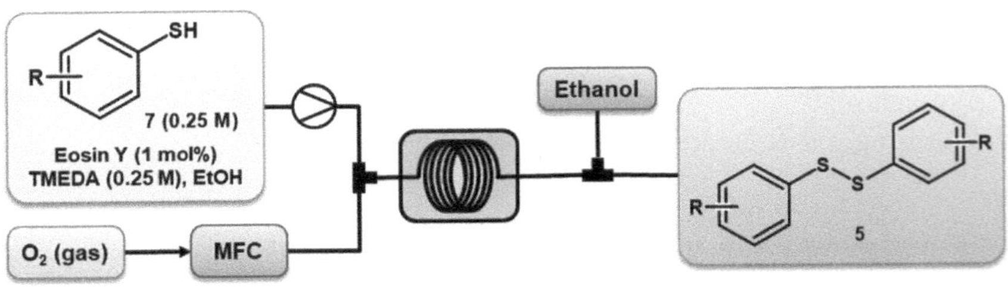

SCHEME 3.13 Continuous flow set-up for the visible-light photocatalytic aerobic oxidation of thiols to disulfides using Eosin Y.

TABLE 3.9 Photocatalytic aerobic oxidation of thiols (0.25 M) to disulfides using Eosin Y (1 mol.%), tetramethylethylenediamine (0.25 M), EtOH, air, 24 W compact fluorescent light.

#	R–SH	R–SS–R	Yield (%)
1	**7a**	**5g**	99
2	**7g**	**5i**	99
3	**7h**	**5m**	99
4	**7i**	**5n**	99
5	**7j**	**5o**	99
6	**7k**	**5p**	93
7	**7l**	**5q**	99
8	**7f**	**5k**	87

Reported yields are those of isolated material on a 1 mmol scale. Reaction time = 20 min.

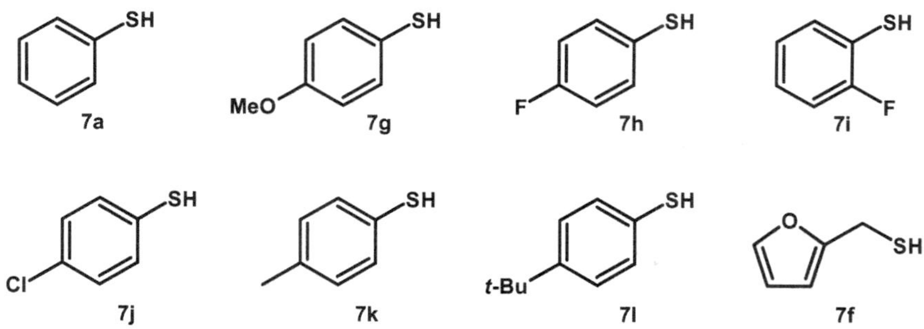

FIGURE 3.10 Set of thiols (7) used in the Photocatalytic aerobic oxidation to disulfides (5).

The adaptability and mildness of this continuous flow protocol allowed the synthesis of complex molecules such as the cyclic peptide hormone, oxytocin,[78] which was obtained with full conversion in water with only 200 s residence time, and the Antabuse drug (disulfiram), used in chronic alcoholism therapy,[79] which was obtained in 45% of isolated yield.

Another field where flow chemistry has actually found important application is the multiphase reaction.[80]

One of the most important factors regulating multiphase reaction is the mass transfer capability of reagents to a reactive site in another phase. This fact implies that the reactivity is often limited by the mass transfer which is directly correlated to the surface contact area.[81] From this perspective, the implementation of continuous flow and microreactor technologies is an effective approach to overcome this issue and to enhance mass transfers, obtaining a very high surface area/volume ratio.[75,82]

The ideal condition is a turbulent flow in which droplets are highly dispersed in an emulsion, but even when the flow rate is low and the residence time not too long, according to the kinetic model proposed by Shimanouchi et al., a slug flow produces local vortex fields able to increase the mass-transfer rates (Fig. 3.11).[83]

Although liquid–liquid reaction in flow can generally carried out with or without a phase-transfer catalyst (PTC), the presence of PTC facilitates the transport of a reagent from one phase to another phase.

On the basis of a promising procedure reported by Sonanave et al. in 2007 for the synthesis of symmetrical disulfides starting from alkyl halides and sodium disulfide under phase-transfer reaction catalyzed by didecyldimethylammonium bromide,[72] Wirth et al. in 2013, developed a mild flow protocol for the synthesis of symmetrical di- and trisulfides using phase-transfer catalysis.[84]

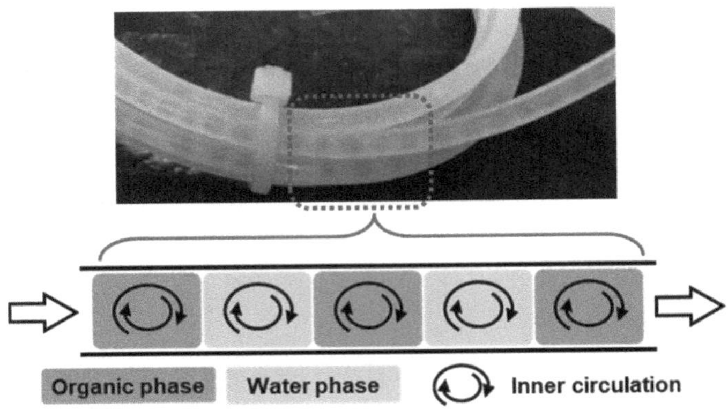

FIGURE 3.11 An ideal turbulent flow condition producing local vortex fields able to increase mass transfer rate.

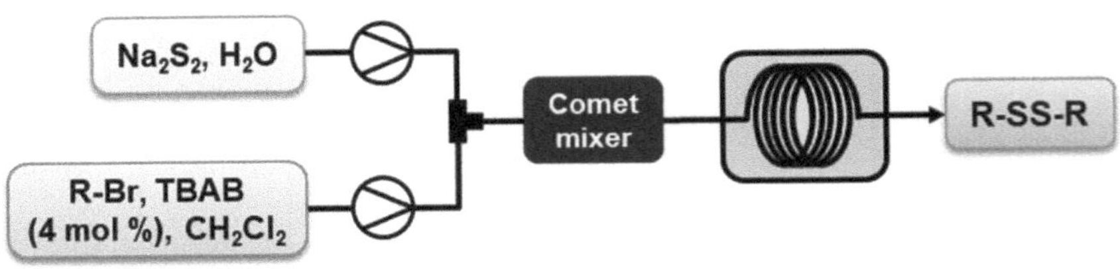

SCHEME 3.14 Flow reactor setup for the synthesis of disulfides.

First, the reaction of allyl bromide (**8a**) was optimized for the synthesis of diallyl disulfide (**5f**), an important volatile component of garlic extract.[85] This study allowed to establish the optimal reaction conditions, such as the flow rate (0.1 mL/min) and the solvent (dichloromethane) at a temperature of 20°C. The process required the mixing (through a micromixer) of the disodium disulfide (Na_2S_2) aqueous solution with the alkyl bromide and tetra-*n*-butylammonium bromide solution by two distinct syringe pumps, as shown in Scheme 3.14.

The reaction takes place in 5 minutes in a reactor coil consisting of a Teflon tubing (0.8 mm diameter, 2 m length, and 1 mL volume). The use of dichloromethane resulted in a biphasic flow system characterized by a segmented flow.[86] Table 3.9 reports the results obtained by this protocol using different alkyl bromides. With the exception of allyl bromide (**8a**) (Table 3.10, #1), all other reagents required an increase of temperature to 30°C–40°C to ensure high yields within the 5-min reaction time.

This procedure was also used for the synthesis of trisulfides. The reaction of elemental sulfur with monosulfide (S^{2-}) required to generate the trisulfide bis-anion (S_3^{2-}) leads to an equilibrium that produce a

TABLE 3.10 Synthesis of dialkyl disulfides (5).

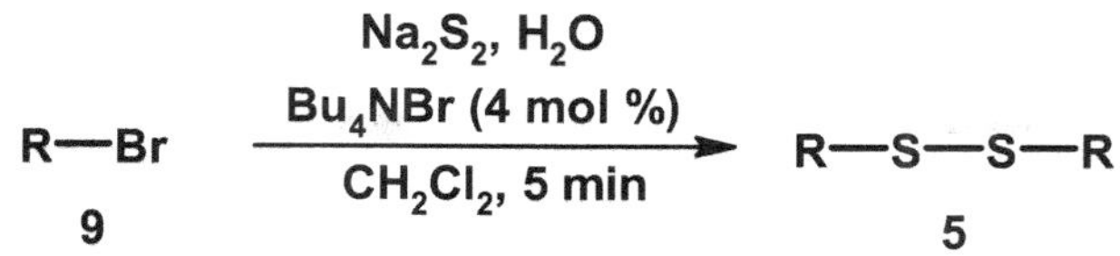

#	R-X (8), R=	R-SS-R (5)	T °C	Yield (%)
1	$CH_2{=}CHCH_2$ (**8a**)	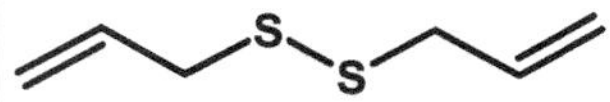 (**5f**)	20	92
2	*n*-C_3H_7 (**8b**)	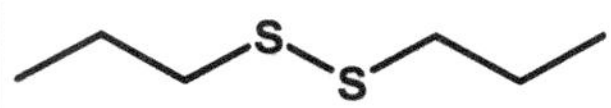 (**5d**)	40	45
3	HO-CH_2CH_2 (**8c**)	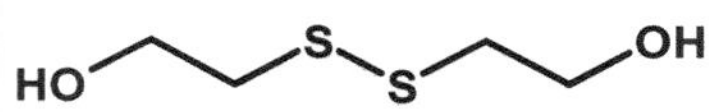 (**5r**)	40	70
4	*n*-C_5H_{11} (**8d**)	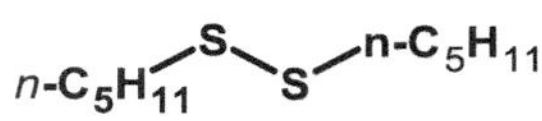 (**5s**)	40	63
5	$PhCH_2$ (**8e**)	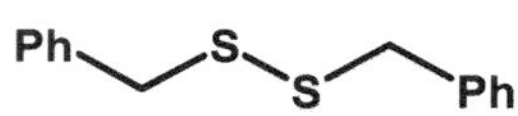 (**5a**)	30	43

Reaction conditions: 4 mol.% TBAB, CH_2Cl_2, 5-min reaction time.

mixture of polysulfide dianions. Thus, the reaction of bromides (**8**), under optimized reaction conditions, resulted in a mixture of alkyl polysulfides (disulsulfides, trisulfides, tetrasulfides, and pentasulfides) in which the disulfide derivatives were the main products. More recently Sancineto et al. applied a similar protocol for the continuous flow synthesis of 2,2′-diselenobis(benzoic acid) and some derivatives by first generating the diazonium salt and its subsequent reaction with a metal diselenide. This is to the best of our knowledge the unique example of in continuous synthesis of diselenides.[87]

In 2011, Santi et al. reported a pioneering example of bioinspired selenium-catalyzed reaction in flow conditions using a biphasic liquid–liquid system.[88] This seminal example deals with the dihydroxylation reaction of methyl cyclohexene using diphenyl diselenide as pre-catalyst and hydrogen peroxide as stoichiometric and green oxidant in 1:1 mixture of EtOAc/water (Scheme 3.15).

The proposed flow methodology presents several advantages in terms of greenest respect to the corresponding batch conditions:[89] reduction of the reaction time from 24 hours to 33 minutes, improvement of the yields

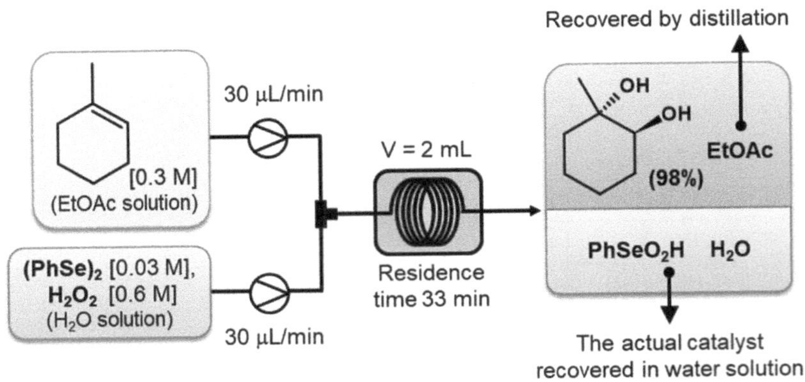

SCHEME 3.15 Dihydroxylation reaction of methyl cyclohexene using diphenyl diselenide as pre-catalyst.

from 76% to 98%, simplification of the workup (consisting in the separation of the two phases).

The under vacuum distillation of organic phase (EtOAc) affords the target compound without further purification. In addition, the actual catalyst ($PhSeO_2H$) can be recovered from the aqueous solution and reused for further reaction cycles.

Some years later, the same author and others reported a general and good methodology for the preparation of hydroxytetrahydrofurane, hydroxytetrahydropyrane and hydroxylactones via the biomimetic selenium-catalyzed oxidative cyclization of alkenols and alkenoic acids, respectively.[90]

Following the seminal result obtained for the dihydroxylation, it was possible to optimize the continuous synthesis of a small library of hydroxylactones as reported in Scheme 3.16.[91]

In order to promote the cyclization of the epoxide intermediate, in the optimized flow setup, a catalytic amount of formic acid (easily removable by evaporation) was added to the aqueous phase, and in order to increase the amount of the processed starting material, 1:1 mixture of acetone/water was used as medium for the catalyst.

These *in continuous* conditions, allowed to reduce the reaction time from 8 hours to 50 minutes obtaining in all the cases excellent results in terms of yields and stereoselectivity. The scale-up of the synthesis of 3-hydroxy-γ-butirrolactone was demonstrated by processing 22 mmol of starting material, which led to a high overall productivity of target compound and purity, without human intervention both in the workup as well as in the purification (Scheme 3.17).

An interesting example of electrochemical process in flow was reported by Wirth et al. in 2018.[92] This process represents the first catalyst- and supporting electrolyte-free method for electrochemical dehydrogenative C–S

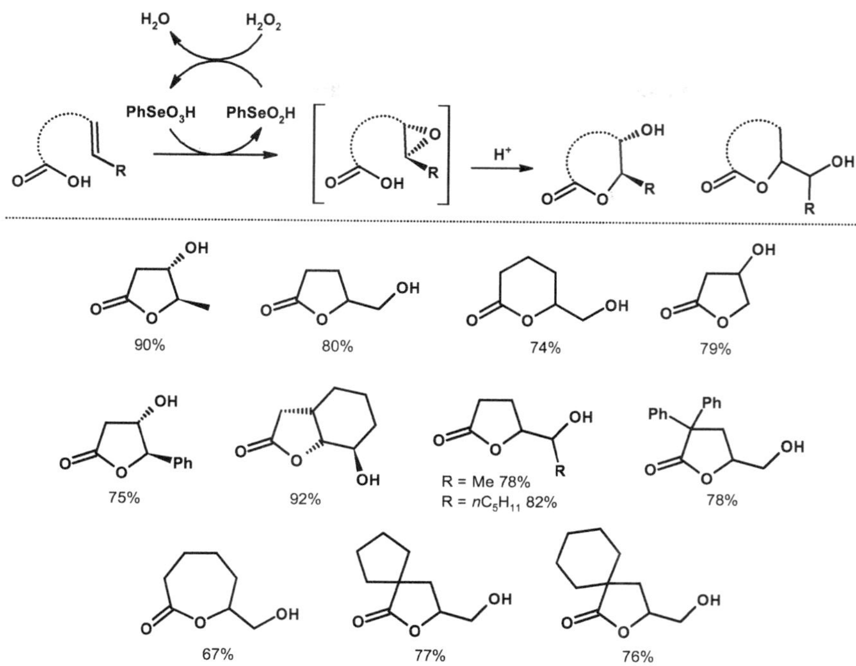

SCHEME 3.16 A small library of hydroxylactones.

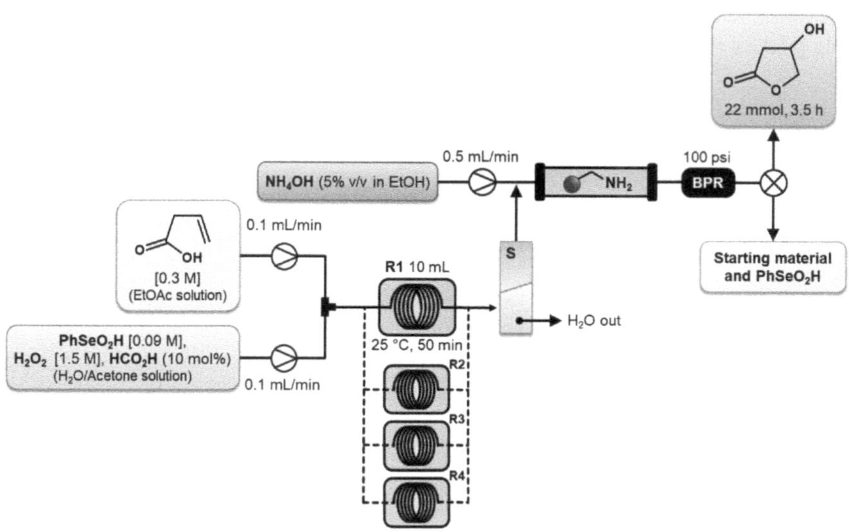

SCHEME 3.17 Setup for the automatic synthesis and purification of 3-hydroxy-γ-butirrolactone.

bond formation in the synthesis of benzothiazoles and thiazolopyridines from *N*-(hetero)arylthioamides using a flow reactor.

The benzothiazole scaffold is part of a large number of bioactive compounds, including antiviral, antitumor, and antiseptic agents, as well as tracers for ß-amyloid plaques in Alzheimer's disease (Fig. 3.12).[93]

The environmental issues and costs have encouraged chemists to seek for innovative strategies for developing new synthetic methodologies such as transition metal-free protocols. This strategy is one of the most interesting tools for the development of innovative synthetic methods of heterocycles.[94] The synthesis of heterocyclic organochalcogen compounds also benefited from this strategy as recently reported by Zeni and coll. that described a three-step one-pot, transition metal-free cyclization for the synthesis of 3-(alkylselanyl) selenazolo indoles.[95]

Benzothiazoles can be synthesized by oxidative cyclization of N-arylthioamides using different reaction conditions.[96–101] Among these procedures, the synthesis of benzothiazole derivatives was also performed by organic electrosynthesis from N-arylthioamides.[102,103] For example, Xu et al., developed a versatile method for synthesis of benzothiazoles and thiazolopyridines through mediated electrochemical oxidation of N-arylthioamides.[104]

The cyclization of N-(quinolin-3-yl)butanethioamide (**9a**) to thiazolo [4,5-c]quinolone (**10a**) (an intermediate in the synthesis of **CL075** – a **TLR8** receptor agonist)[105] was used by Wirth et al. as a model reaction to optimize the electrolysis in flow under conditions similar to those previously employed in batch by Xu.[104] The electrolysis in flow employed platinum as the cathode (Table 3.11), graphite as the anode (working electrode surface: 8.2 cm^2 each), a solvent mixture of acetonitrile and methanol, and a catalytic amount of TEMPO (5 mol.%). It was demonstrated that TEMPO was not required for the electrolysis of thioamide (**9**) in flow

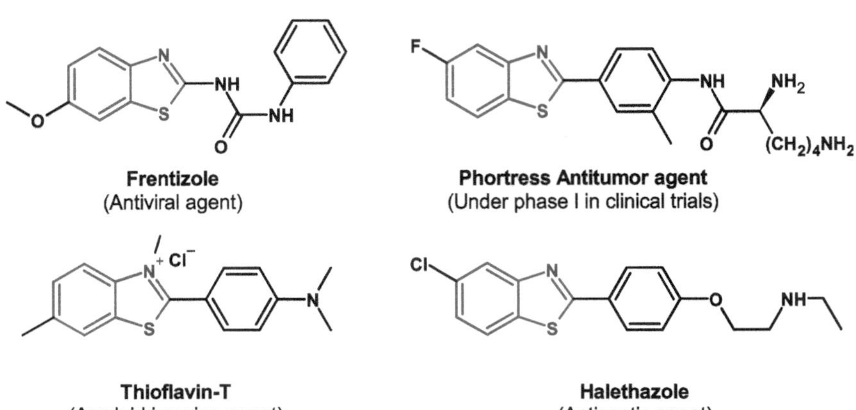

FIGURE 3.12 Benzothiazole-based clinically available drugs.

TABLE 3.11 Optimization of the reaction conditions.

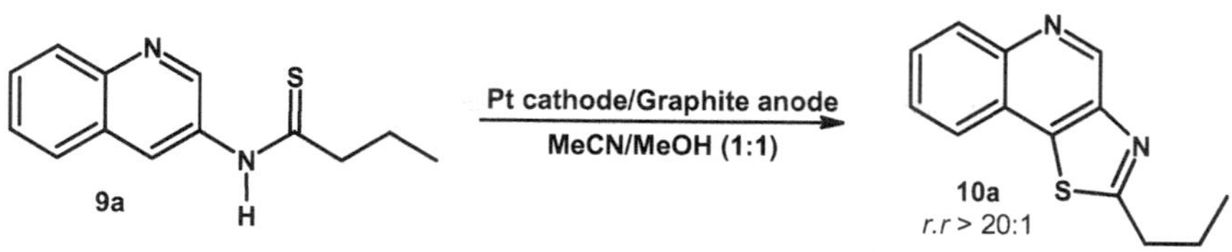

Entry	Concentration (mol/L)	Flow rate (mL/min)	Current density (mA/cm^2)	Current (F/mol)	^{1}H NMR conversion (%)
1[a]	0.025	0.05	0.49	2.0	33
2[a]			0.98	4.0	50
3[a]			1.46	6.0	98
4	0.025	0.05	1.46	6.0	>99
5		0.1	2.93		>99
6		0.2	5.85		>99
7	0.05	0.2	11.71	6.0	>99
8[b]			4.88	2.5	>99
9			3.90	2.0	90

[a] *5 mol.% of TEMPO.*
[b] *97% isolated yield.*

(Table 3.11, entry 4) and higher production rates (conversions >99%) can be obtained by increasing the substrate concentration and flow rate (Table 3.11, entries 5–7). Also the decrease of the electricity amount used in the reaction to 2.5 F/mol did not affect the conversion (Table 3.11, entry 8).

The optimized reaction conditions were applied to investigate the substrate scope (Table 3.12), and the results compared to those of the batch reactor. This electrochemical oxidation process demonstrated good tolerance toward several functional groups and also the possibility to use substrates containing heteroaryl moiety (**9r**, **9s**) or bearing electron-donating groups (**9d**, **9e**, **9g**) and electron-withdrawing groups (**9c**, **9f**) as starting materials. In particular, aminopyridine-derived thioamides (**9r**, **9s**) were cyclized to thiazolopyridines (**10r** and **10s**) with high regioselectivity for the α-position of the pyridyl ring.

The synthetic potential of the method reported by Wirth et al. was also established by a large-scale reaction (13 mmol) performed with the same flow electrochemical reactor (Scheme 3.18). For this purpose, the researchers used a higher flow rate and higher concentration than the one described in the general procedure, in order to decrease the reaction

TABLE 3.12 Substrate scope.

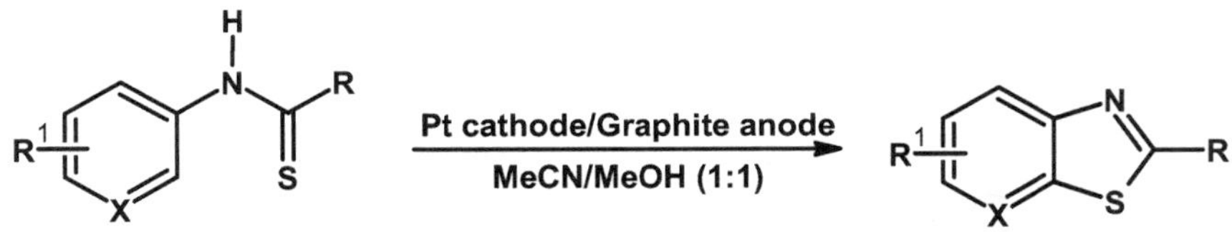

9b-q, X = C or N

10b-q

Product		Flow: Yield [%] (Current [F mol^{-1}])	Batch: Yield [%] (Current [F mol^{-1}])
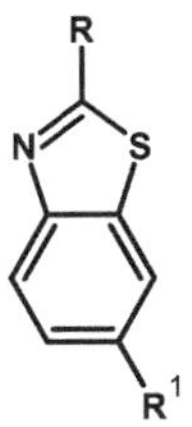	R = Ph	–	–
	10b, R^1 = H	96 (2.6)[a]	95 (3.2)
	10c, R^1 = OCF_3	77 (2.6)[a]	92 (3.3)
	10d, R^1 = SMe	57 (2.9)[a]	68 (4.6)
	10e, R^1 = OMe	92 (4.0)[a]	90 (3.2)
	R = *n*Pr	–	–
	10f, R^1 = CN	58 (2.6)	75 (5.6)
	10g, R^1 = OMe	70 (2.6)	60 (5.3)
	R^1 = H	–	–
	10h, R = Cy	65 (2.0)	82 (5.6)
	10i, R = nBu	81 (2.5)	86 (3.5)
	10j, R = Bn	64 (4.0)	–
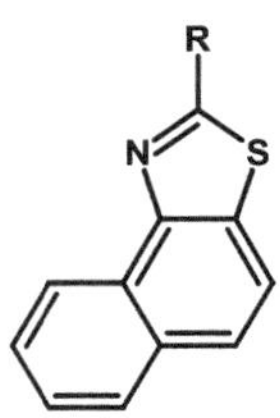	**10k**, R = Ph	97 (2.5)	–
	10m, R = *n*Bu	93 (3.0)	–
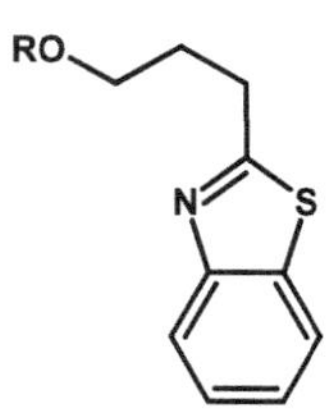	**10n**, R = OH	66 (2.6)	48 (4.0)
	10o, R = OTBDPS	83 (3.0)	80 (3.5)
	10p, R = CO_2tBu	83 (3.0)	92 (3.9)
	10q, R = TsNBoc	36 (3.0)	56 (4.8)
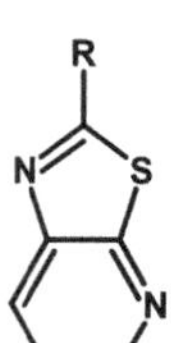	**10r**, R = Ph	81 (3.0) (*r.r.* = 19:1)	–
	10s, R = *n*Bu	87 (3.5) (*r.r.* = 19:1)	–

[a]3% H_2O was added to the solvent mixture. TBDPS = tert-butyldiphenylsilyl, Ts = 4-toluenesulfonyl, Boc = tert-butoxycarbonyl.

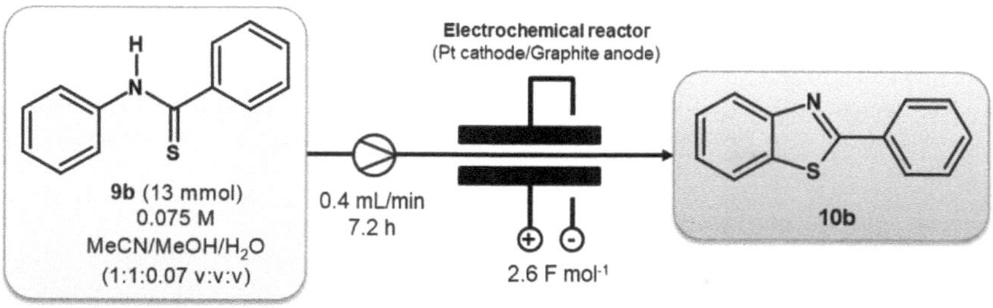

SCHEME 3.18 Large-scale benzothiazole synthesis.

time. This procedure afforded final compound (**10b**) with a yield of 87% instead of 96%. However, the sacrifice of a lower yield is a fair compromise in view of a larger amount of product obtained in shorter time.

3.4 Conclusion

In conclusion, we summarized in this chapter the most recent examples of application of flow chemistry to the organic synthesis involving organochalcogen derivatives as substrates, reactants, and/or catalyst. Even if for this class of compounds, it can be considered still a novel approach, considering that organoselenium and organosulfur derivatives are often considered toxic and potentially dangerous for the environment, the use of in continuous synthesis as well as that of microreactors could represent a promising perspective for the development of new greener procedures. Furthermore, the possibility to setup protocols including also the easy and rapid online purification strongly contribute to reduce the use of organic solvents as well as the wastes production.

References

1. <https://iupac.org/iupac-announces-the-top-ten-emerging-technologies-in-chemistry/>.
2. a) Droescher, M. Chemistry: The Driving Force for Emerging Technologies. *Chem. Int.* **2018,** *40* (4), 14–17.
   b) Kitagawa, S. Chemistry of Small Spaces. *Chem. Int.* **2018,** *40* (4), 4–9.
3. Gomollon-Bel, F. Ten Chemical Innovations that Will Change Our World. *Chem. Int.* **2019,** *41* (2), 12–17.
4. Movsisyan, M.; Delbeke, E. I. P.; Berton, J. K. E. T.; Battilocchio, C.; Leyb, S. V.; Stevens, C. V. Taming Hazardous Chemistry by Continuous Flow Technology. *Chem. Soc. Rev.* **2016,** *45*, 4892–4928.
5. a) Hessel, V. Everything Flows: Continuous Micro-Flow for Pharmaceutical Production. *Chem. Int.* **2018,** *40* (2), 12–16.
   b) Yoshida, J.-I.; Kim, H.; Nagaki, A. Green and Sustainable Chemical Synthesis Using Flow Microreactors. *ChemSusChem* **2011,** *4* (3), 331–340.

6. Wirth, T. Preface. *Bioorg. Med. Chem.* **2017,** *25,* 6179.
7. Plutschack, M. B.; Pieber, B.; Gilmore, K.; Seeberger, P. H. The Hitchhiker's Guide to Flow Chemistry. *Chem. Rev.* **2017,** *117,* 11796–11893.
8. Kirschning, A. *(Ed.) Beilstein J. Org. Chem.,* 5. **2009,** No. 15. Chemistry in Flow Systems and Chemistry in Flow Systems II. Thematic Series in the Open Access Beilstein J. Org. Chem.
9. Mándity, I. M.; Ötvös, S. B.; Fülöp, F. Strategic Application of Residence-Time Control in Continuous-Flow Reactors. *ChemistryOpen* **2015,** *4,* 212–223.
10. Kockmann, N.; Thenée, P.; Fleischer-Trebes, C.; Laudadio, G.; Noël, T. Safety Assessment in Development and Operation of Modular Continuous-Flow Processes. *React. Chem. Eng.* **2017,** *2,* 258–280.
11. Wiles, C.; Watts, P. Recent Advances in Micro Reaction Technology. *Chem. Commun.* **2011,** *47,* 6512–6535.
12. Gutmann, B.; Cantillo, D.; Kappe, C. O. Continuous-Flow Technology. A Tool for the Safe Manufacturing of Active Pharmaceutical Ingredients. *Angew. Chem. Int. (Ed.)* **2015,** *54* (23), 6688–6728.
13. Fitzpatrick, D. E.; Battilocchio, C.; Ley, S. V. A. Novel Internet-Based Reaction Monitoring, Control and Autonomous Self-Optimization Platform for Chemical Synthesis. *Org. Process. Res. Dev.* **2016,** *20* (2), 386–394.
14. Köckinger, M.; Hone, C. A.; Gutmann, B.; Hanselmann, P.; Bersier, M.; Torvisco, A.; Kappe, C. O. Scalable Continuous Flow Process for the Synthesis of Eflornithine Using Fluoroform as Difluoromethyl Source. *Org. Process. Res. Dev.* **2018,** *22,* 1553–1563.
15. Pashkova, A.; Greiner, L. Towards Small-Scale Continuous Chemical Production: Technology Gaps and Challenges. *Chem. Ing. Tech.* **2011,** *83* (9), 1337–1342.
16. Baumann, M.; Moody, T. S.; Smyth, M.; Wharry, S. A Perspective on Continuous Flow Chemistry in the Pharmaceutical Industry. *Org. Process. Res. Dev.* **2020,** *24* (10), 1802–1813.
17. Valikhani, D.; Srivastava, P. L.; Allemann, R. K.; Wirth, T. Immobilised Enzymes for Sesquiterpene Synthesis in Batch and Flow Systems. *ChemCatChem* **2020,** *12,* 2194–2197.
18. Cambié, D.; Noël, T. Solar Photochemistry in Flow. *Top. Curr. Chem.* **2018,** 376.
19. Maljuric, S.; Jud, W.; Kappe, C. O.; Cantillo, D. Translating Batch Electrochemistry to Single-Pass Continuous Flow Conditions: An Organic Chemist's Guide. *J. Flow. Chem.* **2020,** *10,* 181–190.
20. Santi, M.; Seitz, J.; Cicala, R.; Hardwick, T.; Ahmed, N.; Wirth, T. Memory of Chirality in flow Electrochemistry: Fast Optimisation with DoE and Online 2D-HPLC. *Chem. A Eur. J.* **2019,** *25,* 16230–16235.
21. Roy, K. M. *Ullmann's Encyclopedia of Industrial Chemistry;* Wiley-VCH Verlag GmbH & Co. KGaA, 2011.
22. Kagan, H. B.; Chellappan, S. K.; Lattanzi, Al *Encyclopedia of Reagents for Organic Synthesis;* John Wiley & Sons, Ltd, **2014**.
23. Fernández, I.; Khiar, N. Recent Developments in the Synthesis and Utilization of Chiral Sulfoxides. *Chem. Rev.* **2003,** *103* (9), 3651–3705.
24. Johnson, D. S.; Li, J. J. *Modern Drug Synthesis;* John Wiley & Sons, Inc: Hoboken, NJ, 2010.
25. a) Dirikolu, L.; Yohn, R.; Garrett, E. F.; Chakkath, T.; Ferguson, D. C. Detection, Quantifications and Pharmacokinetics of Toltrazuril Sulfone (Ponazuril®) in Cattle. *J. Vet. Pharmacol. Ther.* **2009,** *32,* 280–288.
    b) Zhu, Y. I.; Stiller, M. J. Dapsone and Sulfones in Dermatology: Overview and Update. *J. Am. Acad. Dermatol.* **2001,** *45,* 420–434.
    c) Lind, T.; Rydberg, L.; Kyleback, A.; Jonsson, A.; Andersson, T.; Hasselgren, T.; Holmberg, J.; Rohss, K. Esomeprazole Provides Improved Acid Control vs. Omeprazole in Patients with Symptoms of Gastro-Esophageal Reflux Disease. *Aliment. Pharmacol. Ther.* **2000,** *14,* 861–867.

26. Ilic, D.; Nikolic, V.; Nikolic, L.; Stankovic, M.; Stanojevic, L.; Cakic, M. Allicin and Related Compounds: Biosynthesis, Synthesis and Pharmacological Activity. *Facta Univ.- Ser. Phys. Chem. Technol.* **2011,** *9,* 9–20.
27. Ellman, J. A.; Owens, T. D.; Tang, T. P. N-*tert*-Butanesulfinyl Imines: Versatile Intermediates for the Asymmetric Synthesis of Amines. *Acc. Chem. Res.* **2002,** *35,* 984–995.
    b) Hyean Kim, B.; Curran, D. P. Asymmetric Thermal Reactions with Oppolzer's Camphor Sultam. *Tetrahedron* **1993,** *49,* 293–318.
28. a) Trost, B. M.; Rao, M. Development of Chiral Sulfoxide Ligands for Asymmetric Catalysis. *Angew. Chem. Int. (Ed.)* **2015,** *54,* 5026–5043.
    b) Fernandez, I.; Khiar, N. Recent Developments in the Synthesis and Utilization of Chiral Sulfoxides. *Chem. Rev.* **2003,** *103,* 3651–3706.
29. Stikute, A.; Peipins, V.; Turks, M. Synthesis of Allyl Sulfoxides from Allylsilanes via Silyl Sulfonates. *Tetrahedron Lett.* **2015,** *56* (31), 4578–4581.
30. Turks, M.; Lawrence, A. K.; Vogel, P. The Bora-Ene Reaction of Sulfur Dioxide and Prop-2-ene-1-boronic esters. New Route to Sulfoxides. *Tetrahedron Lett.* **2006,** *47* (16), 2783–2786.
31. Lenstra, D. C.; Vedovato, V.; Ferrer Flegeau, E.; Maydom, J.; Willis, M. C. One-Pot Sulfoxide Synthesis Exploiting a Sulfinyl-Dication Equivalent Generated from a DABSO/Trimethylsilyl Chloride Sequence. *Org. Lett.* **2016,** *18* (9), 2086–2089.
32. Peyronneau, M.; Roques, N.; Mazières, S.; Le Roux, C. Catalytic Lewis Acid Activation of Thionyl Chloride: Application to the Synthesis of Arylsulfinyl Chlorides Catalyzed by Bismuth(iii) Salts. *Synlett* **2003,** *5,* 0631–0634.
33. Frank, R. L.; Seven, R. P. Isoprene Cyclic Sulfone. *Org. Synth.* **1949,** *29,* 59.
34. Répichet, S.; Le Roux, C.; Hernandez, P.; Dubac, J.; Desmurs, J. R. Bismuth(III) Trifluoromethanesulfonate: An Efficient Catalyst for the Sulfonylation of Arenes. *J. Org. Chem.* **1999,** *64* (17), 6479–6482.
35. Ferry, C. W.; Buck, J. S.; Baltzly, R. 4,4′-Diaminodiphenylsulfone. *Org. Synth.* **1942,** *22,* 31.
36. Paquette, L. A.; Carr, R. V. C. Phenyl Vinyl Sulfone and Sulfoxide. *Org. Synth.* **1986,** *64,* 157.
37. a) Trivedi, R.; Lalitha, P. VCl_3-Catalyzed Selective Oxidation of Sulfides to Sulfoxides Using H_2O_2 as Oxidant. *Synth. Commun.* **2006,** *36,* 3777–3782.
    b) Matteucci, M.; Bhalay, G.; Bradley, M. Mild and Highly Chemoselective Oxidation of Thioethers Mediated by $Sc(OTf)_3$. *Org. Lett.* **2003,** *5,* 235–237.
38. Shin, S. S.; Byun, Y.; Lim, K. M.; Choi, J. K.; Lee, K.-w; Moh, J. H.; Kim, J. K.; Jeong, Y. S.; Kim, J. Y.; Choi, Y. H., et al. In Vitro Structure – Activity Relationship and In Vivo Studies for a Novel Class of Cyclooxygenase-2 Inhibitors: 5-Aryl-2,2-dialkyl-4-phenyl-3(2*H*)furanone Derivatives. *J. Med. Chem.* **2004,** *47,* 792–804.
39. Hiskey, R. G.; Harpold, M. A. Chemistry of Aliphatic Disulfides. XIV. The Preparation of Disulfide Sulfoxides by Selective Oxidation. *J. Org. Chem.* **1967,** *32,* 3191–3194.
40. Lai, S.; Lee, D. G. Lewis Acid Assisted Permanganate Oxidations. *Tetrahedron* **2002,** *58,* 9879–9887.
41. Colonna, S.; Gaggero, N. Enantioselective Oxidation of Sulphides by Dioxiranes in the Presence of Bovine Serum Albumin. *Tetrahedron Lett.* **1989,** *30,* 6233–6236.
42. Johnson, C. R.; Keiser, J. E. Methyl Phenyl Sulfoxide. *Org. Synth.* **1966,** *46,* 78.
43. Rayner, C. M. Synthesis of Thiols, Sulfides, Sulfoxides and Sulfones. *Contemp. Org. Synth.* **1995,** *2,* 409–440.
44. Kamata, K.; Hirano, T.; Mizuno, N. Highly Efficient Oxidation of Sulfides with Hydrogen Peroxide Catalyzed by $[SeO_4\{WO(O_2)_2\}_2]^{2-}$. *Chem. Commun.* **2009,** 3958–3960.
45. Griffin, R. J.; Henderson, A.; Curtin, N. J.; Echalier, A.; Endicott, J. A.; Hardcastle, I. R.; Newell, D. R.; Noble, M. E. M.; Wang, L. Z.; Golding, B. T. Searching for Cyclin-Dependent Kinase Inhibitors Using a New Variant of the Cope Elimination. *J. Am. Chem. Soc.* **2006,** *128,* 6012–6013.

46. Varma, R. S.; Naicker, K. P. The Urea – Hydrogen Peroxide Complex: Solid-State Oxidative Protocols for Hydroxylated Aldehydes and Ketones (Dakin Reaction), Nitriles, Sulfides, and Nitrogen Heterocycles. *Org. Lett.* **1999,** *1*, 189–192.
47. Fukuda, N.; Ikemoto, T. Imide-Catalyzed Oxidation System: Sulfides to Sulfoxides and Sulfones. *J. Org. Chem.* **2010,** *75*, 4629–4631.
48. Mirkhani, V.; Tangestaninejad, S.; Moghadamb, M.; Mohammadpoor-Baltork, I.; Kargar, H. Efficient Oxidation of Sulfides with Sodium Periodate Catalyzed by Manganese(III) Schiff Base Complexes. *J. Mol. Catal. A* **2005,** *242*, 251–255.
49. Li, Z.-Z.; Yao, S.-Y.; Ye, B.-H. Enantioselective Oxidation of Thioethers to Sulfoxides by Means of a Structural Template with Chiral-At-Metal Ruthenium Complexes. *ChemPlusChem* **2015,** *80*, 141–150.
50. Hussain, H.; Green, I. R.; Ahmed, I. Journey Describing Applications of Oxone in Synthetic Chemistry. *Chem. Rev.* **2013,** *113*, 3329–3371.
51. Yu, B.; Liu, A.-H.; He, L.-N.; Li, B.; Diao, Z.-F.; Li, Y.-N. Catalyst-Free Approach for Solvent-Dependent Selective Oxidation of Organic Sulfides with Oxone. *Green Chem.* **2012,** *14*, 957–962.
52. Mello, R.; Olmos, A.; Alcalde-Aragonés, A.; Díaz-Rodríguez, A.; González Núñez, M. E.; Asensio, G. Oxidation of Sulfides with a Silica-Supported Peracid in Supercritical Carbon Dioxide under Flow Conditions: Tuning Chemoselectivity with Pressure. *Eur. J. Org. Chem.* **2010,** 6200–6206.
53. a) Leitner, W. Supercritical Carbon Dioxide as a Green Reaction Medium for Catalysis. *ACC. Chem. Res.* **2002,** *35*, 746–756.
    b) Wells, S.; DeSimone, J. M. CO_2 Technology Platform: An Important Tool for Environmental Problem Solving. *Angew. Chem. Int. (Ed.)* **2001,** *40*, 518–527.
    c) McHugh, M.; Krukonis, V. *Supercritical Fluid Extraction*, 2nd ed.; Butterworth-Heinemann: Boston, MA, 1994.
54. a) Salameh, A.; Baudouin, A.; Basset, J.-M.; Copéret, C. Tuning the Selectivity of Alumina-Supported $(CH_3)ReO_3$ by Modifying the Surface Properties of the Support. *Angew. Chem. Int. (Ed.)* **2008,** *47*, 2117–2120.
    b) Berthoud, R.; Baudouin, A.; Fenet, B.; Lukens, W.; Pelzer, K.; Basset, J.-M.; Candy, J.-P.; Copéret, C. Mononuclear Ruthenium Hydride Species vs Ruthenium Nanoparticles: The Effect of Silane Functionalities on Silica Surfaces. *Chem. Eur. J.* **2008,** *14*, 3523–3526.
    c) Ruddy, D. A.; Tilley, T. D. Highly Selective Olefin Epoxidation with Aqueous H_2O_2 Over Surface-Modified TaSBA15 Prepared via the TMP Method. *Chem. Commun.* **2007,** 3350–3352.
    d) Dámore, M. B.; Schwarz, S. Trimethylsilylation of Ordered and Disordered Titanosilicates: Improvements in Epoxidation with Aqueous H_2O_2 from Micro- to Meso-Pores and Beyond. *Chem. Commun.* **1999,** 121–122.
    e) Basset, J.-M.; Lefebvre, F.; Santini, C. Density Functional Theory. A Powerful Tool for Theoretical Studies in Coordination Chemistry. *Coord. Chem. Rev.* **1998,** 178–180.
55. Silva, F.; Baker, A.; Stansall, J.; Michalska, W.; Yusubov, M. S.; Graz, M.; Saunders, R.; Evans, G. J. S.; Wirth, T. Selective Oxidation of Sulfides in Flow Chemistry. *Eur. J. Org. Chem.* **2018,** 2134–2137.
56. Hashimoto, N.; Kanda, A. Practical and Environmentally Friendly Epoxidation of Olefins Using Oxone. *Org. Process. Res. Dev.* **2002,** *6*, 405–406.
57 Maggi, R.; Chitsaz, S.; Loebbecke, S.; Piscopo, C. G.; Sartori, G.; Schwarzer, M. Highly Chemoselective Metal-Free Oxidation of Sulfides with diluted H_2O_2 in a Continuous Flow Reactor. *Green. Chem.* **2011,** *13*, 1121–1123.
58. Mangiavacchi, F.; Crociani, L.; Sancineto, L.; Marini, F.; Santi, C. Continuous Bioinspired Oxidation of Sulphides. *Molecules* **2020,** *25*, 2711.

59. Santoro, S.; Azeredo, J. B.; Nascimento, V.; Sancineto, L.; Braga, A. L.; Santi, C. The Green Side of the Moon: Ecofriendly Aspects of Organoselenium Chemistry. *RSC Adv.* **2014,** *4,* 31521–31535.
60. Santi, C.; Santoro, S.; Battistelli, B.; Testaferri, L.; Tiecco, M. Preparation of the First Bench-Stable Phenyl Selenolate: An Interesting "On Water" Nucleophilic Reagent. *Eur. J. Org. Chem.* **2008,** 5387–5390.
61. Tidei, C.; Piroddi, M.; Galli, F.; Santi, C. Oxidation of Thiols Promoted by PhSeZnCl. *Tetrahedron Lett.* **2012,** *53,* 232–234.
62. Santi, C.; Jacob, R.; Monti, B.; Bagnoli, L.; Sancineto, L.; Lenardão, E. Water and Aqueous Mixtures as Convenient Alternative Media for Organoselenium Chemistry. *Molecules* **2016,** *21,* 1482.
63. Di Schino, L.; Incipini, L.; Dragone, V.; Tidei, C.; Scalera, C.; Santi, C. Green Chemistry for Environmental Sustainability: An Example of "bio-logic" Approach. In *Proceedings of the 1st World Sustainability Forum,* Basel, Switzerland, 1–30 November 2011; MDPI: Basel, Switzerland, 2011; p 675.
64. Młochowski, J.; Wójtowicz-Młochowska, H. Developments in Synthetic Application of Selenium(IV) Oxide and Organoselenium Compounds as Oxygen Donors and Oxygen-Transfer Agents. *Molecules* **2015,** *20,* 10205–10243.
65. a) Drabowicz, J.; Mikołajczyk, M. A Facile and Selective Oxidation of Organic Sulphides to Sulphoxides with Hydrogen Peroxide/Selenium Dioxide System. *Synthesis* **1978,** 758–759.
    b) Drabowicz, J.; łyzwa, P.; Mikołajczyk, M. A New Procedure for Oxidation of Sulfides to Sulfones. *Phosphorus Sulfur Relat. Elem.* **1983,** *17,* 169–172.
66. a) Horn, E. J.; Rosen, B. R.; Baran, P. S. Synthetic Organic Electrochemistry: An Enabling and Innately Sustainable Method. *ACS Cent. Sci.* **2016,** *2,* 302–308.
    b) Waldvogel, S. R.; Janza, B. Renaissance of Electrosynthetic Methods for the Construction of Complex Molecules. *Angew. Chem. Int. (Ed.)* **2014,** *53,* 7122–7123.
    c) Yoshida, J.; Kataoka, K.; Horcajada, R.; Nagaki, A. Modern Strategies in Electroorganic Synthesis. *Chem. Rev.* **2008,** *108,* 2265–2299.
    d) Francke, R.; Little, R. D. Redox Catalysis in Organic Electrosynthesis: Basic Principles and Recent Developments. *Chem. Soc. Rev.* **2014,** *43,* 2492–2521.
    e) Frontana-Uribe, B. A.; Little, R. D.; Ibanez, J. G.; Palma, A.; Vasquez-Medrano, R. Organic Electrosynthesis: A Promising Green Methodology in Organic Chemistry. *Green Chem.* **2010,** *12,* 2099–2119.
67. a) Pletcher, D.; Green, R. A.; Brown, R. C. D. Flow Electrolysis Cells for the Synthetic Organic Chemistry Laboratory. *Chem. Rev.* **2018,** *118* (9), 4573–4591.
    b) Atobe, M.; Tateno, H.; Matsumura, Y. Applications of Flow Microreactors in Electrosynthetic Processes. *Chem. Rev.* **2018,** *118* (9), 4541–4572.
    c) Folgueiras-Amador, A. A.; Wirth, T. Perspectives in Flow Electrochemistry. *J. Flow Chem* **2017,** *7,* 94–95.
    d) Green, R. A.; Brown, R. C. D.; Pletcher, D. Electrosynthesis in Extended Channel Length Microfluidic Electrolysis Cells. *J. Flow Chem.* **2016,** *6,* 191–197.
    e) Green, R. A.; Brown, R. C. D.; Pletcher, D.; Harji, B. A Microflow Electrolysis Cell for Laboratory Synthesis on the Multigram Scale. *Org. Process Res. Dev.* **2015,** *19,* 1424–1427.
    f) Watts, K.; Baker, A.; Wirth, T. Electrochemical Synthesis in Microreactors. *J. Flow Chem.* **2014,** *4,* 2–11.
    g) Attour, A.; Dirrenberger, P.; Rode, S.; Ziogas, A.; Matlosz, M.; Lapicque, F. A High Pressure Single-Pass High-Conversion Electrochemical Cell for Intensification of Organic Electrosynthesis Processes. *Chem. Eng. Sci.* **2011,** *66,* 480–489.
68. Getz, C.; Stenglein, A.; Waldvogel, S. R. Highly Modular Flow Cell for Electroorganic Synthesis. *Org. Process. Res. Dev.* **2017,** *21,* 771–778.

69. Laudadio, G.; Straathof, N. J. W.; Lanting, M. D.; Knoops, B.; Hessel, V.; Noël, T. An Environmentally Benign and Selective Electrochemical Oxidation of Sulfides and Thiols in a Continuous-Flow Microreactor. *Green Chem.* **2017,** *19,* 4061–4066.
70. Bodzansky, M. *Principles of Peptide Synthesis;* Springer: Berlin, **1984,** Chapter 4.
71. Cremlyn, R. J. *An Introduction to Organosulfur Chemistry;* Wiley-VCH: New York, **1996**.
72. Sonavane, S. U.; Chidambaram, M.; Almog, J.; Sasson, Y. Rapid and Efficient Synthesis of Symmetrical Alkyl Disulfides under Phase Transfer Conditions. *Tetrahedron Lett.* **2007,** *48,* 6048–6050.
73. Witt, D. Recent Developments in Disulfide Bond Formation. *Sinthesis* **2008,** 2491–2509.
74. Talla, A.; Driessen, B.; Straathof, N. J. W.; Milroy, L.-G.; Brunsveld, L.; Hessel, V.; Noël, T. Metal-Free Photocatalytic Aerobic Oxidation of Thiols to Disulfides in Batch and Continuous-Flow. *Adv. Synth. Catal.* **2015,** *357,* 2180–2186.
75. a) Noël, T.; Hessel, V. Membrane Microreactors: Gas–Liquid Reactions Made Easy. *ChemSusChem* **2013,** *6,* 405–407.
    b) Sobieszuk, P.; Aubin, J.; Phorecki, R. Hydrodynamics and Mass Transfer in Gas-Liquid Flows in Microreactors. *Chem. Eng. Technol.* **2012,** *35,* 1346–1358.
    c) Gunther, A.; Jensen, K. F. Multiphase Microfluidics: From Flow Characteristics to Chemical and Materials Synthesis. *Lab Chip* **2006,** *6,* 1487–1503.
76. a) Su, Y.; Straathof, N. J. W.; Hessel, V.; Noël, T. Photochemical Transformations Accelerated in Continuous-Flow Reactors: Basic Concepts and Applications. *Chem. Eur. J.* **2014,** *20,* 10562–10589.
    b) Straathof, N. J. W.; Tegelbeckers, B. J. P.; Hessel, V.; Wang, X.; Noël, T. A Mild and Fast Photocatalytic Trifluoromethylation of Thiols in Batch and Continuous-Flow. *Chem. Sci.* **2014,** *5,* 4768–4773.
77. He, Z.; Jamison, T. F. Continuous-Flow Synthesis of Functionalized Phenols by Aerobic Oxidation of Grignard Reagents. *Angew. Chem. Int. (Ed.)* **2014,** *53,* 3353–3357.
78. Carter, C. S. Oxytocin Pathways and the Evolution of Human Behavior. *Annu. Rev. Psychol.* **2014,** *65,* 17–39.
79. Beach, C. A.; Mays, D. C.; Guiler, R. C.; Jacober, C. H.; Gerber, N. Inhibition of Elimination of Caffeine by Disulfiram in Normal Subjects and Recovering Alcoholics. *Clin. Pharmacol. Ther.* **1986,** *39,* 265–270.
80. Weeranoppanant, N. Enabling Tools for Continuous-Flow Biphasic Liquid-Liquid Reaction. *React. Chem. Eng.* **2019,** *4,* 235–243.
81. Wang, K.; Li, L.; Xie, P.; Luo, G. Liquid-Liquid Microflow Reaction Engineering. *React. Chem. Eng.* **2017,** *2,* 611–627.
82. a) Jensen, K. F. Flow Chemistry - Microreaction Technology Comes of Age. *AIChE J.* **2017,** *63,* 858–869.
    b) Hessel, V.; Kralisch, D.; Kockmann, N.; Noël, T.; Wang, Q. Novel Process Windows for Enabling, Accelerating, and Uplifting Flow Chemistry. *ChemSusChem* **2013,** *6,* 746–789.
    c) Nieves-Remacha, M. J.; Kulkarni, A. A.; Jensen, K. F. Hydrodynamics of Liquid-Liquid Dispersion in an Advanced-Flow Reactor. *Ind. Eng. Chem. Res.* **2012,** *51,* 16251–16262.
    d) Lobry, E.; Theron, F.; Gourdon, C.; Le Sauze, N.; Xuereb, C.; Lasuye, T. Turbulent Liquid-Liquid Dispersion in SMV Static Mixer at High Dispersed Phase Concentration. *Chem. Eng. Sci.* **2011,** *66,* 5762–5774.
83. a) Shimanouchi, T.; Kataoka, Y.; Yasukawa, M.; Ono, T.; Kimura, Y. Simplified Model for Extraction of 5-Hydroxymethylfurfural from Fructose: Use of Water/oil Biphasic System under High Temperature and Pressure Conditions. *Solvent Extr. Res. Dev.* **2013,** *20,* 205–212.
    b) Shimanouchi, T.; Kataoka, Y.; Tanifuji, T.; Kimura, Y.; Fujioka, S.; Terasaka, K. Chemical Conversion and Liquid-Liquid Extraction of 5-Hydroxymethylfurfural from Fructose by Slug Flow Microreactor. *AIChE J.* **2016,** *62,* 2135–2143.

c) Ahmed-Omer, B.; Barrow, D.; Wirth, T. Effect of Segmented Fluid Flow, Sonication and Phase Transfer Catalysis on Biphasic Reactions in Capillary Microreactors. *Chem. Eng. J.* **2008,** *135,* S280–S283.

d) Mallia, C. J.; Baxendale, I. R. The Use of Gases in Flow Synthesis. *Org. Process Res. Dev.* **2016,** *20,* 327–360.

84. Baker, A.; Graz, M.; Saunders, R.; Evans, G. J. S.; Kaul, S.; Wirth, T. Flow Synthesis of Symmetrical Di- and Trisulfides Using Phase-Transfer Catalysis. *J. Flow. Chem.* **2013,** *3* (4), 118–121.

85. Block, E.; Ahmad, S.; Catalfamo, J. L.; Jain, M. K.; Apitzcastro, R. Antithrombotic Organosulfur Compounds from Garlic: Structural, Mechanistic, and Synthetic Studies. *J. Am. Chem. Soc.* **1986,** *108,* 7045–7055.

86. Hutchings, M.; Ahmed-Omer, B.; Wirth, T. In *Microreactors in Organic Chemistry and Catalysis*; Wirth, T., Ed.; 2nd ed. Wiley-VCH: Weinheim, **2013**, 197–219.

87. Begini, F.; Krasowska, D.; Jasiak, A. M.; Drabowicz, J.; Santi, C.; Sancineto, L. Continuous Flow Synthesis of 2,2′-Diselenobis(Benzoic Acid) and Derivatives. *React. Chem. Eng.* **2020,** *5,* 641–644.

88. Santi, C.; Scalera, C.; Tidei, C.; Dragone, V.; Incipini, L.; Di Schino, L. *Green Chemistry for Environmental Sustainability: An Example of "Bio-Logic" Approach* **2011**. Available from: https://doi.org/10.3390/wsf-00675.

89. Santoro, S.; Santi, C.; Sabatini, M.; Testaferri, L.; Tiecco, M. Eco-Friendly Olefin Dihydroxylation Catalyzed by Diphenyl Diselenide. *Adv. Synth. Catal.* **2008,** *350,* 2881–2884.

90. Sancineto, L.; Mangiavacchi, F.; Tidei, C.; Bagnoli, L.; Marini, F.; Gioiello, A.; Scianowski, J.; Santi, C. Selenium-Catalyzed Oxacyclization of Alkenoic Acids and Alkenols. *Asian J. Org. Chem.* **2017,** *6,* 988–992.

91. Cerra, B.; Mangiavacchi, F.; Santi, C.; Lozza, A. M.; Gioiello, A. Selective Continuous Flow Synthesis of Hydroxyl Lactones from Alkenoic Acids. *React. Chem. Eng.* **2017,** *2,* 467–471.

92. Folgueiras-Amador, A. A.; Qian, X.-Y.; Xu, H.-C.; Wirth, T. Catalyst- and Supporting-Electrolyte-Free Electrosynthesis of Benzothiazoles and Thiazolopyridines in Continuous Flow. *Chem. Eur. J.* **2018,** *24,* 487–491.

93. Sharma, P. C.; Sinhmar, A.; Sharma, A.; Rajak, H.; Pathak, D. P. Medicinal Significance of Benzothiazole Scaffold: An Insight View. *J. Enzyme Inhib. Med. Chem.* **2013,** *28,* 240–266.

94. Zhang, B.; Armido Studer, A. Recent Advances in the Synthesis of Nitrogen Heterocycles via Radical Cascade Reactions Using Isonitriles as Radical Acceptors. *Chem. Soc. Rev.* **2015,** *44,* 3505–3521.

95. Prochnow, T.; Maroneze, A.; Back, D. F.; Zeni, G. Synthesis of 3-(Organochalcogen) Chalcogenazolo Indoles via Cascade Cyclization of N-Alkynylindoles. *J. Org. Chem.* **2019,** *84,* 2891–2900.

96. Wang, H.; Wang, L.; Shang, J.; Li, X.; Wang, H.; Gui, J.; Lei, A. Fe-Catalysed Oxidative C-H Functionalization/C-S Bond Formation. *Chem. Commun.* **2012,** *48,* 76–78.

97. Rey, V.; Soria-Castro, S. M.; Argeello, J. E.; Peñéñory, A. B. Photochemical Cyclization of Thioformanilides by Chloranil: An Approach to 2-Substituted Benzothiazoles. *Tetrahedron Lett.* **2009,** *50,* 4720–4723.

98. Cheng, Y.; Yang, J.; Qu, Y.; Li, P. Aerobic Visible-Light Photoredox Radical C-H Functionalization: Catalytic Synthesis of 2-Substituted Benzothiazoles. *Org. Lett.* **2012,** *14,* 98–101.

99. Bose, D. S.; Idrees, M. Hypervalent Iodine Mediated Intramolecular Cyclization of Thioformanilides: Expeditious Approach to 2-Substituted Benzothiazoles. *J. Org. Chem.* **2006,** *71,* 8261–8263.

100. Inamoto, K.; Hasegawa, C.; Kawasaki, J.; Hiroya, K.; Doi, T. Use of Molecular Oxygen as a Reoxidant in the Synthesis of 2-Substituted Benzothiazoles via

Palladium-Catalyzed C-H Functionalization/Intramolecular C-S Bond Formation. *Adv. Synth. Catal.* **2010,** *352,* 2643–2655.

101. Zhang, G.; Liu, C.; Yi, H.; Meng, Q.; Bian, C.; Chen, H.; Jian, J. X.; Wu, L. Z.; Lei, A. External Oxidant-Free Oxidative Cross-Coupling: A Photoredox Cobalt-Catalyzed Aromatic C-H Thiolation for Constructing C-S Bonds. *J. Am. Chem. Soc.* **2015,** *137,* 9273–9280.
102. Tabaković, I.; Trkovnik, M.; Batušić, M.; Tabakoviš, K. Electrochemical Synthesis of Heterocyclic Compounds; VII[1]. Anodic Synthesis of 1,3-Thiazole Derivatives. *Synthesis* **1979,** 590–592.
103. Wang, P.; Tang, S.; Lei, A. Electrochemical Intramolecular Dehydrogenative C-S Bond Formation for the Synthesis of Benzothiazoles. *Green Chem.* **2017,** *19,* 2092–2095.
104. Qian, X.-Y.; Li, S.-Q.; Song, J.; Xu, H.-C. TEMPO-Catalyzed Electrochemical C-H Thiolation: Synthesis of Benzothiazoles and Thiazolopyridines from Thioamides. *ACS Catal.* **2017,** *7,* 2730–2734.
105. Kokatla, H. P.; Yoo, E.; Salunke, D. B.; Sil, D.; Ng, C. F.; Balakirshna, R.; Malladi, S. S.; Fox, L. M.; David, S. A. Toll-Like Receptor-8 Agonistic Activities in C2, C4, and C8 Modified Thiazolo[4,5-c]Quinolones. *Org. Biomol. Chem.* **2013,** *11,* 1179–1198.

CHAPTER

4

Advances in photochemical seleno-functionalization of (hetero)arenes

Jamal Rafique[1,2], *Juliano B. Azeredo*[3], *Sumbal Saba*[2], *Paulo H. Schneider*[4] and *Antonio L. Braga*[5]

[1]Institute of Chemistry (INQUI), Federal University of Mato Grosso do Sul – UFMS, Campo Grande, Mato Grosso do Sul (MS), Brazil, [2]Institute of Chemistry (IQ), Federal University of Goiás – UFG, Goiânia, Goiás (GO), Brazil, [3]Department of Pharmacy, Federal University of Pampa, Uruguaiana, Rio Grande do Sul (RS), Brazil, [4]Institute of Chemistry (IQ), Department of Organic Chemistry, Federal University of Rio Grande do Sul – UFRGS, Porto Alegre, Rio Grande do Sul (RS), Brazil, [5]Department of Chemistry, Federal University of Santa Catarina – UFSC, Florianópolis, Santa Catarina (SC), Brazil

4.1 Introduction

In the last few decades, there has been a rising interest in synthetic organoselenium compounds. This is mainly due to the discovery of the presence of the selenium atom in the structure of some proteins in the 1970s.[1] It was observed that these selenoproteins are involved in a large number of physiological processes in mammals, such as thyroid hormone production, antioxidant defense, and immune responses.[1] It is noteworthy to mention that the role of selenium atom in the mode of action of such proteins cannot be performed by its closest relative, sulfur.[2] Furthermore, many therapeutic activities are attributed to

Organochalcogen Compounds
DOI: https://doi.org/10.1016/B978-0-12-819449-2.00008-2

organoselenium compounds like the function as apoptosis inducers, chemopreventors, and antioxidants, in several organs including brain, colon, kidney, liver, lung, prostate, stomach, and skin.[3] Therefore, in recent years, synthetic organoselenium compounds have been extensively studied for numerous biological activities, such as anti-HIV, enzyme inhibition,[4,5] enzyme mimetic,[6] immune response, thyroid hormone production, anticancer, anti-Alzheimer, antihypertensive activities,[7] among others.[3c,8] Furthermore, selenium-containing organic structure have been successfully designed and screened as antimicrobial substances.[9] In relation to current COVID-19 pandemic situation, it has been found that the organoselenium compound, ebselen (**I**) (Fig. 4.1) has antiviral activity against the main protease of SARS-CoV2 virus, which causes the COVID-19.[10]

Consequently, in modern organic synthesis, organoselenium compounds have achieved a new dimension, where several applications have been attributed to these compounds, for example, their use as organocatalysts, synthetic intermediates in total synthesis, ligands, and ionic liquids.[11,12] They have also been used as fluorescent probes in material science.[13,14] Therefore, the design and synthesis of selenylated small molecules constitute a significant research theme in both academic laboratories and chemical/pharmaceutical industries.[15] In this regard, heteroarenes and activated arenes substrates containing organoselenium moieties (Fig. 4.1) have appeared as a very important class of molecules with diverse biological applications.[16]

On the other hand, concern for the preservation of the environment has been growing since the 1990s, and the use of alternative and environmentally benign methods in the area of organoselenium

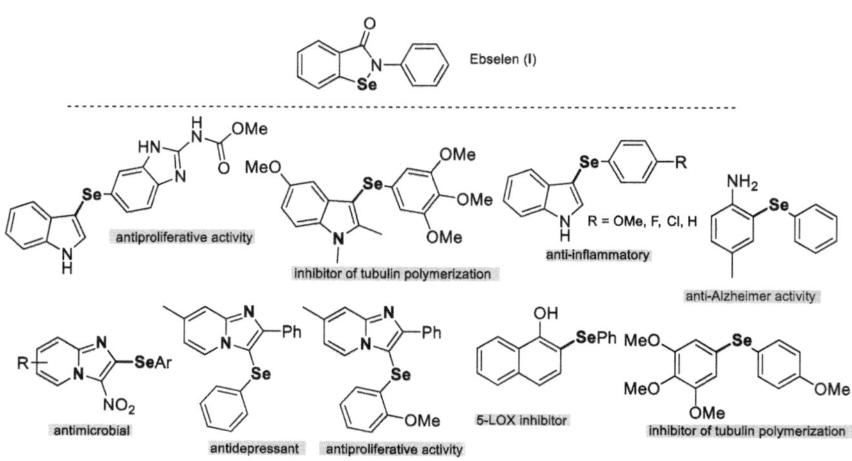

FIGURE 4.1 Biologically relevant selenylated (hetero)arenes.

chemistry is revolutionizing our approach to modern organic synthesis.[17] The use of greener chemistry approaches has a significant increase in the number of alternative protocols for seleno-functionalization of organic structures. This can be observed, for instance, in terms of use of alternative energy sources, renewable solvents, and catalysis.[18]

Among the greener alternative energy sources, photo-induced organic transformations open a new avenue in organic chemistry as an efficient and environmentally benign synthetic strategy, making it desirable as compared to the conventional approaches.[19,20] Under visible light and with the aid of organic dyes (**1-6**) or transition metal salts (**7-9**) as photoredox catalysts (Fig. 4.2), many photocatalytic methodologies

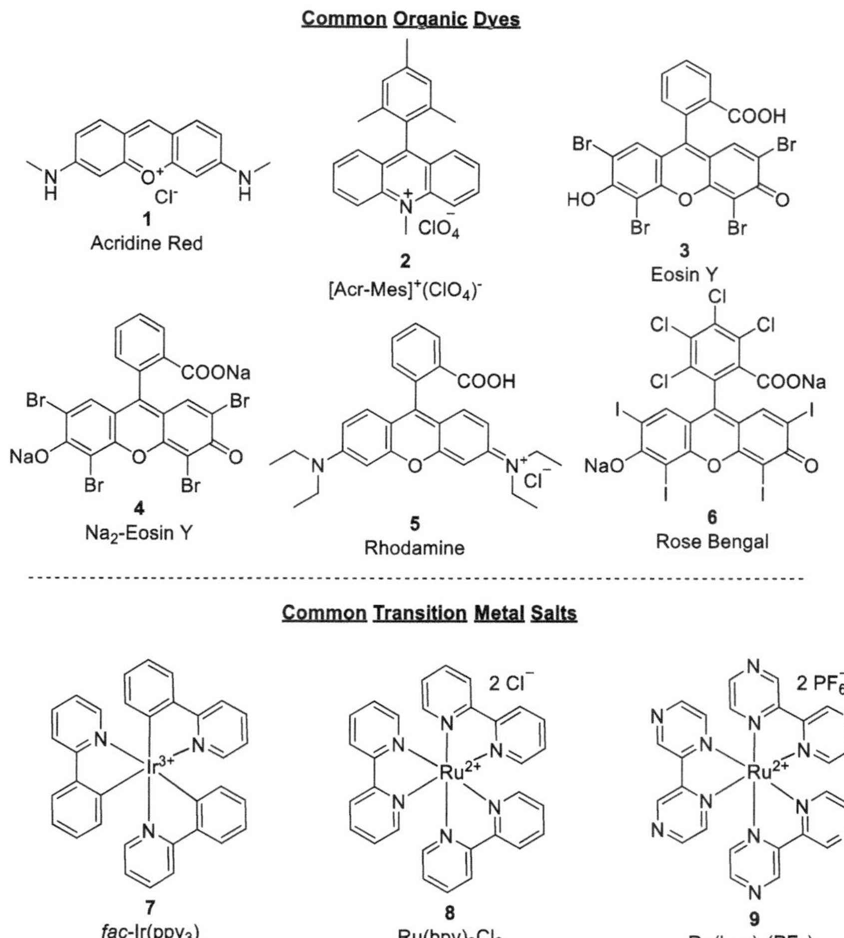

FIGURE 4.2 Common photo-redox catalysts.

have been successfully developed, leading to the achievement of target molecules with high yields, low side-product formation and with high energy efficiency, making these methods interesting from the green chemistry perspective.[19,20] Besides, there are several reports where the photocatalyst is no longer needed.[20] In general, these transformations are accomplished under milder conditions. It is noteworthy to mention that these reactions are easy to set up and are carried out by irradiation with inexpensive light sources, for example, commercial CFL bulbs or light-emitting diodes (LEDs; blue, green, white, violet) or without any special photochemical equipment.[21] Considering these advantages, reports on the light-mediated organic transformations involving seleno-functionalization of heteroarenes as well as activated arenes, among other class of compounds, have been growing considerably in the last few years.[19f] In all these protocols, in general, the substrate, the selenylating reagent, and the photocatalyst, when involved, are stimulated by the visible light irradiation, becoming oxidized or reduced via single electron transfer (SET), generating the radical species that follow in the mechanistic process depending on the reaction medium involved.

In this context, this chapter covers recent scientific literature (excluding the patent literature) addressing the topic of photochemical seleno-functionalization of arenes and heteroarenes. The chemistry of organoselenium compounds that do not involve photo-induced seleno-functionalization of (hetero)arenes has not been included. Consequently, there are a large number of relevant publications reporting very interesting studies, carried out by many groups, which have regrettably been omitted from this chapter.

4.2 Photo-induced selenylation of heteroarenes

The seleno-functionalization of heterocyclic systems is a topic of great interest to the scientific community, owing to their potential therapeutic applications.[22] Among the synthetic methodologies used for this purpose, the direct selenylation of heterocycles through the formation of the C(sp^2)–Se bond is highlighted. In 2018, Kumaraswamy and coworkers, developed a photo-induced direct C(sp^2)–H bond selenylation of indoles in 2-methylpropanol as solvent. The reaction was also extended to imidazopyridines and electron-rich arenes.[23] Under the optimized reaction condition, the authors used 1 molar equivalent of heteroarenes (**10**) in the presence of 2 molar equivalent of LiCl and 0.5 molar equivalent of the diaryl diselenides (**11**) in white LED light source (Scheme 4.1). The reaction operated under air atmosphere and at room temperature without the need for a photocatalyst, resulting the selenylated products (**12**) in 50%–94% yields. The scope of this reaction was explored using variously

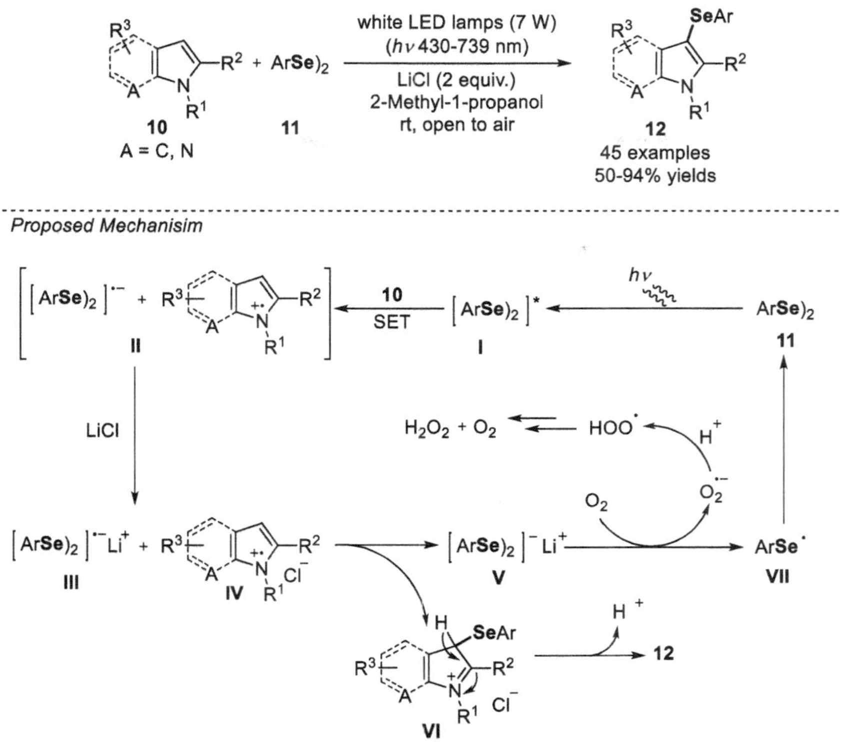

SCHEME 4.1 LiCl mediated, photo-induced selenylation of (hetero)arenes.

substituted heteroarenes (**10**) (with both electron donating group [EDG] and electron withdrawing group [EWG] substituent). The reaction was not effective when *N*-Boc/Ts indoles were tested as substrate. A plausible mechanism for this direct C(sp^2)–H bond selenylation was described by the authors. Under the LED light, diaryl diselenides (**11**) attain a singlet excited-state (**I**), which subsequently reduces the indole via SET in ground state, resulting in exciplex (**II**) formation. The excited complex (**II**) on reaction with LiCl results in solvent separated ions (**III**) and (**IV**). The addition of arylselenyl radical to electron-deficient radical cation of indole leads to the intermediate (**V**) and iminium cation (**VI**). The cation (**VI**) suffers loss of a proton in C-3, resulting into the corresponding selenylated product (**12**). Lastly, (**V**) was oxidized in the presence of O_2 to the radical (**VII**), which in turn dimerizes to form (**11**) (Scheme 4.1).

In 2019, an alternative protocol for the direct C(sp^2)–H bond selenylation of indole and *N*-methyl indole (**13**) was later described by Argüello et al.[24] The authors used ethanol as a greener solvent under blue LEDs in air atmosphere, without the need of any additive or photocatalyst. The selected optimized conditions involve the reaction between

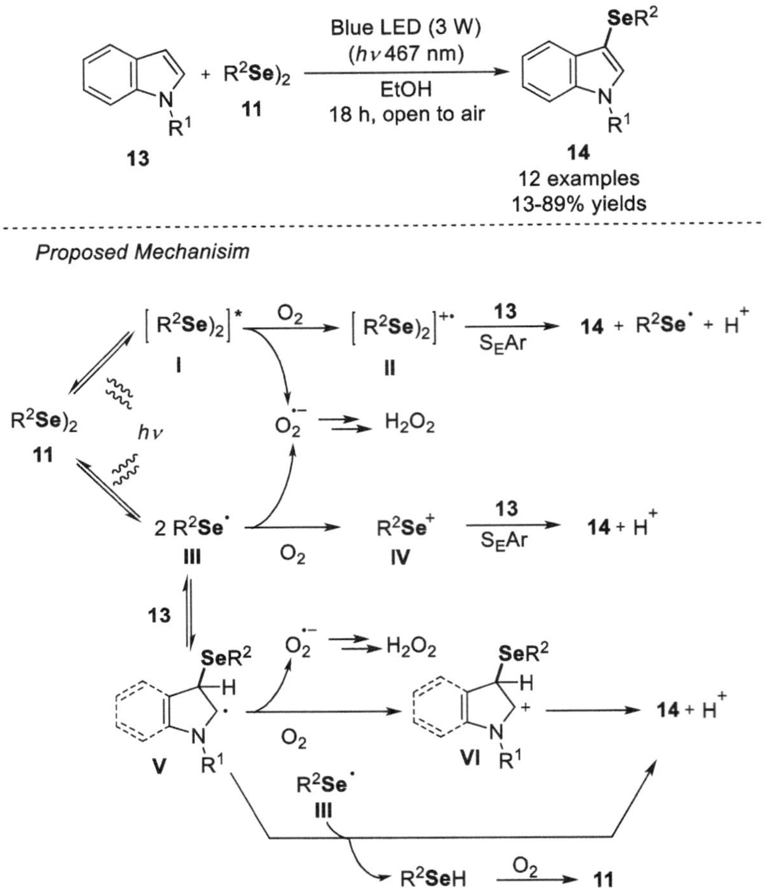

SCHEME 4.2 Additive- and photocatalyst-free selenylation of indoles.

1 molar equivalent of indole (**13**) and 0.5 molar equivalent of the diorganyl diselenides (**11**) under blue LED light for 18 hours (Scheme 4.2). For the scope study, a number of diorganyl diselenides were tested. The reaction gave the selenylated products in satisfactory yields when EWG at para position of diaryl diselenides, dialkyl diselenides, and dibenzyl diselenides were used. However, poor yields were obtained in case of diaryl diselenides with EDG. Furthermore, the reaction was also extended to few activated arenes, that is, *N,N*-dimethylaniline, phloroglucinol, and thiazoles. Based on few control experiments, the authors proposed three possible mechanistic pathways for this transformation (Scheme 4.2). In path **A**, photo-induced excitation of diselenides (**11**) results in species (**I**), which in the presence of oxygen generates electrophilic radical species (**II**). These species on reaction with indole (**13**) via

S_EAr, result in the selenylated products and organoselenyl radicals (**III**). In Path **B**, homolytic cleavage of the excited state diselenide (**11**) results in radical species (**III**), which oxidized in the presence of oxygen to form the electrophilic selenium species (**IV**). The cation species (**IV**) on subsequent reaction with the indole results in the corresponding products (**14**). In Path **C**, the organoselenyl radicals (**III**) could interact directly with the indole (**13**), forming the radical intermediate (**V**). The radical intermediate (**V**) affords the selenylated indole, either via oxidation to form the cationic species (**VI**) or following reaction with the organoselenyl radicals (**III**).

Another alternative approach was developed by Rathore and Kumar, in which they performed the selenylation of indoles under sunlight (5–11 hours) using O_2 as the oxidant. The authors observed that the same reaction can be performed under CFL light; however, a longer time was required with relatively lower yields. Interestingly, the reaction was also effective using diaryl disulfides and ditellurides as chalcogen substrates (Scheme 4.3).[25] The authors observed a limitation when aliphatic dichalcogenide was tested as substrate. Under the optimized reaction conditions, 0.5 molar equivalent of dichalcogenides in acetone reacts with the indoles, resulting the C-3-chalcogenylated products in up to 85% yields. With substituted indoles and dichalcogenides, the reaction tolerated the electronic (EDG and EWG) as well as steric effects.

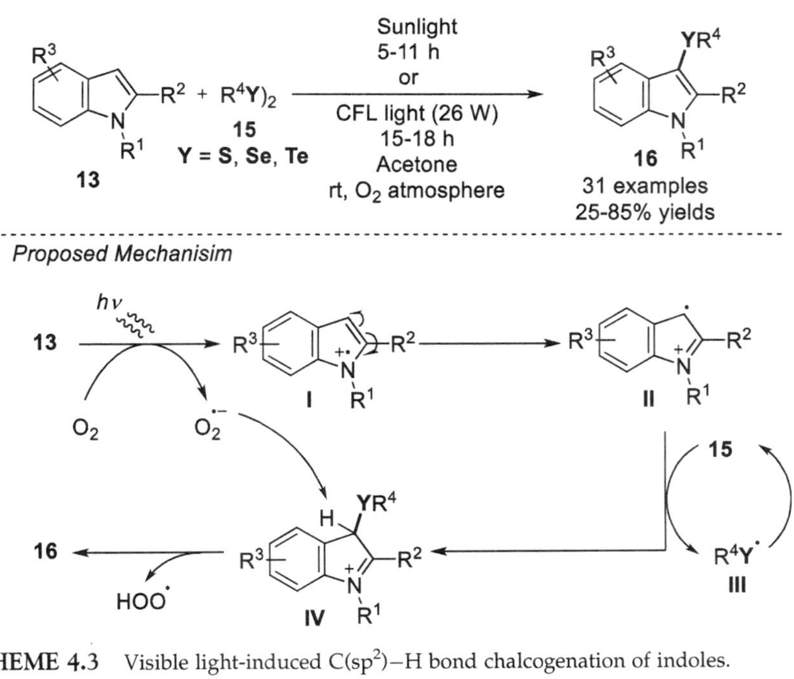

SCHEME 4.3 Visible light-induced C(sp^2)–H bond chalcogenation of indoles.

It is worth noting that the authors also tried thiocyanation reaction with 2 molar equivalent of NH_4SCN as substrate instead of dichalcogenide. After a number of control experiments, the authors proposed a different mechanistic route with photo-induced excitation and where SET generates species (**I**), which could also be represented by the resonance form (**II**). The radical species (**II**) on reaction with the diorganyl dichalcogenides (**15**) forms indole cation (**IV**) and organochalcogenyl radicals (**III**). Lastly, the oxidation of (**IV**) leads to the formation of the desired chalcogenylated products (**16**) (Scheme 4.3).

Recently, Oksdath-Mansilla and coworkers reported a combination of visible light–mediated C(sp^2)–H selenylation of indoles (**13**) in conjunction with continuous-flow systems.[26] The reaction involves reduction of 1 molar equivalent of alkyl and aryl selenocyanates by 0.5 molar equivalent of Rongalite in a mixture of MeOH:H_2O to access corresponding diselenides (**11**) (Scheme 4.4), in 1 minutes of flux. In the second step, these diselenides on photoactivation of the Se – Se bond in MeOH:H_2O (95:5) results in the selenylation of indoles in 2 hours of flux (Scheme 4.4).

In 2017, Liu and coworkers developed FIrPic-catalyzed, photoredox C(sp^2–H) selenylation on indoles (Scheme 4.5).[27] A number of structurally diverse indoles and diorganyl diselenides with different electronic effects were successfully tested under optimized reaction conditions. The protocol was also extended to other heteroarenes, for example, thiophene, pyrrole, azaindole, as well as to activated simple arenes. To elucidate a tentative mechanism, few control experiments were also performed by the authors. The reaction begins with the photo-induced activation of the catalyst, which transfers its energy to the diselenides (**11**), resulting radical species (**I**). These radical species of

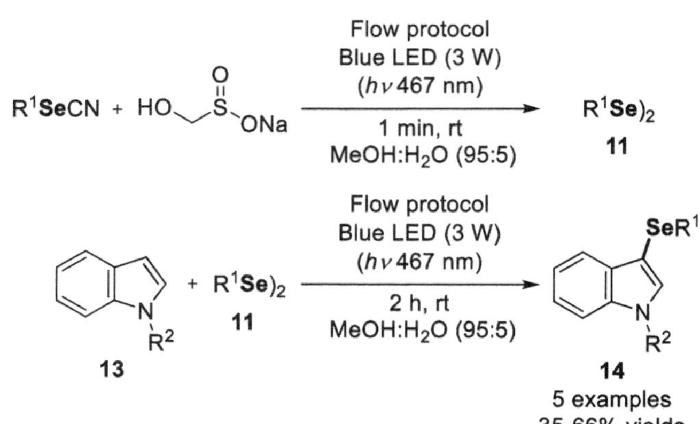

SCHEME 4.4 Visible light-mediated, continuous-flow approach for direct synthesis of unsymmetric selenides.

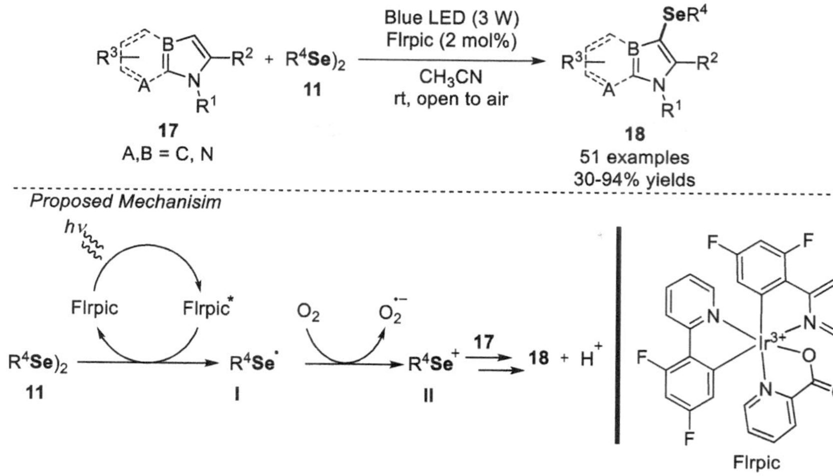

SCHEME 4.5 FIrPic-catalyzed, photo-induced selenylation of (hetero)arenes.

organoselenium oxidize under air atmosphere to electrophilic species R^4Se^+ (**II**), which were subsequently attacked by hetero(arenes) (**17**) to afford the desired selenylated product (**18**) (Scheme 4.5). It is worth mentioning that some of the synthesized selenylated products screened for antiproliferative activity (against HeLa and HepG2 cells) and the results showed that these the tested selenylated compounds exhibit significant antitumor activity.

In the attempt to improve the sustainability of the reaction, Braga's group was able to perform selenylation of (hetero)arenes (**19**) under transition metal–free approach, in the presence of Rose Bengal (**6**) as photocatalyst (Scheme 4.6).[28] They found that 0.5 molar equivalent of diorganoyl diselenides on reaction with indole afforded the selenylated product in good to excellent yields. The reaction tolerated the presence of numerous substituents with different electronic as well as steric effects, on both the substrates. This protocol was also extended to the seleno-functionalization of a number of *N*-heteroarenes, including imidazothiazoles, imidazopyrimidines, imidazopyridines, and indazole, as well as simple activated arenes, for example, anisoles, anilines, naphthols, and naphthylamine. Furthermore, the reaction was also tested successfully at the gram scale. On the basis of few control experiments, two plausible mechanisms have been proposed. The first route operated under oxygen atmosphere (path A), in which heteroarene undergoes SET to photo-excited RB*, resulting radical cation (**I**). This specie via resonance structure (**II**), on reaction with (**11**) forms the selenylated cation (**IV**) and organoselenyl radical (**III**). The cation (**IV**) looses a proton to afford the selenylated product (**20**). Lastly, selenyl

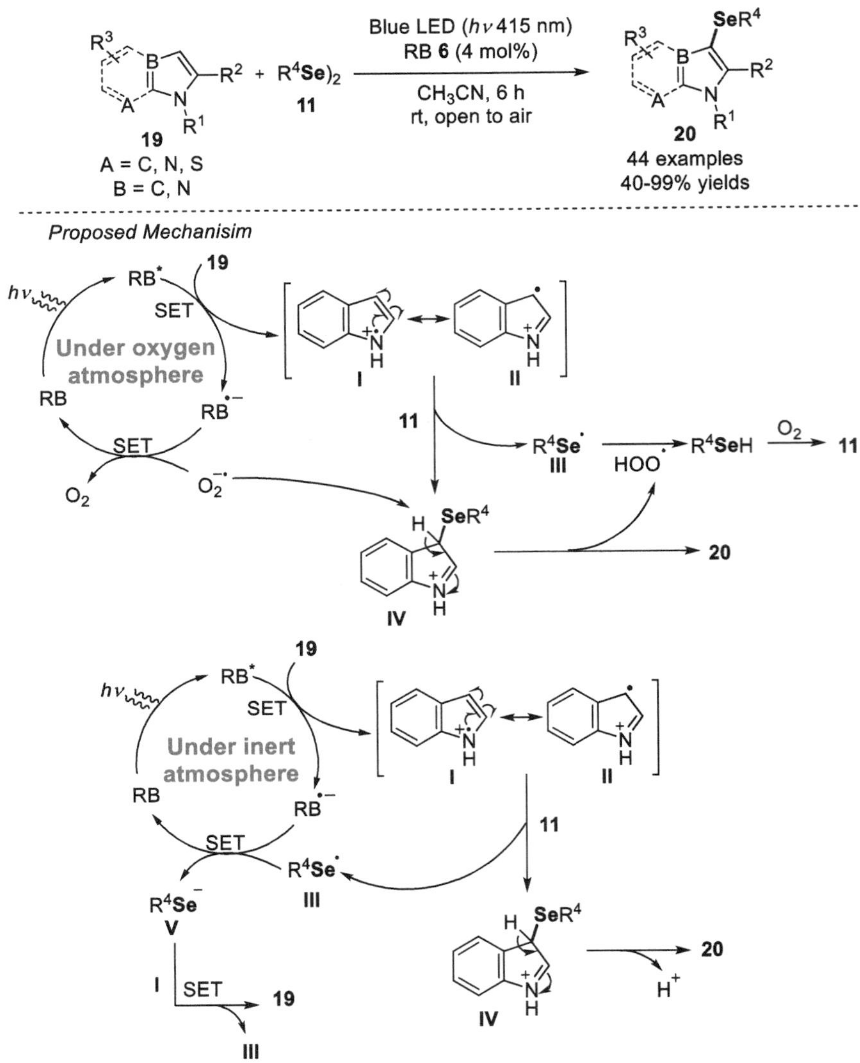

SCHEME 4.6 Rose Bengal-catalyzed, photo-induced direct C(sp^2)-H bond selenylation of (hetero)arenes.

radical (**III**) on oxidation regenerates the diselenide. In another route, which operated under inert atmosphere (pathway B), the organoselenyl radical (**II**) via SET generates nucleophilic selenium species (**V**). These species suffer another SET with the radical cation (**I**), resulting the (hetero)arenes (**19**) and selenyl radical (**III**) (Scheme 4.6).

Another *N*-heteroarene, imidazo[1,2-*a*]pyridine (IP) core constitutes a "privileged scaffold," found in several pharmaceuticals. The IPs are well known because of their applications in material sciences and diverse biological activities, including antitumor, antiviral, antitubercular and antiinflammatory as well as in modern organic syntehsis.[29] In this regards, seleno-functionalization of the IP core results in an interesting hybrid structure with biological properties.[30] Considering the importance of selenium substituted IP, Yasuike and coworkers have described photoredox C(sp^2)−H bond selenylation of IPs (**21**) under aerobic conditions, catalyzed by the ammonium iodide (NH_4I). Under blue LED light, 2-phenylimidazo[1,2-*a*]pyridine and imidazo[1,2-*a*]pyridine were used as substrate in the presence of 10 different diselenides, resulting in the selenylated products (**22**) in up to 93% yields (Scheme 4.7).[31]

Benzothiazole (BT) (**23**) is another sulfur-containing *N*-heteroarene, and its derivatives possess broad spectrum biological activities and are present in several commercially available drugs.[32] The BT nucleus was used in several reports for C−Se bond formation. Ackermann's group has reported a photo-induced multicomponent approach for the selenylation of benzothiazoles (**23**) using Luzchem's LZC-ICH2 photoreactor. The reaction was performed in the presence of copper(I) thiophene-2-carboxylate (CuTc) as catalyst and lithium *tert*-butoxide as base in DMF at ambient temperature for 16 hours. For this multicomponent reaction, the authors used elemental selenium, BTs (**23**) and aryl halides (Br, I) (**24**), resulting the desired products (**25**) in up to 68% yields (Scheme 4.8).[33] The authors also applied their methodology for sulfenylation reactions by using elemental sulfur.

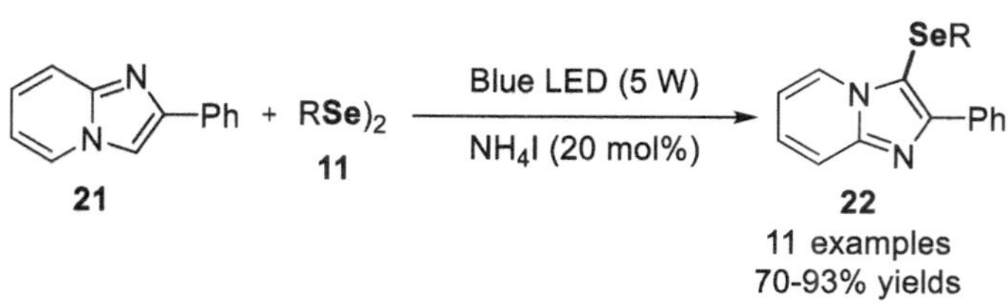

SCHEME 4.7 NH_4I catalyzed selenylation of imidazo[1,2-*a*]pyridine under Blue LEDs.

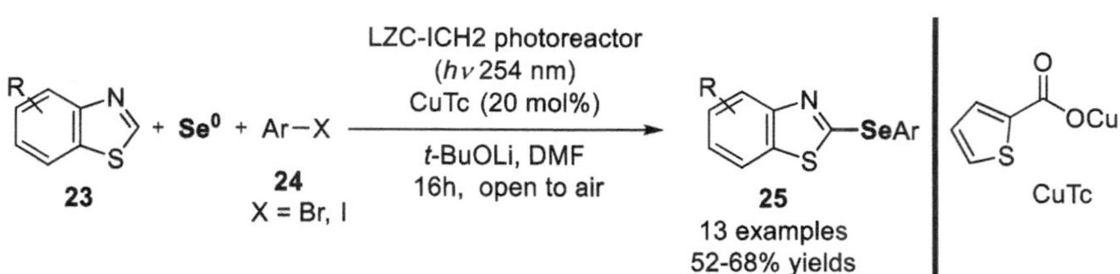

SCHEME 4.8 Photo-induced, CuTC-catalyzed selenylation/sulfenlation of azoles.

Coumarin skeleton is a well-known heteroaryl unit of natural products, and it exhibits a wide range of biological and pharmacological activities.[34] Their application in material science is also an area of interst.[35] Considering the importance of coumarins, Wei and coworkers reported the first protocol of the C(sp^2)–H bond seleno-functionalization of 4-amino substituted coumarins (**26b**) under visible light, in the presence of ammonium persulfate (Scheme 4.9). This methodology provided the synthesis of a series of selenylated compounds (**27**) and (**28**) in 37%–91% yields. Depending on the stoichiometry of diselenide, mono-selenylated and di-selenylated product were achieved. In the presence of 0.75 molar equivalent of diselenides (**11**), mono-selenylated products (**27**) were obtained, while with 1.0 molar

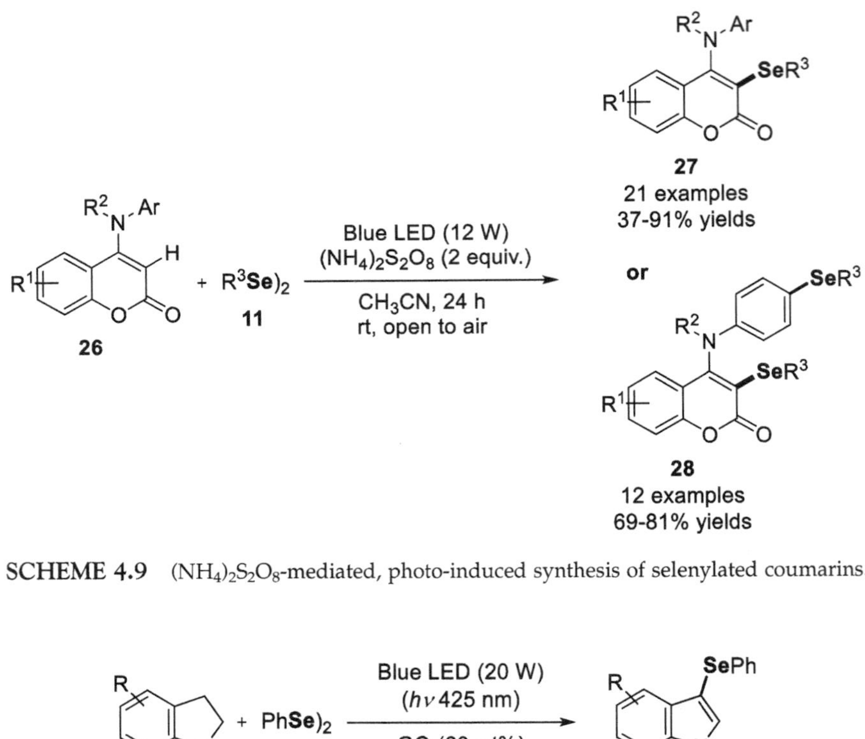

SCHEME 4.9 $(NH_4)_2S_2O_8$-mediated, photo-induced synthesis of selenylated coumarins.

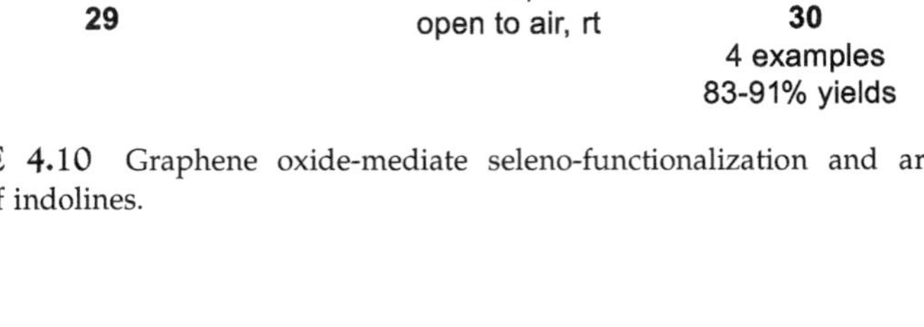

SCHEME 4.10 Graphene oxide-mediate seleno-functionalization and aromatization reaction of indolines.

equivalent of diselenides (**11**), dual selenylation took place (depending on the substrate) (**28**).[36]

Recently, Liu's group reported selenylation of indolines (**29**) using graphene oxide (GO) under blue LEDs (Scheme 4.10). This seleno-functionalization and aromatization reaction of indolines (**29**) with 1.2 molar equivalent of diphenyl diselenide (**11**) resulted in the corresponding 3-selenylindoles (**30**) in good yields. The authors also applied this methodology for sulfenylation and thiocyanation reactions. According to the authors, GO can act as a radical initiator under visible light.[37]

4.3 Photo-induced selenylation of arenes

In the same line of the heterocyclic nuclei, several methodologies have been developed for the photocatalyzed formation of C–Se bond in other arenes systems since selenylated simple arenes are also recognized as biologically promising molecules.[38] In this context, Zhao and coworkers reported a blue LED–driven selenylation of arenes and heteroarenes through the substitution of an azosulfone group (Scheme 4.11). The selenylated products were obtained in up to 91% yields by the reaction of aryl azosulfones (**31**) with diorganoyl diselenides (**11**) in the presence of acetonitrile as solvent. The methodology was successfully applied in the functionalization of several activated arenes allowing even the preparation of organotellurium compounds. According to the authors, the substrate (**31**) is excited in the presence of the light to (**I**) and followed by subsequent decomposition of the azosulfone group generating the aryl radical (**II**). The species (**II**), in the presence of the radical RY•, formed from the dichalcogenides (**11**), afforded the chalcogenylated products (**32**) and organochalcogenyl RY• species (which is converted into dichalcogenides via homodimerization).[39]

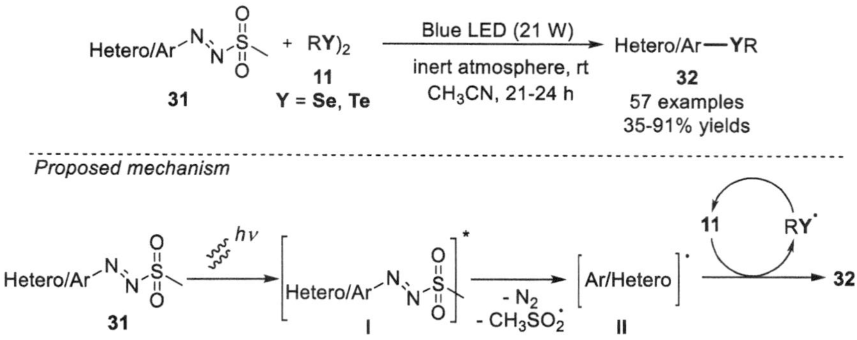

SCHEME 4.11 Photo-induced selenylation of (hetero)aryl azosulfones.

The photo-mediated seleno-functionalization of aryl azosulfones (**33**) was also reported by Kim's group. By using diorganoyl diselenides (**11**) and blue LED irradiation in the presence of DMSO as a solvent, the method allowed the synthesis of 25 diorganoyl selenides (**34**) in up to 89% yield (Scheme 4.12). Furthermore, the authors demonstrated that the methodology is efficient for gram-scale processes.[40]

The blue LED–driven selenylation of aryl amines and heteroarylamines (**35**) via the formation of aryl diazonium salts (**I**) was reported by Ranu and coworkers. In this procedure, the organic dye Eosin Y was used as photosensitizer in the presence of diorganyl diselenides (**11**) and *t*-BuONO, such that 32 compounds could be prepared in yields that ranged from 72% to 92% (Scheme 4.13). The protocol also allowed the carbon–tellurium bond formation by using diorganoyl ditellurides. The mechanism proposed by the authors indicates that the (hetero)aryl

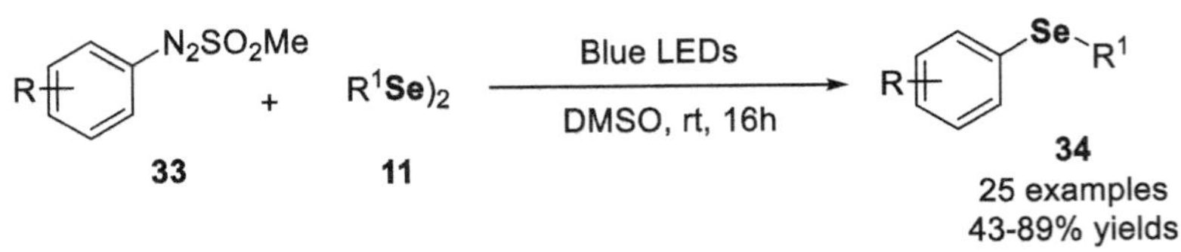

SCHEME 4.12 Visible-light-mediated seleno-functionalization of aryl azosulfones.

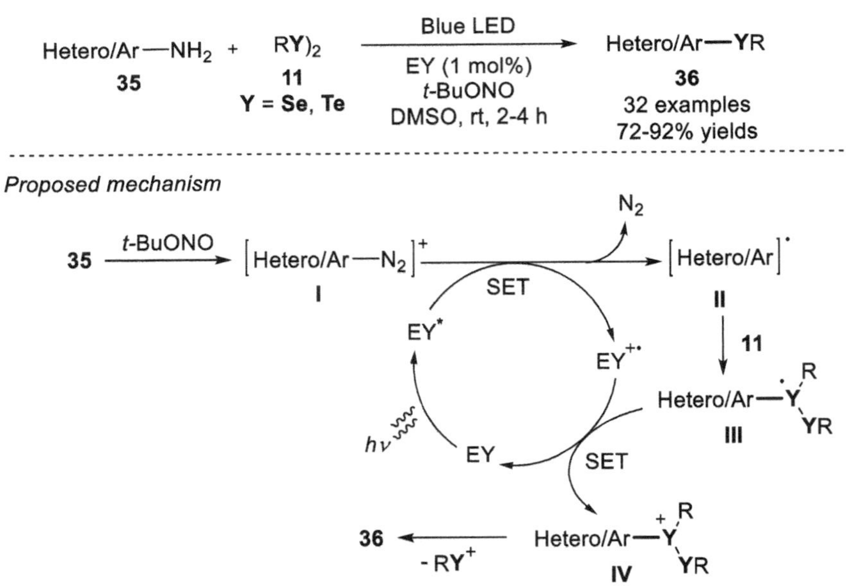

SCHEME 4.13 Eosin Y-catalyzed photo-induced synthesis of selenylated (hetero)arenes from anilines.

diazonium salts (**I**) can be generated in situ from the amines (**35**). The photo-excited sensitizer (Eosin Y*) promotes, via SET, the formation of the (hetero)aryl radical (**II**), which attacks the dichalcogenide (**11**) to generate the intermediate (**III**). Subsequently, a single electron oxidation forms the intermediate (**IV**), which is converted to the product (**36**) by cleavage.[41]

The seleno-functionalization of aryl amines (**37**) mediated by visible light was also developed by Sarkar and coworkers, employing nickel(II) tetraphenylporphyrin (NiTPP) as photocatalyst via generation of aryl diazonium salts (Scheme 4.14).[42] Such methodology allowed the synthesis of 32 selenylated and sulfenylated compounds (**38**) in high yields, demonstrating the efficiency of the catalyst used.

The nucleophilic attack of selenolate anions to aryl halides is an important method for preparing organo-selenides. In this context, the photo-induced generation of arylselenolate anions (**III**) from the reaction of selenobenzamide (**39**), selenourea (**40**), and selenocyanate anions (**41**) with aryl halides, in the presence of bases like *tert*-butoxide or 2-naphthoxide ions, was reported by Argüello and coworkers (Scheme 4.15). The arylselenolate anions formed reacts with MeI in an SN_2

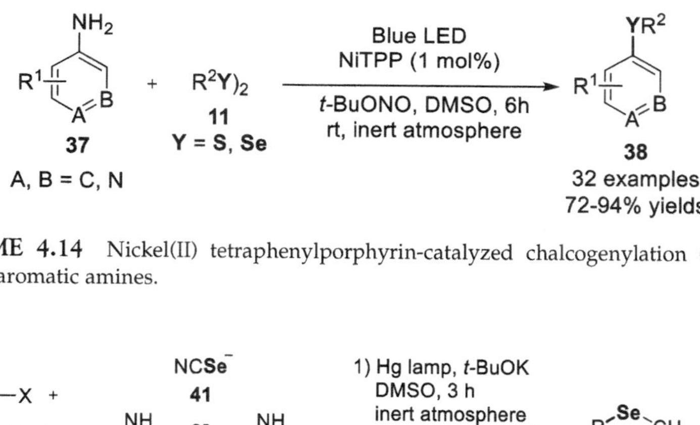

SCHEME 4.14 Nickel(II) tetraphenylporphyrin-catalyzed chalcogenylation (Se, S) of (hetero)aromatic amines.

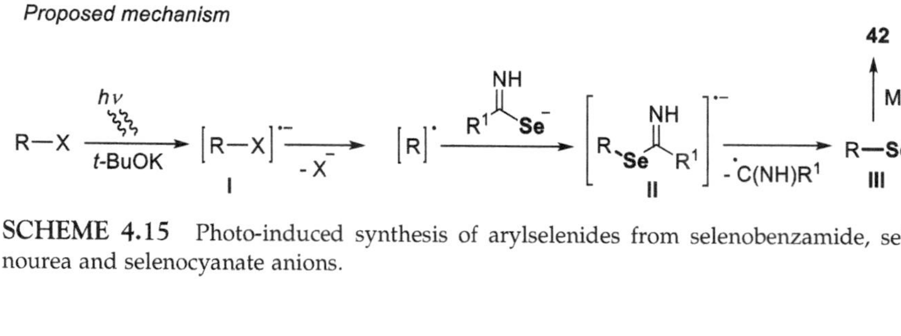

SCHEME 4.15 Photo-induced synthesis of arylselenides from selenobenzamide, selenourea and selenocyanate anions.

process, leading to the formation of ArSeMe (**42**) as a major product (Scheme 4.15). The role of the *t*-butoxide anion is to act as radical initiator to form the radical anion (**I**), which in the presence of the selenolates, produces the intermediate (**II**). After this specie suffers decomposition, the selenolate (**III**) is generated and reacts with MeI leading to the formation of the product (**42**).[43]

Ogawa and coworkers developed a photochemical method for the direct selenylation of triarylbismutines (**43**) with diorganoyl dichalcogenides (**11**) (Scheme 4.16). For this protocol, the authors used Xenon lamp irradiation and obtained 12 examples of arylselenides (**44**) in yields that ranged from 45% to 86%.[44]

In 2020, Huang and coworkers reported the synthesis of deuterated aryl sulfides and selenides (**47**) from the substitution of aryl diazonium salts (**45**) with *S*-methyl-d_3 sulfonothioate or *Se*-methyl-d_3 selenium sulfonate reagents (**46**) (Scheme 4.17). In the presence of Eosin Y as a

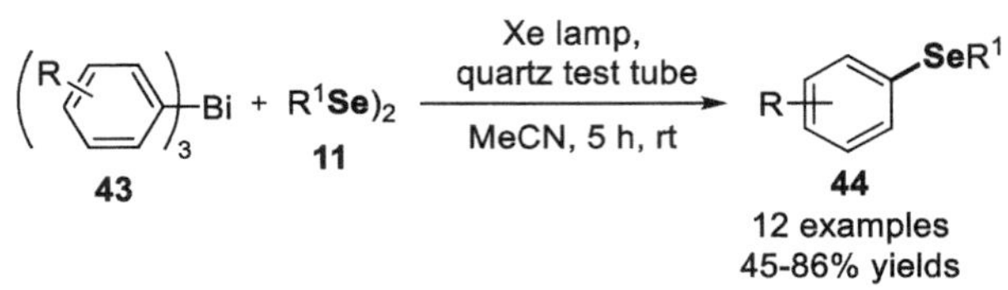

SCHEME 4.16 Photoinduced synthesis of diaryl selenides from triarylbismuthines and diaryl diselenides.

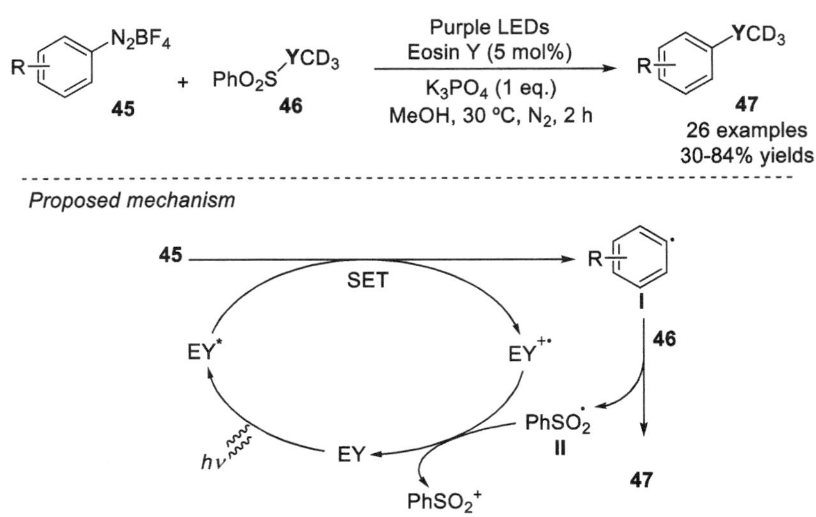

SCHEME 4.17 Eosin Y-catalyzed, visible-light-mediated synthesis of trideuteromethylated sulfides, sulfoxides, and selenides.

photosensitizer and under purple light irradiation, the methodology afforded the synthesis of 26 aryl chalcogenides (**47**) in good yields. The authors suggest that the excited photocatalyst in the presence of aryl diazonium salt, generates the radical (**I**) and the radical cation EY^{+}. The specie (**I**) reacts with the deuterated reagent (**46**) to form the corresponding product (**47**) and eliminate sulfoxide radical (**II**), which is oxidized and regenerates the photocatalyst in the cycle.[45]

4.4 Conclusions and perspectives

In the last decade, there has been a growing interest in the visible light photoredox catalysis, either employing transition metal–free photocatalyst (organic dyes and other transition metal–free organic compounds) or by using transition metal complexes, has emerged as an efficient and powerful tool for seleno-functionalization. Specifically, in this chapter, we discussed the recent advances in seleno-functionalization of hetero(arenes) via $C(sp^2)$–H bond functionalization. A diverse range of biologically relevant selenylated compounds were prepared via this approach, by using indole, imidazopyridine, coumarin, activated arenes, and other related structures as substrates. In general, photoredox transformations are mild and selective, which result into a facile route to access radical species that can be used to incorporate organoselenium moiety into several interesting structures. These reactions are easy to set up and are performed under visible light sources, for example, commercial CFL bulbs, LEDs (white, blue, green), or even in sunlight, without the need for a specialized photochemical reactor. Another feature of these transformation is their sustainability aspect and that they are considered greener and environmentally benign. In comparison with conventional approaches, photo-induced selenylation of hetero(arenes) usually avoids the use of stoichiometric or large amounts of reagents, long reaction times, and high temperature, higher catalytic loading, as well as harsh reaction conditions. Nevertheless, the development of new selenylated products is still highly desirable, and the future development of visible light-promoted seleno-functionalization will lead to the construction of fine and useful chemicals.

The authors of this chapter believe that the progress and perspective described in this chapter will stimulate further efforts from researchers all across the organoselenium community and related fields. The content of this chapter will help future research on light-mediated C–Se bond formation, providing new synthetic routes and allowing improvements to the previous methodologies. This approach could also encourage the development of hybrid systems, with the use of flow or electrochemical process, for the seleno-functionalization (hetero)arenes with biological interest.

Acknowledgments

We gratefully acknowledge CAPES (Finance Code 001), CNPq, INCT-Catálise/CNPq/FAPESC, CERSusChem (grant 2014/50249–8), FAPESP (grant 2014/50249–8), GSK (grant 2014/50249–8) and FAPERGS. J.R., S.S., and A.L.B. would like to acknowledge CNPq (433896/2018–3, 315399/2020–1, 422645/2021-4 and 403210/2021-6).

Conflict of interest

There are no conflicts to declare.

Credit author statement

Jamal Rafique, Juliano B. Azeredo, Sumbal Saba: Methodology and Writing, Original draft preparation; Jamal Rafique, Juliano B. Azeredo, Paulo H. Schneider: Writing, Reviewing, and Editing. Antonio L. Braga: Supervision and Writing, Reviewing, and Editing.

References

1. (a) Rotruck, J. T.; Pope, A. L.; Ganther, H. E.; Swanson, A. B.; Hafeman, D. G.; Hoekstra, W. G. *Science* **1973,** *179*, 588–590. Available from: https://doi.org/10.1126/science.179.4073.588.
   (b) Flohé, L.; Günzler, E. A.; Schock, H. H. *FEBS Lett.* **1973,** *32*, 132–134. Available from: https://doi.org/10.1016/0014-5793(73)80755-0.
   (c) Flohé, L.; Andreesen, J. R.; Brigelius-Flohé, R.; Maiorino, M.; Ursini, F. *IUBMB Life* **2000,** *49*, 411–420. Available from: https://doi.org/10.1080/152165400410263.
   (d) Klayman, D.L.; Günter, H.H. Organoselenium Compounds: Their Chemistry and Biology Wiley, New York, 1973.
   (e) Shamberger, R. J. *Biochemistry of Selenium;* Plenum Press: New York, 1983.
2. Wessjohann, L. A.; Schneider, A.; Abbas, M.; Brandt, W. *Biol. Chem.* **2007,** *388*, 997–1006. Available from: https://doi.org/10.1515/bc.2007.138.
3. (a) May, S. W.; Pollock, S. H. *Drugs* **1998,** *56*, 959–964. Available from: https://doi.org/10.2165/00003495-199856060-00001.
   (b) Mugesh, G.; du Mont, W. W.; Sies, H. *Chem. Rev.* **2001,** *101*, 2125–2179. Available from: https://doi.org/10.1021/cr000426w.
   (c) Nogueira, C. W.; Zeni, G.; Rocha, J. B. T. *Chem. Rev.* **2004,** 6255–6285. Available from: https://doi.org/10.1021/cr0406559.
   (d) Braga, A. L.; Rafique, J. *Synthesis of Biologically Relevant Small Molecules Containing Selenium. Part C. Miscellaneous Biological Activities.* In *The Chemistry of Organic Selenium and Tellurium Compounds;* Rappoport, Z., Ed.; Vol. 4; John Wiley & Sons, Ltd: Chichester, 2013; pp 1119–1174. Available from: https://doi.org/10.1002/9780470682531.pat0727.
4. (a) Zhan, P.; Liu, X.; Fang, Z.; Panneouque, C.; Clercq, E. D. *Bioorg. Med. Chem.* **2009,** *17*, 6374–6379. Available from: https://doi.org/10.1016/j.bmc.2009.07.027.

(b) Alexander, V.; Choi, W. J.; Chun, J.; Kim, H. O.; Jeon, J. H.; Tosh, D. K.; Lee, H. W.; Chandra, G.; Choi, J.; Jeong, L. S. *Org. Lett.* **2010,** *12,* 2242–2245. Available from: https://doi.org/10.1021/ol1005906.

5. (a) Braga, A. L.; Rafique, J. *Synthesis of Biologically Relevant Small Molecules Containing Selenium. Part B. Anti-Infective and Anticancer Compounds.* In *The Chemistry of Organic Selenium and Tellurium Compounds;* Rappoport, Z., Ed.; Vol. 4; John Wiley & Sons, Ltd: Chichester, 2013; pp 1053–1117. Available from: https://doi.org/10.1002/9780470682531.pat0728.
(b) Braga, H. C.; Stefani, H. A.; Paixão, M. W.; Santos, F. W.; Lüdtke, D. S. *Tetrahedron* **2010,** *66,* 3441–3446. Available from: https://doi.org/10.1016/j.tet.2010.03.033.

6. (a) Braga, A. L.; Rafique, J. *Synthesis of Biologically Relevant Small Molecules Containing selenium. Part A. Antioxidant Compounds.* In *The Chemistry of Organic Selenium and Tellurium Compounds;* Rappoport, Z., Ed.; Vol. 4; John Wiley & Sons, Ltd: Chichester, UK, 2013; pp 989–1052. Available from: https://doi.org/10.1002/9780470682531.pat0721.
(b) Botteselle, G. V.; Elias, W. C.; Bettanin, L.; Canto, R. F. S.; Salin, D. N. O.; Barbosa, F. A.; Saba, S.; Gallardo, H.; Ciancaleoni, G.; Domingos, J. B.; Rafique, J.; Braga, A. L. *Molecules* **2021,** *26,* 4446. Available from: https://doi.org/10.3390/molecules26154446.

7. May, S. W.; Wang, L.; Gill-Woznichak, M. M.; Browner, R. F.; Ogonowski, A. A.; Smith, J. B.; Pollock, S. H. *J. Pharmacol. Exp. Ther.* **1997,** *283,* 470–477. Available from: https://jpet.aspetjournals.org/content/283/2/470.

8. (a) Jacob, C.; Giles, G. I.; Giles, N. M.; Sies, H. *Angew. Chem. Int. (Ed.)* **2003,** *42,* 4742–4758. Available from: https://doi.org/10.1002/anie.200300573.
(b) Martinez-Ramos, F.; Salgado-Zamora, H.; Campos-Aldrete, M. E.; Melendez-Camargo, Y.; Márquez-Flores, Y.; Soriano-García, M. *Eur. J. Med. Chem.* **2008,** *43,* 1432–1437. Available from: https://doi.org/10.1016/j.ejmech.2007.09.029.
(c) Rafique, J.; Farias, G.; Saba, S.; Zapp, E.; Casagrande, I. B.; Salla, C. A. M.; Bechtold, I. H.; Scheide, M. R.; Neto, J. S. S.; Souza, D. M., Jr.; Braga, H. C.; Ribeiro, L. F. B.; Gastaldon, F.; Pich, C. T.; Frizon, T. E. A. *Dyes Pigm.* **2020,** *180,* 108519. Available from: https://doi.org/10.1016/j.dyepig.2020.108519.
(d) Galant, L. S.; Rafique, J.; Braga, A. L.; Braga, F. C.; Saba, S.; Radi, R.; Rocha, J. B. T.; SantI, C.; Monsalve, M.; Farina, M.; Bem, A. F. *Neurochem. Res.* **2021,** *46,* 120–130. Available from: https://doi.org/10.1007/s11064-020-03026-x.
(e) Frizon, T. E. A.; Cararo, J. H.; Saba, S.; Dal-Pont, G.; Michels, M.; Braga, H. C.; Pimentel, T.; Dal-Pizzol, F.; Valvassori, S. S.; Rafique, J. *Oxid. Med. Cell. Longev.* **2020,** *2020,* 5417024. Available from: https://doi.org/10.1155/2020/5417024.

9. (a) Deidda, D.; Lampis, G.; Maullu, C.; Pompei, R.; Isaia, F.; Lippolis, V.; Verani, G. *Pharm. Res.* **1997,** *36,* 193–197. Available from: https://doi.org/10.1006/phrs.1997.0224.
(b) Hafez, S. H. A. *Russ. J. Org. Chem.* **2005,** *41,* 396–401. Available from: https://doi.org/10.1007/s11178-005-0177-z.
(c) Foley, J. W.; Song, X.; Demidova, T. N.; Jilal, F.; Hamblin, M. R. *J. Med. Chem.* **2006,** *49,* 5291–5299. Available from: https://doi.org/10.1021/jm060153i.
(d) Hafez, S. H. A. *Eur. J. Med. Chem.* **2008,** *43,* 1971–1977. Available from: https://doi.org/10.1016/j.ejmech.2007.12.006.
(e) Victoria, F. N.; Radatz, C. S.; Sachini, M.; Jacob, R. G.; Perin, G.; da Silva, W. P.; Lenardão, E. J. *Tetrahedron Lett.* **2009,** *50,* 6761–6763. Available from: https://doi.org/10.1016/j.tetlet.2009.09.093.

10. Jin, Z.; Du, X.; Xu, Y.; Deng, Y.; Liu, M.; Zhao, B.; Li, X.; Zhang, P.; Peng, C.; Duan, Y.; Yu, J.; Wang, L.; Yang, K.; Liu, F.; Jiang, R.; Yanh, X.; You, T.; Liu, X.; Yanh, X.; Bai, F.; Liu, H.; Liu, X.; Gudat, L. W.; Xu, W.; Xiao, G.; Qin, C.; Shi, H.; Jiang.; Jiang, H.; Rao,

Z.; Yang, H. *Nature* **2020,** *582,* 289–293. Available from: https://doi.org/10.1038/s41586-020-2223-y.

11. (a) Shao, L.; Li, Y.; Lu, J.; Jiang, X. *Org. Chem. Front.* **2019,** *6,* 2999–3041. Available from: https://doi.org/10.1039/C9QO00620F.
    (b) Tungen, J. E.; Kristianlund, R.; Vik, A.; Hansen, T. V. *J. Org. Chem.* **2019,** *84,* 11373–11381. Available from: https://doi.org/10.1021/acs.joc.9b01294.
    (c) Singh, F. V.; Wirth, T. *Catal. Sci. Technol.* **2019,** *9,* 1073–1091. Available from: https://doi.org/10.1039/C8CY02274G.
12. (a) Cao, H.; Qian, R.; Yu, L. *Catal. Sci. Technol.* **2020,** *10,* 3113–3121. Available from: https://doi.org/10.1039/D0CY00400F.
    (b) Kim, B.; Lee, M.; Kim, M. J.; Kim, S.; Kim, J.; Koh, M.; Park, S. B.; Shin, K. J. *J. Am. Chem. Soc.* **2008,** *130,* 16807–16811. Available from: https://doi.org/10.1021/ja806304s.
    (c) Sohn, T. I.; Kim, J.; Kim, D. *J. Am. Chem. Soc.* **2010,** *132,* 12226–12227. Available from: https://doi.org/10.1021/ja106116v.
13. (a) Godoi, M.; Paixão, M. W.; Braga, A. L. *Dalton Trans.* **2011,** *40,* 11347–11355. Available from: https://doi.org/10.1039/C1DT11022E.
    (b) Panda, S.; Panda, A.; Zade, S. S. *Coord. Chem. Rev.* **2015,** *300,* 86–100. Available from: https://doi.org/10.1016/j.ccr.2015.04.006.
    (c) Jain, V. K. *Coord. Chem. Rev.* **2021,** *427,* 213456–213478. Available from: https://doi.org/10.1016/j.ccr.2020.213546.
14. (a) Manjare, S. T.; Kim, Y.; Churchill, D. G. *Acc. Chem. Res.* **2014,** *47,* 2985–2998. Available from: https://doi.org/10.1021/ar500187v.
    (b) Frizon, T. E.; Rafique, J.; Saba, S.; Bechtold, I. H.; Gallardo, H.; Braga, A. L. *Eur. J. Org. Chem.* **2015,** 3470–3476. Available from: https://doi.org/10.1002/ejoc.201500124.
    (c) Hoover, G. C.; Seferos, D. S. *Chem. Sci.* **2019,** *10,* 9182–9188. Available from: https://doi.org/10.1039/C9SC04279B.
    (d) Ho, P. C.; Rafique, J.; Lee, J.; Lee, L. M.; Jenkins, H. A.; Britten, J. F.; Braga, A. L.; Vargas-Baca, I. *Dalton Trans.* **2017,** *46,* 6570–6579. Available from: https://doi.org/10.1039/C7DT00612H.
    (e) Li, Q.; Zhang, Y.; Chen, Z.; Pan, X.; Zhang, A.; Zhu, J.; Zhu, X. *Org. Chem. Front.* **2020,** *7,* 2815–2841. Available from: https://doi.org/10.1039/D0QO00640H.
15. (a) Garcia, M. S. *Curr. Med. Chem.* **2004,** *11,* 1657–1669. Available from: https://doi.org/10.2174/0929867043365053.
    (b) Reilly, C. *Selenium in Food and Health,* 1st ed.; Blackie Academic and Professional: London, 1997.
16. (a) Braga, A. L.; Rafique, J. *The Chemistry of Organic Selenium and Tellurium Compounds,* Vol. 4. Wiley & Sons, Ltd: Chichester, 2014989–1174Ch. 13–15.
    (b) Reich, H. J.; Hondal, R. J. *ACS Chem. Biol.* **2016,** *11,* 821–841. Available from: https://doi.org/10.1021/acschembio.6b00031.
    (c) Jacob, C.; Giles, G. I.; Giles, N. M.; Sies, H. *Angew. Chem. Int. (Ed.)* **2003,** *42,* 4742. Available from: https://doi.org/10.1002/anie.200300573 4258.
    (d) Almeida, G. M.; Rafique, J.; Saba, S.; Siminski, T.; Mota, N. S. R. S.; Gilho, D. W.; Braga, A. L.; Pedrosa, R. C.; Ourique, F. *Biochem. Biophys. Res. Commun.* **2018,** *503,* 1291–1297. Available from: https://doi.org/10.1016/j.bbrc.2018.07.039.
    (e) Veloso, I. C.; Delanogare, E.; Machado, A. E.; Braga, S. P.; Rosa, G. K.; Bem, A. F.; Rafique, J.; Saba, S.; Trindade, R. N.; Galetto, F. Z.; Moreira, E. L. G. *J. Pharm. Pharmacol.* **2021,** *73,* 673–681. Available from: https://doi.org/10.1093/jpp/rgaa070.
    (f) Rodrigues, J.; Saba, S.; Joussef, A. C.; Rafique, J.; Braga, A. L. *Asian J. Org. Chem.* **2018,** *7,* 1819–1824. Available from: https://doi.org/10.1002/ajoc.201800346.

17. (a) Guan, Z.; Wang, Y.; Huang, Y.; Wang, S.; Tang, H.; Zhang, H.; Lei, A. *Green Chem.* **2019,** *21,* 4976–4980. Available from: https://doi.org/10.1039/C9GC02665G.
(b) Scheide, M. R.; Peterle, M. M.; Saba, S.; Neto, J. S. S.; Lenz, G. F.; Cezar, R. D.; Felix, J. F.; Botteselle, G. V.; Schneider, R.; Rafique, J.; Braga, A. L. *Sci. Rep.* **2020,** *10,* 15233–15244. Available from: https://doi.org/10.1038/s41598-020-72129-w.
(c) Peterle, M. M.; Scheide, M. R.; Silva, L. T.; Saba, S.; Rafique, J.; Braga, A. L. *ChemistrySelect* **2018,** *3,* 13191–13196. Available from: https://doi.org/10.1002/slct.201801213.
(d) Wu, G.; Min, L.; Li, H.; Gao, W.; Ding, J.; Huang, X.; Liu, M.; Wu, H. *Green Chem.* **2018,** *20,* 1560–1563. Available from: https://doi.org/10.1039/C8GC00341F.
(e) Meirinho, A. G.; Pereira, V. F.; Martins, G. M.; Saba, S.; Rafique, J.; Braga, A. L.; Mendes, S. R. *Euro. J. Org. Chem.* **2019,** 6465–6469. Available from: https://doi.org/10.1002/ejoc.201900992.

18. (a) Rafique, J.; Saba, S.; Franco, M. S.; Bettanin, L.; Schneider, A. R.; Silva, L. T.; Braga, A. L. *Chem. Eur. J.* **2018,** *24,* 4173–4180. Available from: https://doi.org/10.1002/chem.201705404.
(b) Rafique, J.; Saba, S.; Franco, M. S.; Bettanin, L.; Schneider, A. R.; Silva, L. T.; Braga, A. L. *Chem. Eur. J.* **2018,** *24,* 4173–4180. Available from: https://doi.org/10.1002/chem.201705404.
(b) Scheide, M. R.; Schneider, A. R.; Jardim, G. A. M.; Martins, G. M.; Durigon, D. C.; Saba, S.; Rafique, J.; Braga, A. L. *Org. Biomol. Chem.* **2020,** *18,* 4916–4921. Available from: https://doi.org/10.1039/D0OB00629G.

19. (a) Crisenza, G. E. M.; Melchiorre, P. *Nat. Commun.* **2020,** *11,* 803–807. Available from: https://doi.org/10.1038/s41467-019-13887-8.
(b) Sun, K.; Lv, X. L.; Chen, X. L.; Qu, L. B.; Yu, B. *Green Chem.* **2021,** *23,* 232–248. Available from: https://doi.org/10.1039/D0GC03447A.
(c) Yoon, T. P.; Ischay, M. A.; Du, J. *Nat. Chem.* **2010,** *2,* 527–532. Available from: https://doi.org/10.1038/nchem.687.
(d) Williams, J. D.; Kappe, C. O. *Curr. Opin. Green Sustain. Chem.* **2020,** *25,* 100351–100358. Available from: https://doi.org/10.1016/j.cogsc.2020.05.001.
(e) Oelgemöller, M. *Chem. Rev.* **2016,** *116,* 9664–9682. Available from: https://doi.org/10.1021/acs.chemrev.5b00720.
(f) Rafique, J.; Rampon, D. S.; Azerdedo, J. B.; Coelho, F. L.; Schneider, P. H.; Braga, A. L. *Chem. Rec.* **2021,** *21,* 2739–2761. Available from: https://doi.org/10.1002/tcr.202100006.

20. (a) Prier, C. K.; Rankic, D. A.; MacMillan, D. W. C. *Chem. Rev.* **2013,** *113,* 5322–5363. Available from: https://doi.org/10.1021/cr300503r.
(b) Romero, N. A.; Nicewicz, D. A. *Chem. Rev.* **2016,** *116,* 10075–10166. Available from: https://doi.org/10.1021/acs.chemrev.6b00057.
(c) Hari, D. P.; König, B. *Chem. Commun.* **2014,** *50,* 6688–6699. Available from: https://doi.org/10.1039/C4CC00751D.
(d) Wang, C. S.; Dixneuf, P. H.; Soulé, J. F. *Chem. Rev.* **2018,** *118,* 7532–7585. Available from: https://doi.org/10.1021/acs.chemrev.8b00077.
(e) Glaser, F.; Wenger, O. S. *Coord. Chem. Rev.* **2020,** *405,* 213129–213143. Available from: https://doi.org/10.1016/j.ccr.2019.213129.
(f) Twilton, J.; Le, C.; Zhang, P.; Shaw, M. H.; Evans, R. W.; MacMillan, D. W. C. *Nat. Rev. Chem.* **2017,** *1,* 52–70. Available from: https://doi.org/10.1038/s41570-017-0052.

21. (a) Laishram, R. D.; Chen, J.; Fan, B. *Chem. Rec.* **2021,** *21,* 69–86. Available from: https://doi.org/10.1002/tcr0.202000094.
(b) Liu, Q.; Wu, L. Z. *Natl. Sci. Rev.* **2017,** *4,* 359–380. Available from: https://doi.org/10.1093/nsr/nwx039.

(c) Parasram, M.; Gevorgyan, V. *Chem. Soc. Rev.* **2019,** *46,* 6227–6240. Available from: https://doi.org/10.1039/C7CS00226B.

22. (a) Thanikachalam, P. V.; Maurya, R. K.; Garg, V.; Monga, V. *Eur. J. Med. Chem.* **2019,** *180,* 562–612. Available from: https://doi.org/10.1016/j.ejmech.2019.07.019.
   (b) Festa, A. A.; Voskressensky, L. G.; der Eycken, E. V. V. *Chem. Soc. Rev.* **2019,** *48,* 4401–4423. Available from: https://doi.org/10.1039/C8CS00790J.
   (c) Matzkeit, Y. H.; Tornquist, B. L.; Manarin, F.; Botteselle, G. V.; Rafique, J.; Saba, S.; Braga, A. L.; Felix, J. F.; Schneider, R. *J. Non Cryst. Solids* **2018,** *498,* 153–159. Available from: https://doi.org/10.1016/j.jnoncrysol.2018.06.020.
   (d) Tronquist, B. L.; Bueno, G. P.; Wiling, M.; Oliveira, I.; Stefani, H.; Rafique, J.; Saba, S.; Iglesias, B. A.; Botteselle, G. V.; Manarin, F. *ChemistrySelect* **2018,** *23,* 6358–6363. Available from: https://doi.org/10.1002/slct.201800673.
23. Kumaraswamy, G.; Ramesh, V.; Gangadhar, M.; Vijaykumar, S. *Asian J. Org. Chem.* **2018,** *7,* 1689–1697. Available from: https://doi.org/10.1002/ajoc.201800332.
24. Lemir, I. D.; Castro-Godoy, W. D.; Heredia, A. A.; Schmidt, L. C.; Argüello, J. E. *RSC Adv.* **2019,** *9,* 22685–22694. Available from: https://doi.org/10.1039/C9RA03642C.
25. Rathore, V.; Kumar, S. *Green. Chem.* **2019,** *21,* 2670–2676. Available from: https://doi.org/10.1039/C9GC00007K.
26. Heredia, A. A.; Soria-Castro, S. M.; Castro-Godoy, W. D.; Lemir, I. D.; Lopez-Vidal, M.; Bisogno, F. R.; Agüello, J. E.; Oksdath-Mansilla, G. *Org. Process. Res. Dev.* **2020,** *24,* 540–545. Available from: https://doi.org/10.1021/acs.oprd.9b00548.
27. Zhang, Q. B.; Ban, Y. L.; Yuan, P. F.; Peng, S. J.; Fang, J. G.; Wu, L. Z.; Liu, Q. *Green Chem.* **2017,** *19,* 5559–5563. Available from: https://doi.org/10.1039/C7GC02803B.
28. Saba, S.; Rafique, J.; Franco, M. S.; Schneider, A. R.; Espíndola, L.; Silva, D. O.; Braga, A. L. *Org. Biomol. Chem.* **2018,** *16,* 880–885. Available from: https://doi.org/10.1039/C7OB03177G.
29. (a) Bagdi, A. K.; Santra, S.; Monir, K.; Hajra, A. *Chem. Commun.* **2015,** *51,* 1555–1575. Available from: https://doi.org/10.1039/C4CC08495K.
   (b) Saba, S.; dos Santos, C. R.; Zavarise, B. R.; Naujorks, A. A. S.; Franco, M. S.; Schneider, A. R.; Affeldt, R. F.; Rafique, J.; Braga, A. L. *Chem. Eur. J.* **2020,** *26,* 4461–4466. Available from: https://doi.org/10.1002/chem.201905308.
   (c) Neto, J. S. S.; Balaguez, R. A.; Franco, M. S.; de Sá Machado, V. C.; Saba, S.; Rafique, J.; Galetto, F. Z.; Braga, A. L. *Green Chem.* **2020,** *22,* 3410–3415. Available from: https://doi.org/10.1039/D0GC00137F.
   (d) Franco, M. S.; Saba, S.; Rafique, J.; Braga, A. L. *Angew. Chem. Int. (Ed.)* **2021,** *133,* 18602–18608. Available from: https://doi.org/10.1002/ange.202104503.
30. dos Santos, D. C.; Rafique, J.; Saba, S.; Almeida, G. M.; Siminski, T.; Filho, D. W.; Zamoner, A.; Braga, A. L.; Pedrosa, R. C.; Ourique, F. *J. Biochem. Mol. Toxicol.* **2021,** *35,* e22663. Available from: https://doi.org/10.1002/jbt.22663.
31. Murata, Y.; Kanasaki, K.; Kondo, K.; Nakusawa, N.; Matsumura, M.; Yasuike, S. *HeteroCycles* **2019,** *99,* 596–603.
32. (a) Rafique, J.; Saba, S.; Frizon, T. E. A.; Braga, A. L. *ChemistrySelect* **2018,** *3,* 12387–12393. Available from: https://doi.org/10.1002/slct.201702623.
   (b) Sharma, P. C.; Sinhmar, A.; Sharma, A.; Rajak, H.; Pathak, D. P. *J. Enzyme Inhib. Med.* **2013,** *28,* 240–266. Available from: https://doi.org/10.3109/14756366.2012.720572.
33. Grandeepan, P.; Mo, J.; Ackermann, L. *Chem. Commun.* **2017,** *53,* 5906–5909. Available from: https://doi.org/10.1039/C7CC03107F.
34. (a) Thakur, A.; Singla, R.; Jaitak, V. *Eur. J. Med. Chem.* **2015,** *101,* 476–495. Available from: https://doi.org/10.1016/j.ejmech.2015.07.010.

(b) Li, H.; Yao, L.; Li, L. *J. Pharm. Pharmacol.* **2017,** *69,* 1253–1264. Available from: https://doi.org/10.1111/jphp.12774.

35. Cuevas, J. M.; Seoane-Rivero, R.; Navarro, R.; Marcos-Fernandez, A. *Polymers* **2020,** *12,* 630–656. Available from: https://doi.org/10.3390/polym12030630.
36. Yang, D.; Li, G.; Xing, C.; Cui, W.; Li, K.; Wei, W. *Org. Chem. Front.* **2018,** *5,* 2974–2979. Available from: https://doi.org/10.1039/C8QO00899J.
37. Liu, C.; Peng, X.; Hu, D.; Shi, F.; Huang, P.; Luo, J.; Liu, Q.; Liu, L. *New. J. Chem.* **2020,** *44,* 17245–17251. Available from: https://doi.org/10.1039/D0NJ00747A.
38. Liu, J.; Tian, M.; Li, Y.; Shan, X.; Li, A.; Lu, K.; Fagnoni, M.; Protti, S.; Zhao, X. *Eur. J. Org. Chem.* **2020,** 7358–7367. Available from: https://doi.org/10.1002/ejoc.202001386.
39. (a) Saba, S.; Rafique, J.; Braga, A. L. *Catal. Sci. Technol.* **2016,** *6,* 3087–3098. Available from: https://doi.org/10.1039/C5CY01503K.

    (b) Silva, L. T.; Azeredo, J. B.; Saba, S.; Rafique, J.; Bortoluzzi, A. J.; Braga, A. L. *Euro. J. Org. Chem.* **2017,** 4740–4748. Available from: https://doi.org/10.1002/ejoc.201700744.
40. Jang, J.; Kim, R.; Kim, D. Y. *Synth. Commun.* **2020,** 720–726. Available from: https://doi.org/10.1080/00397911.2020.1850796.
41. Kundu, D.; Ahammed, S.; Ranu, B. C. *Org. Lett.* **2014,** *16,* 1814–1817. Available from: https://doi.org/10.1021/ol500567t.
42. Mandal, T.; Das, S.; Sakar, S. D. *Adv. Synth. Catal.* **2019,** *361,* 3200–3209. Available from: https://doi.org/10.1002/adsc.201801737.
43. Bouchet, L. M.; Peñéñory, A. B.; Argüello, J. E. *Tetrahedron Lett.* **2011,** *52,* 969–972. Available from: https://doi.org/10.1016/j.tetlet.2010.11.115.
44. Kobiki, Y.; Kawaguchi, S.; Ohe, T.; Ogawa, A. *Beilstein J. Org. Chem.* **2013,** *9,* 1141–1147. Available from: https://doi.org/10.3762/bjoc.9.127.
45. Huang, C. M.; Li, J.; Ai, J. J.; Liu, X. Y.; Rao, W.; Wang, S. Y. *Org. Lett.* **2020,** *22,* 9128–9132. Available from: https://doi.org/10.1021/acs.orglett.0c03562.

CHAPTER

5

Synthesis of organochalcogens: use of nonconventional solvents/reaction media

Francesca Marini[1], *Luana Bagnoli*[1] *and Martina Palomba*[2]

[1]**Department of Pharmaceutical Sciences (Group of Catalysis, Synthesis and Organic Green Chemistry), University of Perugia, Perugia, Italy,**
[2]**Fondazione Istituto Italiano di Tecnologia, Genova, Italy**

5.1 Introduction

Over the last two decades, the synthesis of organochalcogen compounds containing S, Se, or Te has attracted massive interest not only for their versatility as synthetic intermediates, ligands, or catalysts,[1] but also for their applications in material science, biology, and medicinal chemistry.[1,2] In the last few years, as an answer to environmental concerns, organochalcogen synthetic strategies have moved toward more sustainable and eco-friendly methods. Compounds characterized by chemical and thermal stabilities, nonflammability, low, or negligible vapor pressure, low cost, ease of recovery and recycling have emerged as attractive alternatives to nonrenewable, volatile, toxic, or hazardous solvents. Water, polymers, biomass-derived compounds, ionic liquids (IL), and deep eutectic solvents (DES) have replaced traditional solvents. In many cases, these compounds can even improve the efficiency of the process by increasing the rate of the reaction, facilitating the workup and product isolation, or enhancing regio- and stereo-selectivity. Bearing in mind that no solvent is a priori green and that the evaluation of the *greenness* of a reaction requires a careful analysis of the single process, the proposal of

Organochalcogen Compounds
DOI: https://doi.org/10.1016/B978-0-12-819449-2.00005-7

this chapter is to illustrate synthetic approaches to organochalcogen compounds in alternative solvents/reaction media. This topic has been the object of excellent reviews.[3] For this reason, the selected examples will primarily cover the literature of the last decade. This chapter is arranged in six sections according to the structure of the organochalcogen compound: diorganic dichalcogenides, chalcogen ethers, alkenyl and alkynyl chalcogenides, chalcogen-containing carbo- and heterocycles, chalcogenoesters, and compounds containing chalcogens at higher oxidation states. Synthetic procedures in pure water, water/organic biphasic systems, glycerol, ethyl lactate, polyethylene glycol, IL, DES, and solvent-free reactions will be reported and discussed.

5.2 Synthesis of diorganic dichalcogenides

Organic dichalcogenides are stable, easy to handle and store, and poorly volatile compounds. They are important synthetic building blocks commonly employed for the generation of electrophilic, nucleophilic, or radical chalcogen species through oxidative, reductive, or homolytic cleavage. In recent years, their biological and pharmaceutical properties have also been evaluated.[1,2] A classical and simple method for the preparation of disulfides involves the oxidative coupling of thiols. Such conversion has been investigated in solvent-free conditions,[4] ILs,[5] and polyethylene glycol (PEG).[6] In Scheme 5.1, recent examples of

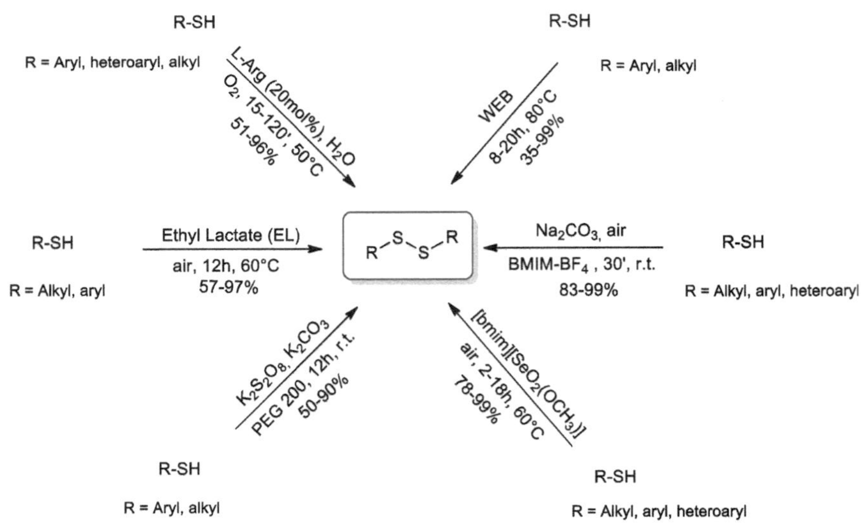

SCHEME 5.1 Synthesis of disulfides from thiols.

such protocols are reported. In 2010, Braga[5a] developed a metal-free air oxidation of alkyl, aryl, and heteroaryl thiols in [bmim]BF_4 in the presence of Na_2CO_3. The oxidation of biologically active thiols such as Captopryl, Penicillamine, and *L*-cysteine was also described. An efficient base-free oxidation protocol of alkyl and aryl thiols was described by Lenardão[5b] using a selenium-containing IL, the 1-*n*-butyl-3-methylimidazolium methylselenite, [bmim][$SeO_2(OCH_3)$], under gentle heating. In both cases, the recycling of the solvent was studied and only a reasonable loss of efficiency was observed after six runs. The same conversion was carried out in PEG-200 in the presence of potassium persulfate.[6] Interestingly, unsymmetrical disulfides were also isolated. Ethyl lactate, a biomass-derived fully biodegradable solvent, was also effective for the synthesis of dialkyl or diaryl disulfides from thiols without the addition of any catalyst or additive.[7] The aerobic oxidative coupling of aromatic thiols, *L*-cysteine, or *L*-glutathione was carried out in water using a substoichiometric amount of an aminoacidic catalyst.[8] Water is a cheap, readily available, nontoxic, and noninflammable solvent with a central role in biological reactions as well as amino acids. The *L*-arginine (20 mol%) gave the best yields as the result of a hydrogen-bonding network involving the basic amino acid, two molecules of thiophenol, and the oxygen. The method has been adapted to the synthesis of vinyl sulfides by hydrothiolation of terminal alkynes. Water extracts of agro-waste ashes have recently attracted great interest for environmentally safe and cost-effective organic transformations. Agro-waste ashes contain mixtures of lithium, potassium, sodium, calcium, and magnesium hydroxides and carbonates and have found useful applications as basic catalysts.

Water extract of banana peel ash (WEB) was used for the oxidative dimerization of thiols in the absence of other catalysts or promoters.[9] Control experiments with radical scavengers such as TEMPO or hydroquinone demonstrated that the reaction occurs *via* anionic pathway. Thiols and selenols are volatile, unstable, toxic, and persistently bad smelling. For this reason, their replacement with halides is very attractive. Metal-promoted or catalyzed coupling reactions of alkyl or aryl halides have provided disulfides and diselenides in water or PEG/water mixtures. As the first example, symmetric disulfides were obtained from the corresponding halides using thiourea as the odorless sulfur source. The basic hydrolysis of the generated S-alkylisothiouronium intermediates afforded thiolates that suffer oxidation by stoichiometric MnO_2[10] or catalytic amounts of Europhtal (8020), a cobalt organometal species.[11] The choice of the oxidant was crucial to avoid the decomposition of thiourea or the formation of consistent amounts of byproducts. In the case of Europhtal, the catalytic species is regenerated by atmospheric oxygen as reported in Scheme 5.2. The reaction has been adapted to the synthesis of diaryl disulfides, by addition of CuI, which activates aryl halides to the substitution. Very recently, the tetrathionate anion, produced from $Na_2S_2O_3 \cdot 5H_2O/I_2$, was used to catalyze a similar aerobic oxidation of in situ–generated alkyl thiols.[12]

Interestingly, the thiourea-based method was also adapted to the synthesis of unsymmetrical dialkyl disulfides. Benzylthiocyanates[13] or S-alkylthiosulfates, known as Bunte salts,[14] were used as electrophilic sulfur reagents able to react with the thiourea-generated thiolates. Except that with tertiary halides, products were recovered in acceptable-to-good yields. Unfortunately, the former method generates highly toxic cyanide as the byproduct, and the latter affords good results only with cyano-containing Bunte salts. Protocols of copper-catalyzed dichalcogenation of aryl halides with elemental sulfur or selenium have also been developed. Reactions were carried out with aryl iodides with sulfur or selenium powder at high temperatures in water in the presence of Cs_2CO_3, $CuCl_2$ as catalyst, and 1,10 phenanthroline as chelating ligand (Scheme 5.3, route a). Diaryl disulfides and diselenides were obtained in good-to-excellent yields.[15] Moreover, a crystalline and nanoporous copper-based metal–organic framework, MOF-199, was recently synthesized, characterized, and employed as catalyst in similar conversions involving halides or tosylates in an alkaline PEG/H_2O mixture (Scheme 5.3, route b).[16] Diaryl or diheteroaryl chalcogenides were

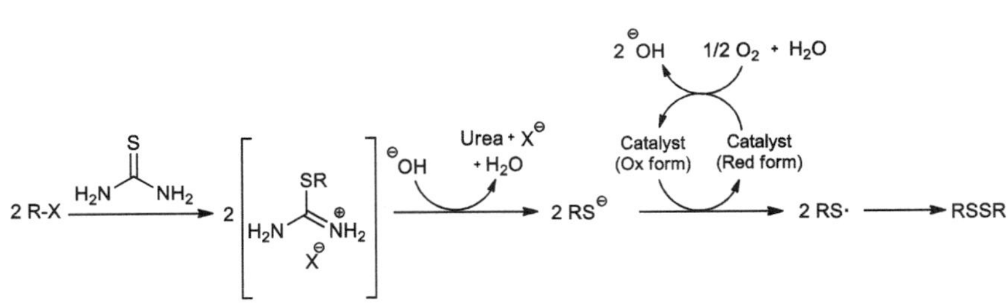

SCHEME 5.2 One-pot synthesis of disulfides from halides and thiourea.

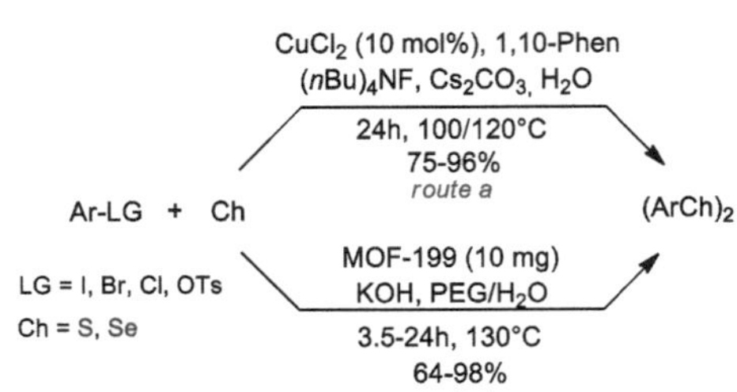

SCHEME 5.3 Copper-catalyzed syntheses of dichalcogenides from aryl halides and elemental chalcogens.

obtained in good-to-excellent yields and good functional group tolerance. The heterogeneous catalyst was easily recovered by simple filtration and recycled without a significant loss of activity.

Diorganic ditellurides[17] were formed in good yields by oxidative coupling of sodium alkyl tellurolate generated in situ by nucleophilic ring-opening of epoxides or aziridines with Na_2Te (Scheme 5.4). Na_2Te was generated from elemental tellurium, NaOH, and Rongalite as the reducing agent at 90°C in water. Diluted conditions were crucial to control the reaction and avoid the formation of the corresponding tellurides. The scheme shows reactions from the *L*-valine-derived aziridine and benzyl glycidyl ether. Reactions occurred with excellent regioselectivity and stereospecificity.

Finally, a practical method to prepare symmetrical dichalcogenides involves the alkylation of metal dichalcogenides with alkyl or aryl halides or tosylates.[3b,18] Alkali metal diselenides were prepared safely in alkaline aqueous conditions and used directly to synthesize symmetric dialkyl- or diaryl-diselenides in high yields under normal heating or ultrasonic activation as reported in Scheme 5.5.[19] The method has been applied to the synthesis of diselenobis(benzoic acid) (up to 7 g), a widely employed building block for the synthesis of ebselen and other organoselenium compounds with antitumor, antiviral, and glutathione-peroxidase-like activities.[1f,20]

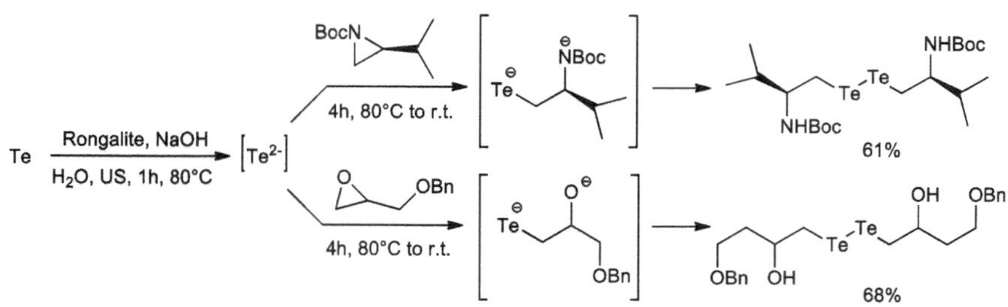

SCHEME 5.4 On water synthesis of β−hydroxy- and β−aminoditellurides.

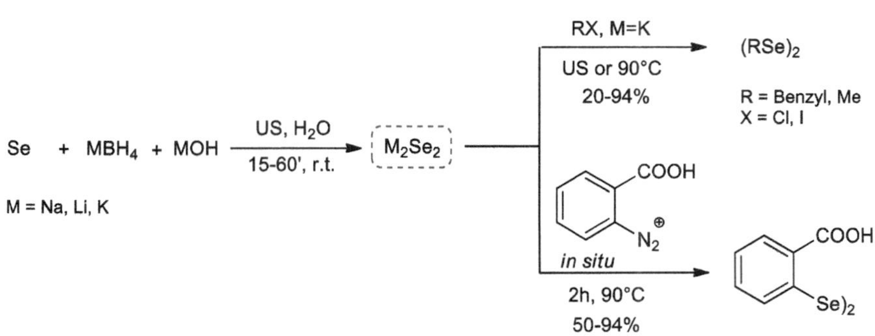

SCHEME 5.5 Ultrasound-assisted, gram-scale synthesis of diselenides.

5.3 Synthesis of chalcogen ethers

In the last two decades, a number of vicinally functionalized chalcogen ethers have been synthesized by ring-opening or addition to alkenes of chalcogenated reagents in alternative solvents. The addition of chalcogen nucleophiles to epoxides and aziridines is an important strategy in the preparation of oxygen- or nitrogen-functionalized chalcogen ethers. These compounds are versatile intermediates in the synthesis of a variety of heterocycles and natural products, owing to the easy manipulation of carbon–chalcogen bond. Starting from the first decade of the 2000s, several chemo-, regio-, and stereoselective thiolysis of epoxides in water,[21] ILs,[22] or solvent-free conditions[23] were developed (Scheme 5.6). Usually, reactions proceed under neutral or basic conditions through a S_N2 mechanism with a nucleophilic *anti*-attack, generally at the less hindered C atom, with inversion of configuration. Under Brønsted or Lewis acid activation, the ring-opening usually proceeds with opposite regioselectivity, as a consequence of the formation of an incipient carbocation. Interestingly, desymmetrizations of *meso* epoxides with thiophenols in water were carried out with a good level of enantioselectivity using a scandium-surfactant combined catalyst and chiral bipyridine ligands.[24]

Recently, DES were employed in the thiolysis of epoxides. DES are a class of polar substances, obtained by simple mixing and heating of two compounds, having a melting point lower than those of the single components. Classic examples of DES are mixtures of choline chloride (CCl) with urea and CCl with $ZnCl_2$ or $SnCl_2$. DES are usually liquids at room temperature and have physical and chemical properties similar to ILs. However, they have several advantages, such as cheapness, easy preparation, and often biocompatibility and biodegradability. The first example of thiolysis in DES was reported by Azizi in 2012.[25] The DES, obtained by gentle stirring of CCl and $SnCl_2$ at 130°C in air, was used for the regioselective ring-opening of unsymmetrical epoxides with sterically and electronically different alkyl and aryl thiols at room temperature. Except for styrene oxide, the major products derived from the attack of the thiol to the less-substituted carbon atom. The recyclability of the DES was investigated, and after precipitation by addition of diethyl ether, the viscous liquid could be recovered and reused for four cycles. Subsequently, the same research group studied the applicability of choline hydroxide as biodegradable catalyst for the thiolysis of

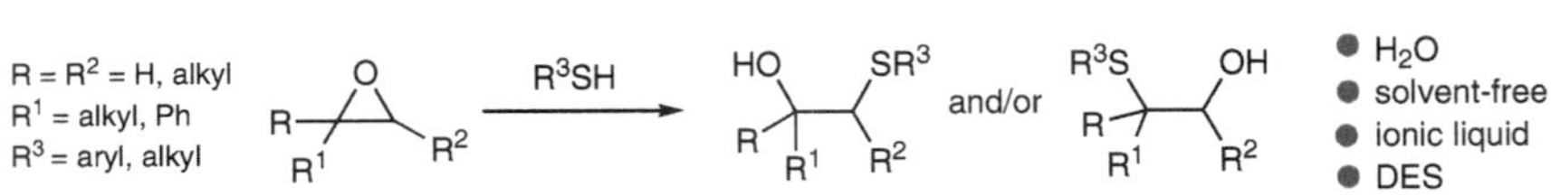

SCHEME 5.6 Thiolysis of epoxides.

epoxides.[26] A three-component one-pot approach for the synthesis of β-hydroxy sulfides was performed, employing alkyl halides, unsymmetrical epoxides, and thiourea in CCl-urea-based DES.[27] As previously mentioned, thiourea acts as nucleophilic sulfur source—generating urea, the component of the choline-urea DES, as the sole side-product. In the same article, the authors also described a simple procedure for the conversion of the epoxides into the corresponding thiiranes. Ring-openings of epoxides with benzenselenol,[28] PhSeZnCl[29] or in situ—generated selenolates and tellurates[30] have been developed in alternative reaction media. As in the example described by Santi,[29] the bench-stable PhSeZnCl was easily prepared from the PhSeCl through an oxidative insertion of Zn into the selenium-chloride bond and used in efficient ring-opening of epoxides in water suspension at room temperature. Moreover as shown in Scheme 5.7, the alkyl and aryl selenolates and tellurates generated in situ by reduction of the corresponding dichalcogenides with $NaBH_4$, were applied in the ring-opening of epoxides, β-lactones and unprotected aziridines.[30] Reactions were performed in the presence of Al_2O_3 at 50°C under solvent-free conditions, using an excess of the electrophile. PEG-400 was the solvent of choice for reactions carried out with aziridines, affording β-chalcogen amines in good yields and short reaction times. Nucleophilic selenium species usually gave better results than tellurium ones.

Moreover, symmetrically oxygen- or nitrogen-functionalized tellurides were efficiently formed *via* ring-opening of epoxides and aziridines by Na_2Te by reaction of the intermediate tellurates described in Scheme 5.4 with an excess of the electrophile.[17] Chiral aziridines are key precursors for the construction of nitrogen-containing valuable compounds such as amino acids, heterocycles, and peptides in a stereo-controlled fashion. The thiolysis of protected aziridines in water was developed by Rao under supramolecular catalysis conditions with β-cyclodextrin.[31] The catalyst accelerated the reaction by hydrogen-bonding activation of the aziridines. The formation of the supramolecular complex, depending on size, shape, and hydrophobicity of the guest, beneficially influenced the regioselectivity of the reaction. A similar protocol was explored with

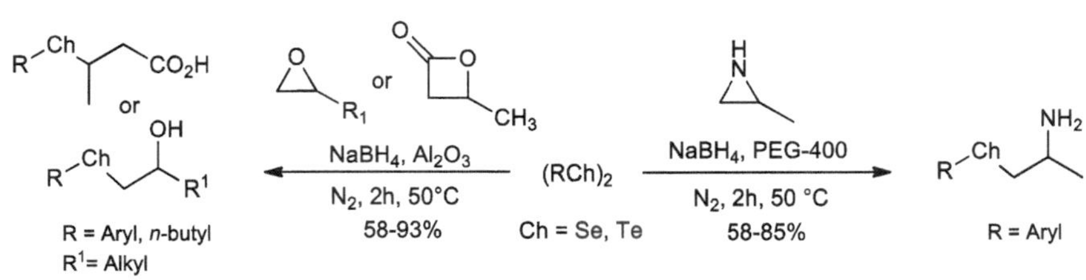

SCHEME 5.7 Ring-opening of epoxides, β-lactones, and unprotected aziridines by chalcogenolates.

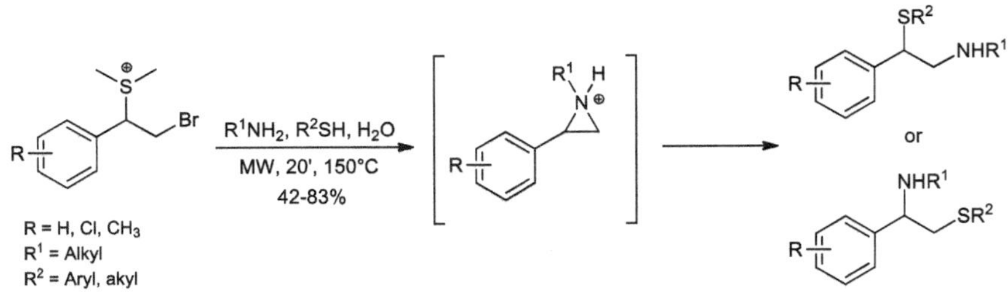

SCHEME 5.8 Multicomponent strategy for the synthesis of β-aminosulfides.

potassium thiocyanate (KSCN) as the nucleophile.[32] Ring-openings with thiols of protected and unprotected aziridines were also performed in PEG[33] and solvent-free conditions,[34] respectively. Recently, the thiolysis of functionalized bicyclic aziridines not only with aryl and alkyl thiols but also with biologically interesting esters of cysteine, *N*-acetyl cysteine, and glutathione, was successfully investigated under physiological conditions (buffer phosphate, pH 7.4, 37°C).[35] A good chemoselectivity for sulfur-based nucleophiles *versus*amines or other nucleophiles was observed in competition experiments. The method has been successfully applied in sequential reduction-bioconjugation processes using salmon calcitonin, a disulfide bridge containing protein. In 2019, a microwave-assisted multicomponent approach to achieve β-aminosulfides was described.[36] As reported in Scheme 5.8, reactions were performed starting from styrene sulfonium bromide derivatives, alkyl amines, and alkyl or aryl thiols in water. No acid/base catalyst or protecting group was needed. In most cases, the nucleophilic attack of the thiol occurred regioselectively at the most substituted carbon. In selected cases, enzymatic kinetic resolutions of the aminosulfides with monoamine oxidase MAO-N led to enantio-enriched compounds.

Enantiopure selenoamines were prepared by ring-opening of chiral aziridines with PhSeZnBr[37] or in situ–generated zinc[38] selenolates in the IL [Bmim]BF_4. The reduction of dialkyl and diaryl diselenides with Zn in a biphasic Et_2O/acidic aqueous system was a successful strategy for the opening of unprotected aziridines.[39] Michael-type addition is one of the most general, fast, and clean strategy to achieve 1,2-functionalized chalcogen ethers. A large number of protocols have been reported in literature, in particular regarding the formation of β-carbonyl sulfides starting from alkyl and aryl thiols and α,β-unsaturated ketones, aldehydes, or esters. Reactions were performed in different reaction media such as ILs, solvent-free conditions, PEG, and water in good chemical yields.[40] Nitroalkenes,[40c] α,β-unsaturated nitriles,[40a,c,e,f,g] and poorly reactive α,β-unsaturated amides[40f,g,h] were also competent substrates. As an example, unsubstituted α,β-unsaturated amides reacted in water at room temperature in the presence of 10 mol% of an *N*-methyl morpholine–based IL as catalyst, whereas the substituted ones gave the best results under solvent-free conditions at 50°C.[41] Quite recently, a practical multicomponent strategy afforded β-carbonyl sulfides starting from aromatic aldehydes, acetone, and aryl thiols via a cascade aldol/Knoevenagel–thia–Michael/Michael process (Scheme 5.9, route a).[42] Reactions were catalyzed by bovine serum albumin (BSA) and the neutral IL 1-butyl-3-methylimidazolium bromide ([bmim] Br). The bio- and the organocatalyst synergistically cooperate to improve the efficiency of the three-component process, as supported by NMR and computational studies. The same research group proposed a similar decarboxylative Knoevenagel-thia-Michael cascade to afford β-sulfanyl esters. In this case, the easy-to-handle *L*-arginine was used as cocatalyst (Scheme 5.9, route b).[43] In both cases, the catalytic system was recovered and reused up to five cycles.

Enantioselective versions of the thio-Michael addition have been developed. Although most reactions employed conventional organic solvents, some examples in water are available (Scheme 5.10).

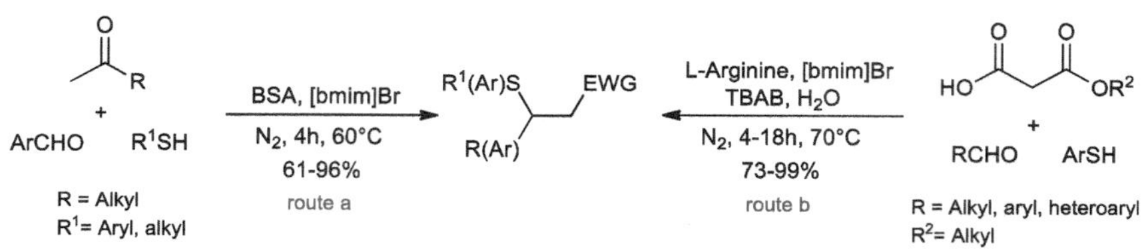

SCHEME 5.9 Three-component approaches for the synthesis of β-carbonyl sulfides.

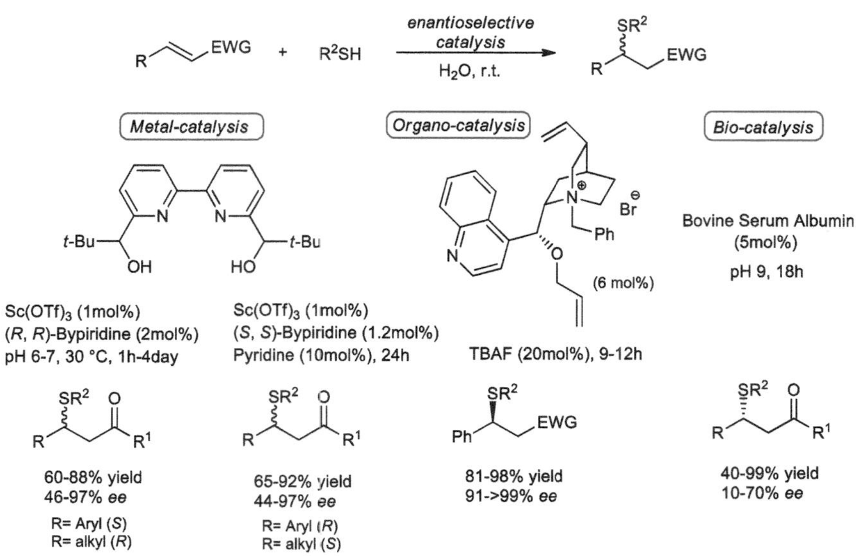

SCHEME 5.10 Enantioselective thia-Michael additions.

Lewis acid–catalyzed enantioselective sulfa-Michael additions using $Sc(OTf)_3$/bipyridine as the catalytic system were developed independently by Vaccaro[44] and Kobayashi[45] in 2011. In both cases, products were obtained in good yields and high enantioselectivity, despite the fact that noncovalent interactions between reactants, chiral ligands, and metal ions should be weakened in water. The protocol was extended to α-substituted enones and gave access to chiral products as well with two consecutive stereocenters through a challenging enantioselective protonation.[46] Organocatalysts were also tested to control the enantioselectivity of the addition in water. First experiments were carried out with chalcones and thiols using per-6-amino-β-cyclodextrin as chiral catalyst, but products were obtained with very poor enantioselectivities.[47] A cinchona-derived phase-transfer catalyst gave good results in the asymmetric Michael addition of thiols to α,β-unsaturated esters, acid, and ketones in combination with tetra-*n*-butylammonium fluoride as additive.[48] Finally, a biomimetic approach with BSA in aqueous borate buffer (pH 9) at room temperature gave enantio-enriched Michael adducts in poor to moderate enantioselectivities.[49]

Thia-Michael reactions were also performed with in situ–generated nucleophilic species. In fact, in the last decade, the use of odorless surrogates of thiols such as elemental sulfur, S-alkylisothiouronium salts or disulfides, has attracted great attention. As an example, Gholinejad and Firouzabadi developed a one-pot thia–Michael reaction catalyzed by copper ferrite nanoparticles (NPs) in PEG-200 with elemental sulfur, aryl halides, or boronic acids, and α,β-unsaturated esters, nitriles and ketones.[50] Preformed or in situ–generated S-alkylisothiouronium salts gave successful results in both water[51] and DES.[52] Interestingly, these alternative strategies afforded structurally diverse thia–Michael adducts not accessible by previous methods, because of the limited commercial availability of thiols. S–S bond cleavage of disulfides in thio-Michael additions was carried out following two protocols. One method involves $NaBH_4$ in PEG-400[53] and the other one PPh_3 in an IL.[54] Both reagents were successfully employed for the reductive cleavage of aryl and alkyl ditellurides and diselenides, also. Finally, Scheme 5.11 shows the results of some seleno Michael additions carried out with benzenselenol,[55] PhSeZnCl[56] in water, or in situ–generated $[(PhSe)_2Zn]$ (in equilibrium with PhSeH) in a EtOAc/aqueous HCl biphasic system.[57] Products were usually isolated in good-to-excellent yields, complete regioselectivity, and moderate stereoselectivity.

The addition of a thiol to a double bond (thiol-ene reaction or hydrothiolation) is a well-known procedure for the synthesis of thioethers (Scheme 5.12). Several studies have been reported in literature owing to its interesting features such as mild reaction conditions, high atom economy, minimal waste generation, high yielding, and solvent and

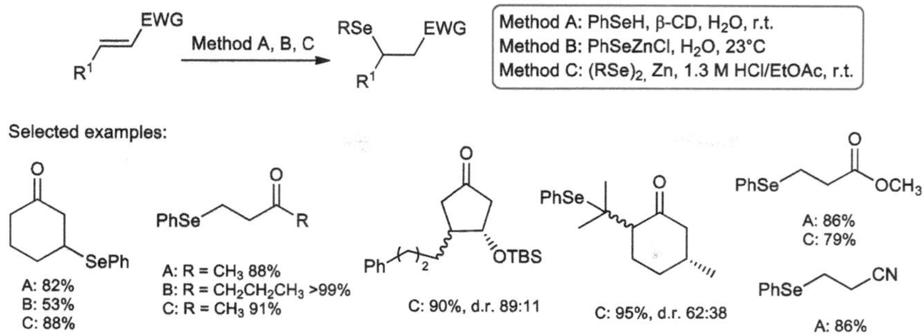

SCHEME 5.11 Seleno-Michael additions.

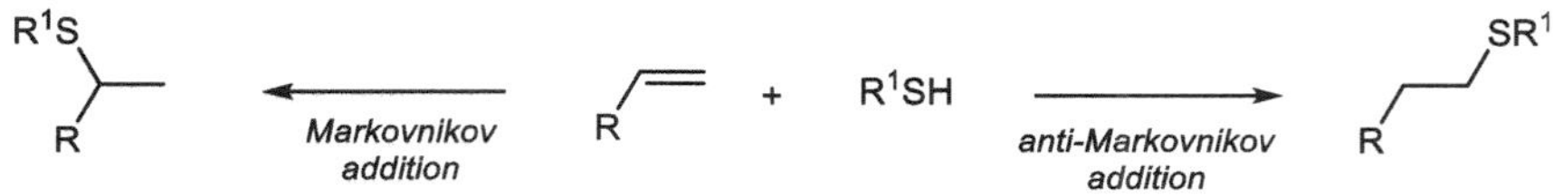

SCHEME 5.12 Thiol-ene reaction.

functional group tolerance. Mechanistically, the reactions can proceed through ionic or radical pathways leading, in principle, to Markovnikov or anti-Markovnikov products.

Several Brønsted[58] or Lewis acid[59] and native silica NP[60] catalyzed hydrothiolations. Also, catalyst-free[61] procedures were developed under solvent-free conditions. Protocols were carried out with unactivated alkenes, enol ethers or α,β-unsaturated esters, and aryl or alkyl thiols. In all cases, anti-Markovnikov selectivity was observed. Hydrothiolation processes were also carried out in neutral ILs[62] and glycerol.[63] In both cases, the solvent drove the reactions toward an anti-regioselectivity. A valuable photoinitiated thiol-ene–mediated trialkoxysilane functionalization of alkenes has been described by Garrell.[64] Such compounds are extensively employed in surface coatings with potential applications in polymer and material sciences, synthesis of dendrimers, and functionalization of biomolecules. Reactions were carried out in solvent-free conditions using UV light. Alkenes decorated with ligands, organocatalysts, and various functional groups have been functionalized in almost quantitative yields without the need of any chromatographic purification.

Although the ring-opening of epoxides is the most straightforward strategy to achieve β-hydroxysulfides, examples of hydroxysulfenylation of alkenes with the simultaneous C–O and C–S bond formation have been developed (Scheme 5.13). Hydroxysulfenylations of terminal alkenes with aryl thiols and aerial oxygen in water in the presence of β-cyclodextrin has been described by Rao.[65] The supramolecular catalyst assists through hydrogen bonds the oxidation and the nucleophilic

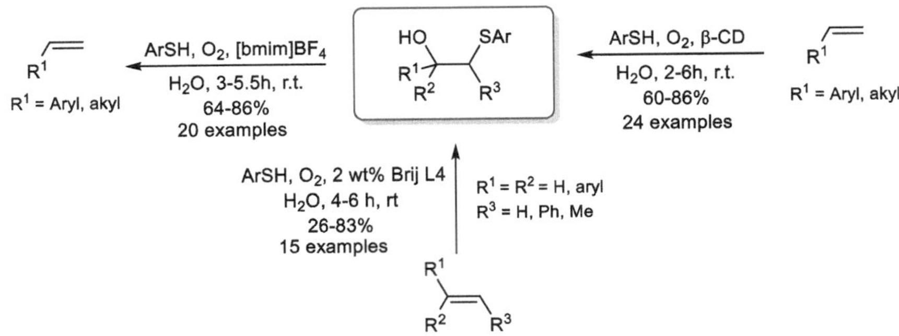

SCHEME 5.13 Hydroxysulfenylation reactions of alkenes.

addition of thiols to alkenes. Similar processes were carried out in a surfactant/water system. In this case, the authors suggested a radical pathway in which O_2 acts as radical initiator for the generation of thiyl radicals. Products were obtained in good yields and high chemoselectivity starting from terminal and internal alkenes. Single regioisomers derived from the attack of thiols at the less-substituted carbon were obtained with both procedures. An alternative one-pot hydroxylation process was reported in a mixture of IL and water without acid or base catalysts.[66]

The addition of electrophilic selenium species to double bonds is a well-established approach to construct functionalized organoselenium compounds; however, poor examples are described in nonconventional reaction media. In 2015, Braga and coworkers developed a solvent and metal-free strategy for the alkoxyselenylation of terminal alkenes under microwave irradiation.[67] Reactions were performed using diselenides, dimethylsulfoxide (DMSO) as oxidant and I_2 as catalyst to generate in situ the electrophilic selenium species. The procedure was successfully extended to disulfides and diphenyl ditelluride. Reaction and plausible mechanism are reported in Scheme 5.14.

5.4 Synthesis of alkenyl and alkynyl chalcogenides

Alkenyl chalcogenides are easily accessible by hydrochalcogenation of terminal or internal alkynes. These processes have been deeply investigated in the presence of alternative solvents or under solvent-free conditions. Scheme 5.15 shows some representative examples on nonactivated terminal alkynes.

Hydrothiolations of alkynes have been developed in solvent-free conditions using solid-supported catalysts, such as KF/Al_2O_3[68] or native silica NPs[60] and in the presence of catalytic amounts of a selenium-based

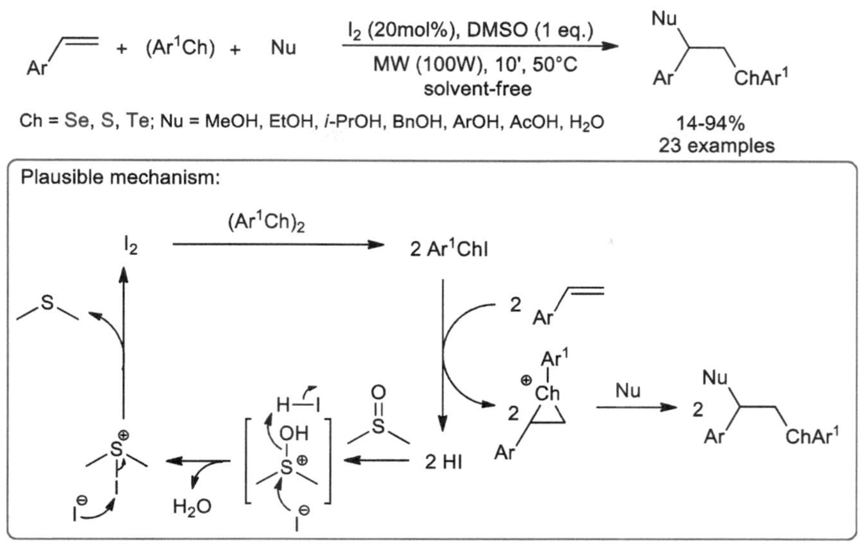

SCHEME 5.14 Microwave-assisted chalcogen (S/Se/Te) functionalization of olefins.

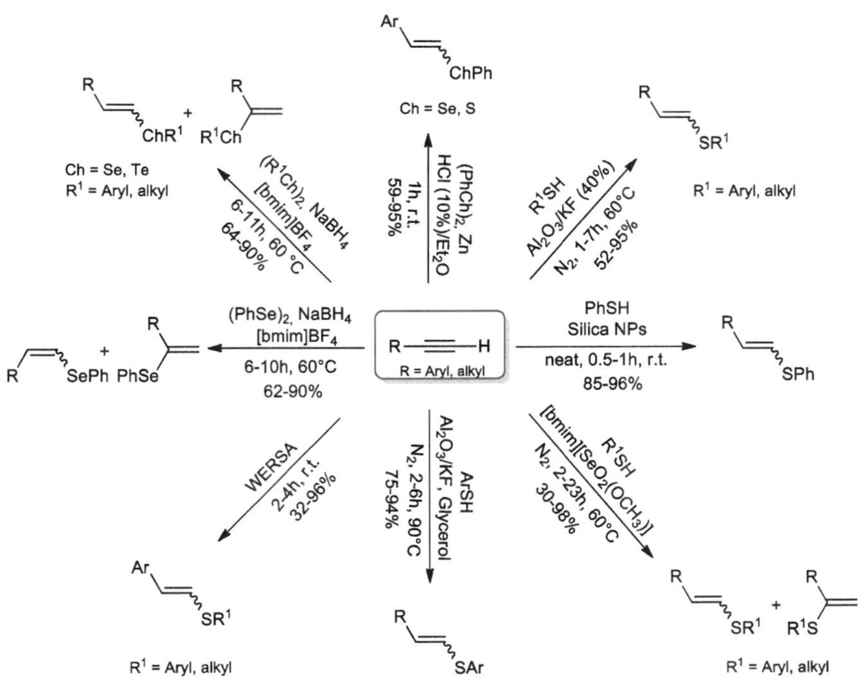

SCHEME 5.15 Synthesis of alkenyl chalcogenides.

IL.[69] Anti-Markovnikov products were mainly generated, even if poor regio- and Z/E selectivity were sometimes observed. Similar results were obtained with KF/Al_2O_3 in glycerol[70] as the recyclable solvent. The water extract of rice straw ash was recently used for the hydrothiolation of terminal aryl alkynes with aryl and alkyl thiols.[71] Reactions proceed at room temperature in good yields, high regioselectivity, and moderate-to-good Z/E selectivity. Reactions with alkyl-substituted alkynes were unsuccessful under these conditions. Several alkenyl selenides and tellurides were efficiently prepared replacing the foul-smelling chalcogenols with dichalcogenides as the nucleophile precursors. $NaBH_4$ in [bmim]BF_4 was the reducing system for the cleavage of the Se–Se and Te–Te bonds and in situ generation of the chalcogenols.[72] Alternatively, zinc in a biphasic Et_2O/acidic aqueous medium was employed for hydrothiolations and hydroselenations of terminal and internal alkynes.[73] Vinyl selenides and sulfides were obtained with moderate to good Z-selectivity. Higher selectivities were observed with selenium. The synthesis of the phenyl styryl selenide was scaled up to the gram-scale by recycling the mixture of aqueous HCl and unreacted zinc for five cycles.

Interestingly also sodium aryl sulfinates gave hydrothiolation of terminal alkynes in place of foul thiols.[74] In both cases, reactions were carried out in water, in the presence of PPh_3 as the reducing agent (Scheme 5.16). Although the detailed mechanisms of these reactions have not been fully clarified, the formation of sulfenyl radicals has been proposed.

Divinyl selenides and tellurides were obtained by regio- and stereoselective addition of in situ–generated sodium selenide to terminal aryl alkynes in PEG-400 (Scheme 5.17).[75] The Z,Z isomers were preferentially isolated. Few protocols for the access to these compounds are disposable in the literature.

Scheme 5.18 shows the recent synthesis of Z-alkenyl monoselenides or (E)-1,2-bis-selenides using the metal-free reducing agent H_3PO_2 (50 wt.% in water) at high temperature.[76] The selective formation of

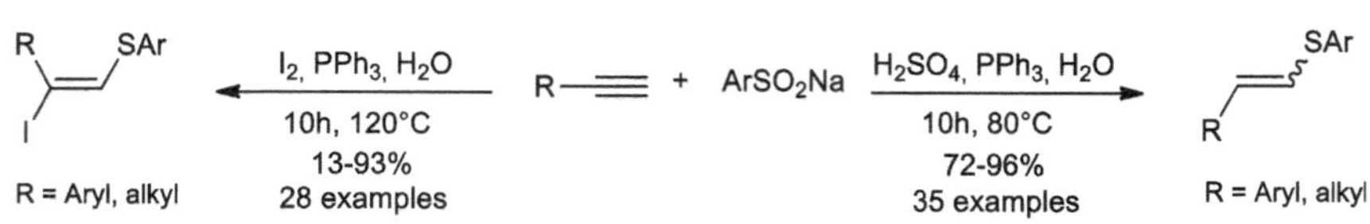

SCHEME 5.16 Hydrothiolation of terminal alkynes by sodium aryl sulfinates.

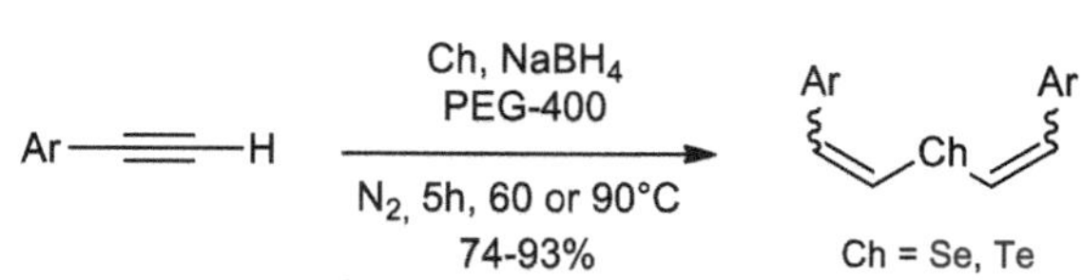

SCHEME 5.17 Synthesis of divinyl selenides and tellurides.

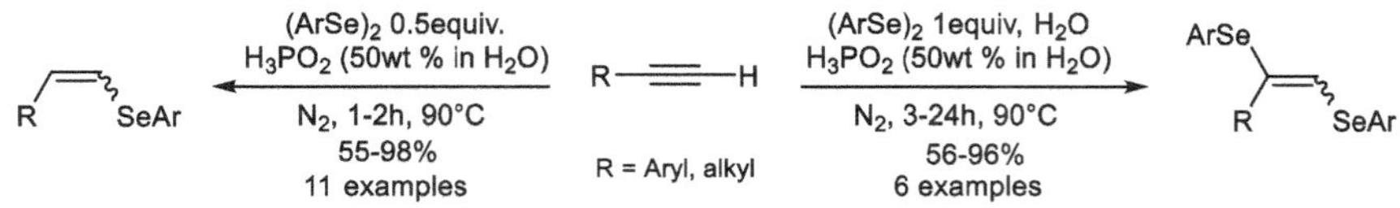

SCHEME 5.18 Selective synthesis of mono or bis-vinyl selenides.

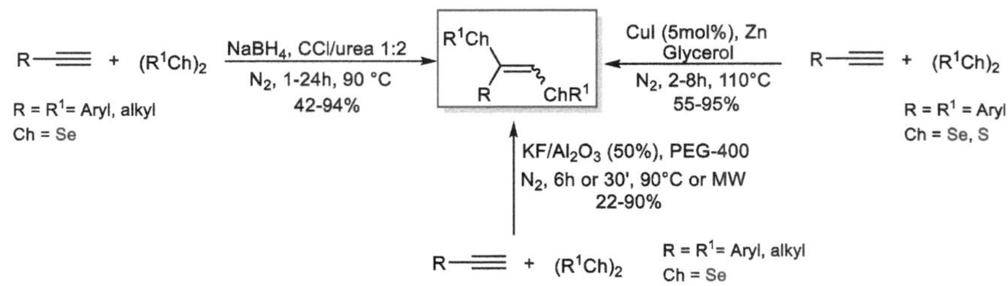

SCHEME 5.19 Synthesis of (*E*) or (*Z*)-1,2-bis-chalcogen alkenes from terminal alkynes.

mono or bis-adducts was controlled by just carefully modulating the stoichiometry and the water content. In fact, the use of a 1:1 diselenide/alkyne ratio and additional amounts of water were crucial for the selective formation of the bis-adducts.

A similar double selenylation of terminal alkynes has been reported using $NaBH_4$ in CCl/urea (1:2) (Scheme 5.19).[77] The concomitant addition of the selenolate to the terminal carbon and the capture of another PhSe group by the incipient negative charge is responsible of the prevalent formation of the (*E*)-bis-adducts. Monoselenylation products were often recovered as byproducts. The scheme also shows the formation of (*E*)-1,2-bis-chalcogen alkenes from terminal alkynes and diaryl diselenides and disulfides. The reactive ArChCu(I) species was generated by a CuI/Zn catalytic system in glycerol at 110°C.[78] Selenium reagents gave products with better selectivities. (*Z*)-1, 2-bis-arylselenyl alkenes were accessible in good yields by one-pot reaction of terminal alkynes with diaryl diselenides in the presence of KF/Al_2O_3 in PEG-400 under conventional heating or microwave irradiation.[79] The formation of a phenylselenoalkyne intermediate that suffers a formal PEG-assisted *anti*-addition of arylselenol is responsible of the observed stereoselectivity.

(*Z*)-Bis-phenylchalcogen alkenes were alternatively obtained from phenylselenoalkynes by reduction with $NaBH_4$ of S, Se or Te dichalcogenides in PEG-400 at 60°C.[80] 1,2-Bisorganoseleno alkenes as well as mixed S/Se or Te/Se 1,2-organochalcogen alkenes were accessible under these conditions. Several examples of hydrochalcogenation of electron-poor alkynes in alternative media or under solvent-free conditions have been described.[3b] Scheme 5.20 illustrates examples from the literature of the last decade. Aromatic and aliphatic thiols reacted with dialkyl acetylene dicarboxylate in water at room temperature in the

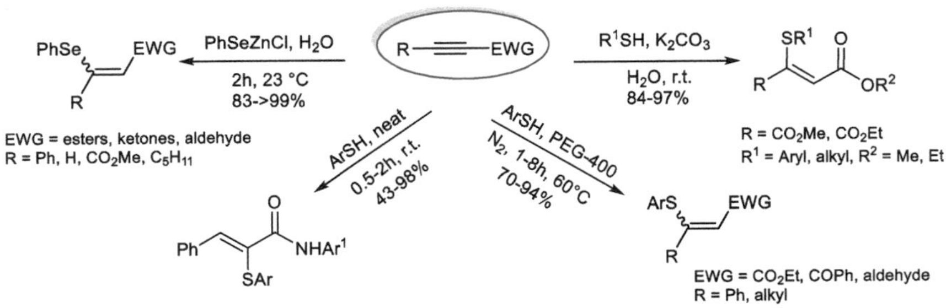

SCHEME 5.20 Synthesis of α-β-unsaturated carbonyl compounds.

presence of K_2CO_3.[81] Moreover, simple alkynoates reacted with thiols under base-free conditions in PEG-400 at 60°C as recyclable solvent.[82] *Z*-β-arylthio adducts were mainly generated. Reactions of alkynyl-*N*-arylamides with thiophenols under solvent-free conditions gave products with opposite regioselectivity.[83] The α-arylthio products were formed exclusively as a consequence of an hydrogen bond between the sulfur atom and the amidic proton. This interaction assists the attack on the triple bond with favorable effects on regioselectivity. Finally, on-water Michael–type additions of the commercial and bench-stable PhSeZnCl to alkynones or alkynoates gave β-alkenyl selenides in high yields and good to excellent *Z*-selectivity.[56] A rate acceleration was observed under "on water" conditions in respect to the corresponding reaction in THF.

Transition-metal catalyzed cross-coupling reactions of alkenyl halides or boronic acids by nucleophilic chalcogen species is a popular method to access to vinyl chalcogenides with an excellent level of stereocontrol. These reactions usually occur with complete retention of configuration. Cu-catalyzed couplings with thiols or chalcogenides have been investigated in water, ILs or glycerol.[84] Scheme 5.21 shows five recent examples using copper NPs or supported catalysts.

The first reaction is a copper(0) NP catalyzed phenylselenylation of vinyl bromides in the presence of zinc in water.[85] $CuFe_2O_4$ NPs were employed in the conversion of (*E*)- and (*Z*)-styrenyl bromides or iodides with aryl/heteroaryl thiols in water in the presence of tetrabutylammonium bromide (TBAB) and K_3PO_4.[86] The method has a wide scope and can be used also for the chalcogenation of aryl/heteroaryl halides. Similar results were obtained with the same catalyst, organoboronic acids and diselenides or ditellurides PEG-400.[87] The catalyst is magnetically separable and recyclable without any appreciable loss of catalytic activity.

Alumina-supported Cu(II) efficiently catalyzed the selenylation of styryl bromides in refluxing water, KOH and TBAB.[88] The reaction of

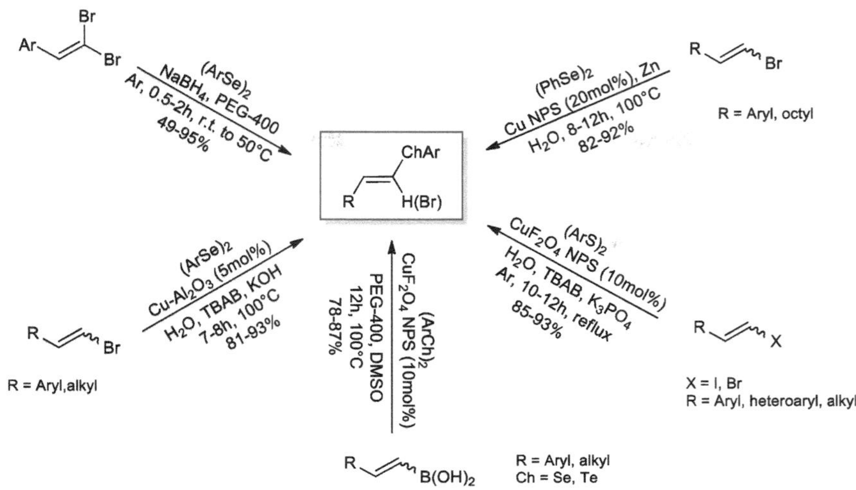

SCHEME 5.21 Synthesis of vinyl chalcogenides from vinyl halides or boronic acids.

(*E*)-alkenyl halides occurred with retention of configuration, while (*Z*) halides gave mixtures of geometric isomers. Finally, a practical transition metal–free strategy in PEG-400 under heating gave access to (*E*)-1-bromo-1-seleno alkenes or ketene selenoacetals starting from easily accessible 1,1-dibromoalkenes, diaryl diselenides, and $NaBH_4$.[89] Factors controlling the selectivity of the process are the stoichiometry of the reactants and the reaction temperature. Activated alkynes have been quite recently employed in eco-friendly hydrothio and hydroselenocyanation protocols affording multifunctionalized vinyl chalcogenocyanates. These compounds serve as versatile building-blocks for organic synthesis, medicinal chemistry, and material sciences. The conventional synthetic approach involves the nucleophilic substitution of vinyl halides by chalcogen cyanate salts. These protocols suffer from certain drawbacks such as the use of toxic metal catalysts, poor *Z*/*E* selectivities, and/or harsh reaction conditions. In the recent protocols reported in Scheme 5.22, reactions were carried out in Brønsted acid IL, DES, or lactic acid as alternative recyclable solvents using water as the proton source.[90] The thermodynamically unfavorable *Z*-vinyl compounds were readily recovered in high yields and stereoselectivity. Similar mechanisms were suggested for all the processes. The solvent plays a dual role serving as catalyst and reaction medium. The enhanced polarization of the carbonyl group through the formation of hydrogen bonds with the solvent promotes the formation of the intermediate allenolate. Then, the attack at the β-carbon by the thio or selenocyanate followed by protonation at the less hindered face gave access to the product.

Radical additions of the thiocyanyl radical, produced by oxidation of KSCN with $PhI(OAc)_2$ in water gave thiocyanation of enaminones in the α-position with broad functional and structural scope.[91] According to the structure of the starting material, thiocyano enaminones or chromenones were efficiently synthesized (Scheme 5.23).

(Z)-β-iodo vinylthiocyanates were prepared starting from variously substituted terminal and internal alkynes using KSCN, ammonium persulfate, iodine, and acetic acid (Scheme 5.24). Reactions proceed through a iodine-mediated electrophilic addition of the SCN anion to the alkyne.[92]

Scheme 5.25 shows two recent strategies for the multistep one-pot access to alkynyl selenides in alternative solvents under transition

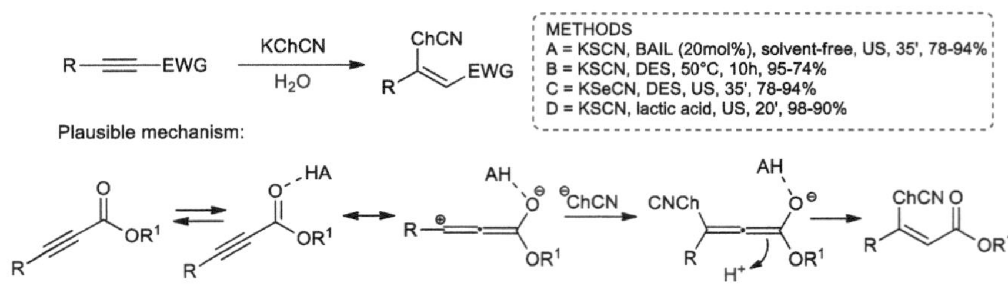

SCHEME 5.22 Thio and selenocyanation of electron-deficient alkynes in water.

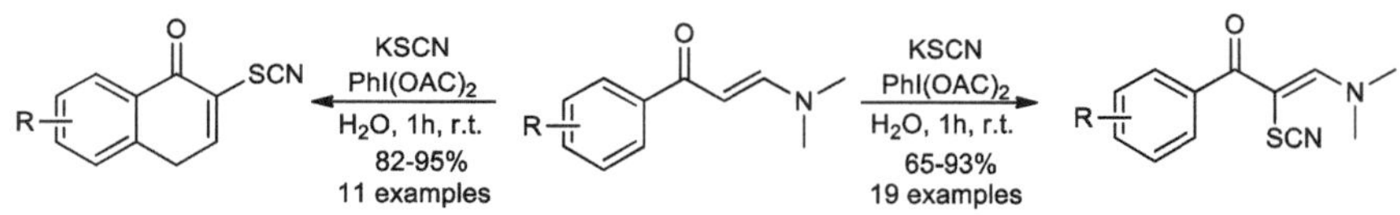

SCHEME 5.23 Synthesis of thiocyano enaminones or chromenones.

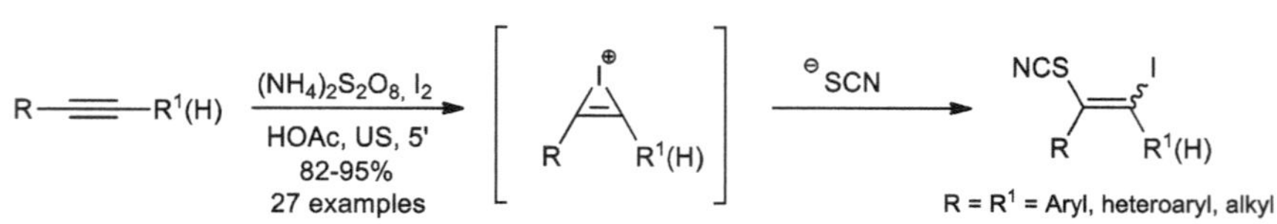

SCHEME 5.24 Multicomponent synthesis of (Z)-β-iodo vinylthiocyanates.

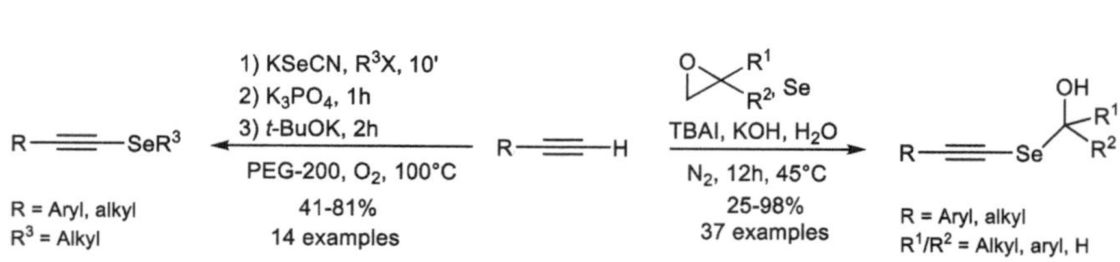

SCHEME 5.25 Synthesis of alkynyl selenides.

metal–free conditions. In the first reaction, diselenides were formed in situ from alkyl bromides and KSeCN in PEG-200 at high temperatures.[93] The reaction of the diselenide with terminal alkynes in the presence of *t*-BuOK generated the corresponding alkyl alkynyl selenides in acceptable-to-good yields. In the second cascade protocol epoxides, elemental selenium and terminal alkenes reacted in pure water to generate alkynyl alkyl selenides with high regioselectivity.[94]

The reaction started with the formation of diselenides through ring-opening of the epoxide with selenide anions formed by disproportion of elemental selenium, followed by oxidative coupling. Then, the selenation of the terminal alkynes generated the alkynyl alkyl selenides. A good functional group tolerance was observed.

5.5 Synthesis of chalcogen-containing carbocycles and heterocycles

Chalcogen-containing hetero- and carbocyclic compounds are prevalent in natural products, synthetic drugs, and organic materials. In the last few years, massive attention has been devoted to developing economical and practical processes for C-chalcogen bond formation. Numerous direct C(sp2)-H chalcogenations of carbo- and heteroarenes have been performed employing an ionic pathway through environment-friendly protocols. In this context, reactions catalyzed by molecular iodine and iodine salts have emerged as versatile approaches (Scheme 5.26).

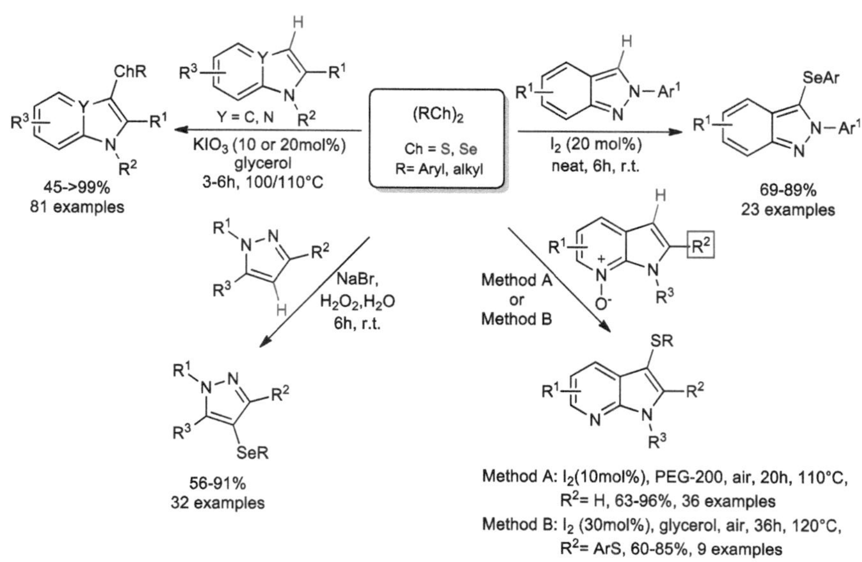

SCHEME 5.26 Chalcogenations of electron-rich *N*-heteroarenes with dichalcogenides.

These species possess reactivity similar to that of transition metals but offer several advantages such as atom economy and milder reaction conditions. Synthesis of 3-Se/S-indoles and imidazo[1,2-*a*]pyridines has been carried out using a KIO_3/glycerol catalytic system with different diorganyl dichalcogenides as the sulfur and selenium source.[95] This robust procedure, employing a nontoxic and easy-to-handle catalyst, has shown tolerance for various substituents with different electronic and steric effects. Moreover, a number of 3-(phenylselanyl)-2*H*-indazoles were synthesized in good yields at room temperature in solvent-free conditions using readily available diaryl diselenides and a catalytic amount of I_2.[96] Both of these methodologies proceeded regioselectively at the C3 position of heteroarenes without the exclusion of air and moisture. Reactions were readily scaled up on gram-scale. A general and sustainable approach for the chalcogenation of 7-azindoles from the corresponding *N*-oxides using the I_2/PEG as an efficient and reusable catalytic system has been also developed.[97] Controlled mono or dichalcogenylations were obtained by variation of the reaction parameters. In this transformation, 7-azindole *N*-oxides served both as substrates and as internal oxidants and various alkyl, aryl, and heteroaryl chalcogenating reagents could be employed with success. Using H_2O_2 and halide salts as catalyst, variously substituted 4-selanyl pyrazoles were synthesized in good yield.[98]

The scope of the C(sp^2)-H chalcogenation of carbo- and heteroarenes was extended to the use of thiols, heterothiols, and thiones as the sulfur source (Scheme 5.27) in water under mild reaction conditions. An environment-friendly arylthiolation of imidazo heterocycles using

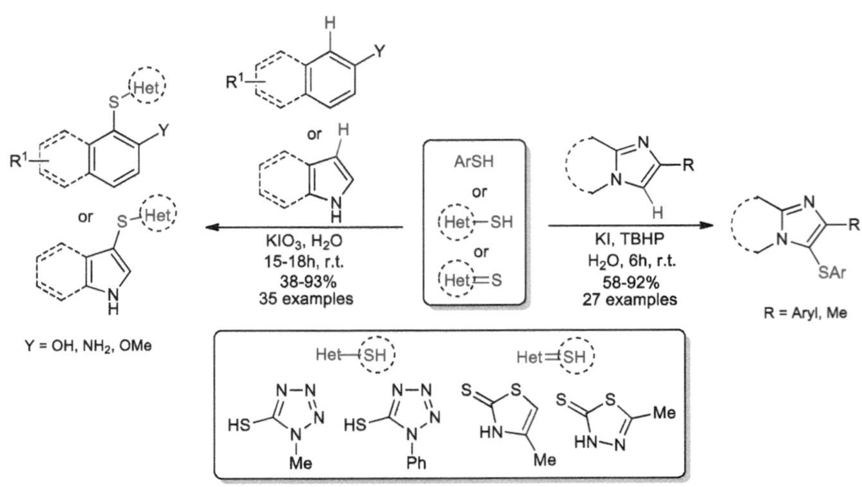

SCHEME 5.27 Chalcogenations of electron-rich arenes and *N*-heteroarenes with thiols or thiones.

potassium iodide (KI) and*tert*-butyl hydroperoxide (TBHP) as oxidant (ratio 1:3) at room temperature was reported.[99] This metal-free process was carried out with different imidazo[1,2-*a*]pyridines and imidazo[2, 1-*b*]thiazoles, demonstrating the generality of this methodology. A KIO_3-catalyzed direct regioselective sulfenylation of a variety of phenols, arylamines, pyrrole, and indole derivatives with heterocyclic thiols or thiones has also been performed.[100]

Many other thiolations of arenes were carried out in water (Scheme 5.28). Piperidine promoted coupling reactions between thiophenols and 2-naphtol derivatives in the presence of DMSO as oxidant.[101] The authors provided a plausible mechanism based on the initial conversion of thiol in disulfide under oxidative conditions. Piperidine interacted with the disulfide to generate an electrophilic sulfenylating agent, which reacted with the enolate of the naphtol. The synthesis of substituted aryl sulfides *via* a practical copper-catalyzed three-component approach (arenes, aryl, or alkyl iodides and elemental sulfur) has been also investigated.[102]

Different functional groups like halogen, nitro, cyano, aldehyde, acetyl, and amino were well tolerated. Sodium aryl sulfinates were alternative sulfur sources for the metal-free synthesis of aryl sulfides at the C-1 position of 2-naphtols.[103] Probably, the addition of iodine induced the reduction of sulfinates and generated the electrophilic intermediate ArSI. Phenols, 2-alkoxy naphthalenes, and *N,N* dimethylaniline also reacted with ArSI to provide the corresponding aryl sulfides. Aryl sulfonyl hydrazides, sulfonyl chlorides or bromides as well as the above-mentioned aryl sulfinates were also employed for thiolations of hetero and carboarenes in different sustainable solvents (Scheme 5.29). A variety of 3-sulfenyl indoles was synthesized in good-to-excellent yields employing sulfonyl hydrazides without any addition of ligand or additive.[104] The generated ArS^+ intermediate gave a Friedel–Crafts process with the formation of nitrogen and water as the only byproducts.

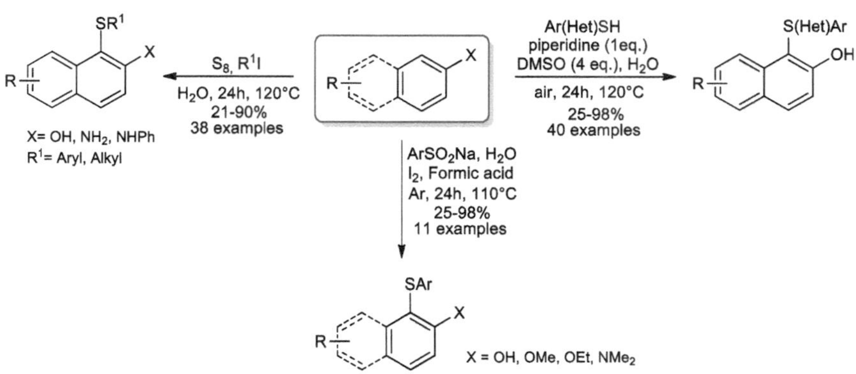

SCHEME 5.28 Thiolations of electron-rich arenes.

Sulfonyl hydrazides were also employed for the iodine-mediated regioselective sulfenylation of coumarins.[105] Several 3-arylthio indoles and 2-halo-3-arylthioindoles were synthesized from sulfonyl chlorides or bromides and indole derivatives in hexylmethylimidazolium bromide [hmim]Br in the presence of a catalytic amount of water.[106] Tunable reactions proceeded at room temperature or at 50°C via a selective C3–S bond formation or a cascade C3–S and C2–Z (Z = Cl and Br) bond formation, respectively. The same IL [hmim]Br was used as both the solvent and the reducer in the acid-promoted chemoselective arylthiolations of electron-rich arenes with sodium aryl sulfinates.[107]

A regioselective aromatic nucleophilic substitution was developed for the synthesis of 4-arylselanyl-7-chloroquinolines (Scheme 5.30).[108] This method involves the reaction of the 4,7-dichloroquinoline with organylselenols, generated in situ by the reaction of diorganyl diselenides with aqueous H_3PO_2 at 60°C under nitrogen atmosphere. The Se-functionalized quinolines were successfully screened for in vitro acetylcholinesterase activity.

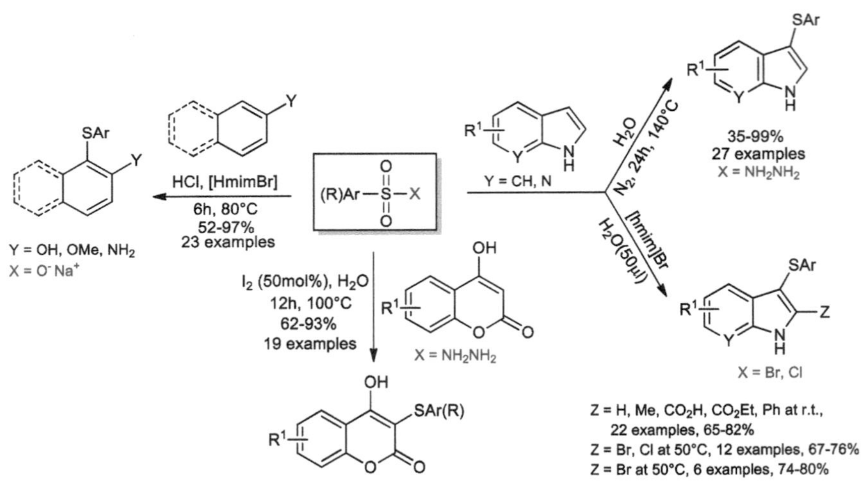

SCHEME 5.29 Thiolations of heterocycles and carbocycles by sulfonyl derivatives.

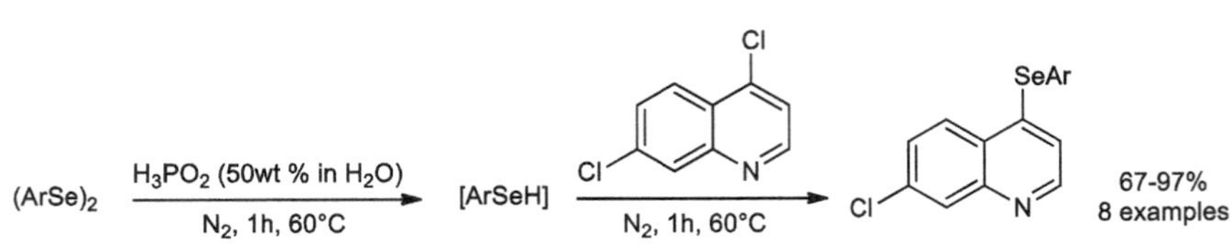

SCHEME 5.30 Synthesis of 4-arylselanyl-7-chloroquinolines via a nucleophilic aromatic substitution.

A radical coupling of imidazopyridines with ethers or alkanes and selenium powder gave 3-alkylselanyl-imidazopyridines in acceptable-to-good yields (Scheme 5.31).[109]

In a plausible reaction mechanism the dicumyl peroxide (DCP) generated alkyl radicals that reacted with the selenium powder. The formed selenyl radicals added at the C-3 position of the imidazopyridines and gave a homolytic aromatic substitution through a radical addition/DCP oxidation/proton elimination sequence. An important process to obtain chalcogen-functionalized hetero and carbocyclic compounds is the cyclofunctionalization reaction. The cyclization of alkenes containing an internal nucleophile promoted by electrophilic chalcogen species has attracted considerable attention in organic synthesis, even if environment-friendly approaches remain poorly investigated. An efficient and simple method for selenoetherifications and lactonizations employs diphenyl diselenide and I_2/DMSO as the catalytic oxidative system, under microwave irradiation (Scheme 5.32).

Biologically interesting Lapachol and C-allyl-Lawsone selenium derivatives were obtained in 10 minutes and in good yields through a 6-*endo-trig* or 5-*exo-trig*-cyclization, respectively.[67] The low concentration of iodide anions is crucial to avoid the formation of side products by competition with the internal nucleophile.

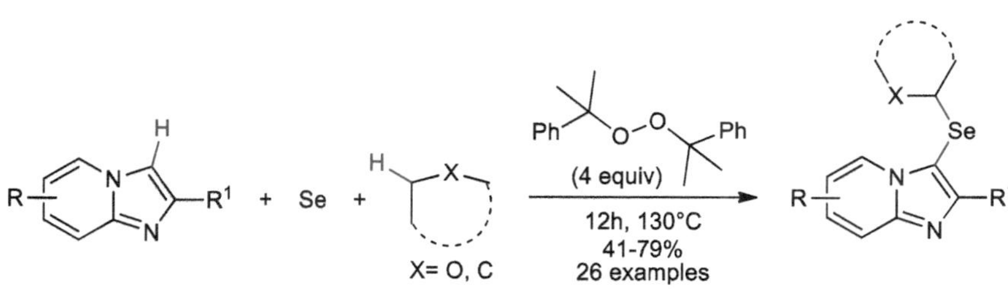

SCHEME 5.31 Oxidative dual C−H selenylation of imidazopyridines.

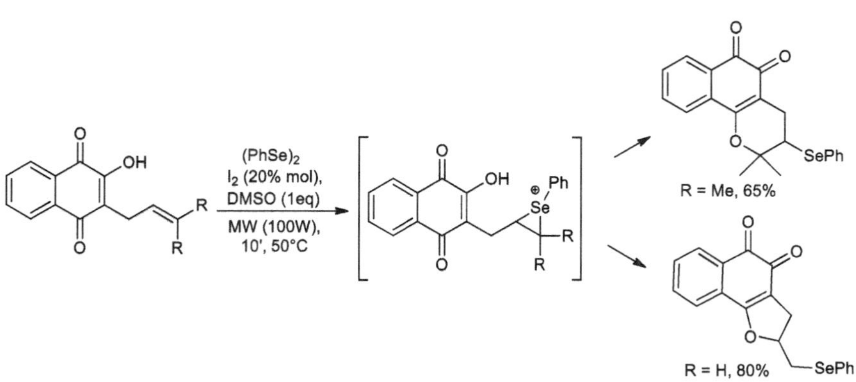

SCHEME 5.32 Selenocyclizations of Lapachol and C-allyl-Lawsone.

An eco-friendly methodology to synthesize 2-organoselenyl–naphthalenes based on the reaction of alkynols with diaryl diselenides was described using Oxone in aqueous medium (Scheme 5.33).[110] The reactions proceeded under ultrasonic irradiation in short reaction times and good yields. The plausible mechanism involved an oxidative cleavage of the Se–Se bond by Oxone. The formed electrophilic selenium species reacted with the carbon–carbon triple bond to produce the seleniranium intermediate.

Then, an intramolecular carbon *6-endo-dig* cyclization occurred and after deprotonation and water elimination the aromaticity of the system was restored to afford the 2-organoselenyl-naphthalenes.

Polymer–bound selenyl bromides with an amide linker were prepared from polystyrene resins and used in the intramolecular oxyselenylation of olefins in water (Scheme 5.34).[111,112] The polymer supported seleno-heterocycles were detached *via* oxidative deselenylation by treatment with H_2O_2 and recovered by simple filtration. The immobilization of the selenium reagents on the polymer was crucial to carry out the reaction in aqueous medium.

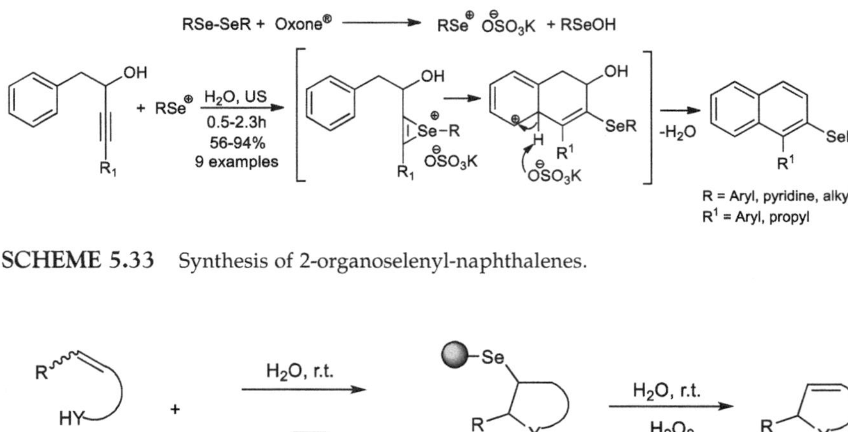

SCHEME 5.33 Synthesis of 2-organoselenyl-naphthalenes.

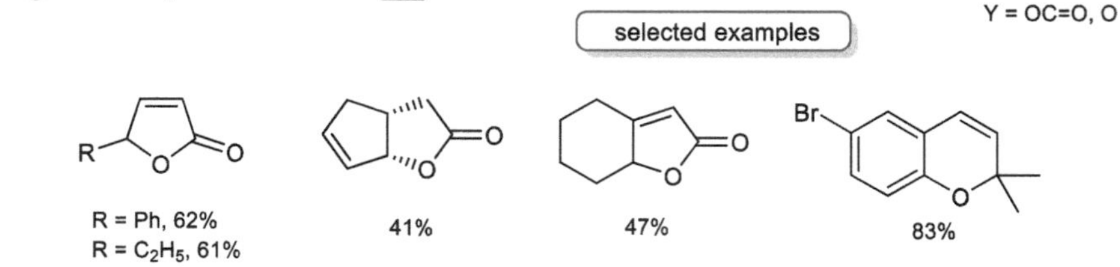

SCHEME 5.34 Selenocyclization-deselenylation reactions with polymer-supported selenyl bromides.

Tin-free group-transfer radical cyclizations in environmentally benign solvent were developed by Engman (Scheme 5.35).[113] Reactions were performed with primary and secondary alkyl tellurides at 250°C in ethylene glycol or at 180°C in water under microwave irradiation. Tetrahydrofurans and cyclopentanes were formed after few minutes in good yields, but with poor diastereoselectivity, probably as a consequence of the high temperatures reached in the microwave-assisted reactions.

Radical cyclizations of eneyne derivatives were developed with diphenyl disulfide and tripropylamine at 140°C in the presence of water (Scheme 5.36).[114] The plausible mechanism involves a single electron transfer reaction of tripropylamine to diphenyl disulfide. The formed benzenethiyl radical reacted with several eneyne compounds in a 5-*exo-trig* manner to afford cyclopentanes or heterocyclic five membered products. The experimental procedures were found to be extremely simple and safe and employed inexpensive and environment-friendly reagents.

In the last decades selenium-containing rings, particularly selenium–nitrogen heterocycles, have received great attention because of their remarkable biological properties.[115] Despite the growing interest, few

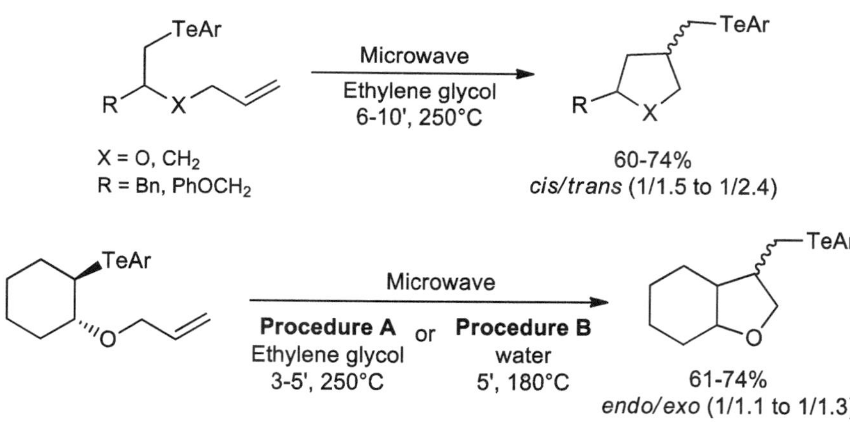

SCHEME 5.35 Microwave-assisted group transfer cyclizations.

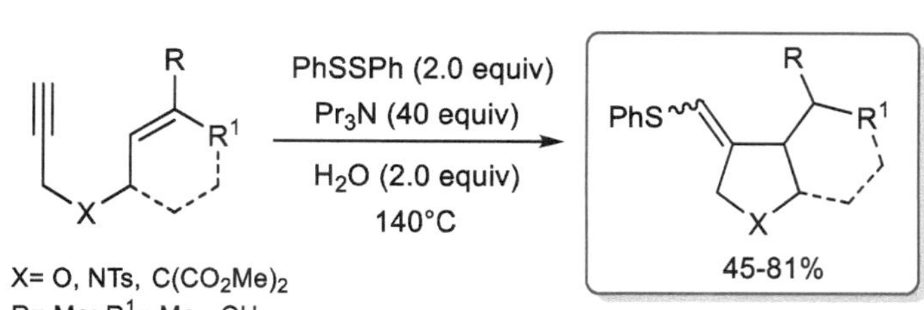

SCHEME 5.36 Disulfide-mediated radical cyclizations of eneyne derivatives.

synthetic strategies in alternative reaction media have appeared in literature. Isoselenocyanates were used as efficient building-blocks for the synthesis of five- and six-membered heterocycles (Scheme 5.37). In 2010, a solvent- and catalyst-free approach was developed to obtain 3,1-benzoselenazines. Isoselenocyanates and *o*-ethynylanilines gave a tandem addition-6-*exo-dig*-cyclization. Microwave-assisted processes improved reaction times and yields.[116] Metal- or organo-catalyzed, on-water [3 + 2] cycloaddition reactions with aziridines and isoselenocyanates have also been described. Reactions catalyzed by an iron salt[117] proceeded on the water/oil interface through a concerted mechanism. On the contrary, a stepwise mechanism via urea-type intermediates was proposed for pyrrolidine-catalyzed cycloadditions.[118] Products were obtained in good-to-excellent yields and protocols were found suitable for gram-scale syntheses.

Finally, a synthetic strategy to afford benzoselenoazoles and benzoselenazolines in glycerol is shown in Scheme 5.38.[119] The addition to dicarbonyl compounds of 2-amino-benzenselenol, generated in situ from the corresponding diselenide and aqueous H_3PO_2, led to iminium intermediates that successively underwent the 5-*endo-trig* cyclization. The regio- and the chemo-selectivity of the process were influenced by the nature of dicarbonyl compounds. When β-diketones were used, a retro-Mannich type fragmentation took place after cyclization to afford the benzoselenazoles.

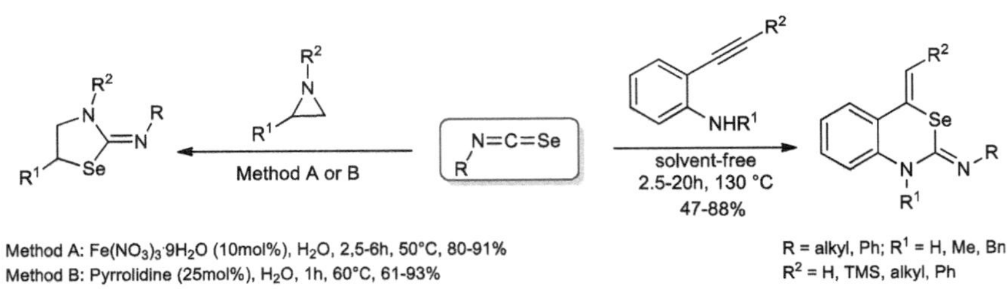

SCHEME 5.37 Isoselenocyanates for the synthesis of five- and six-membered heterocycles.

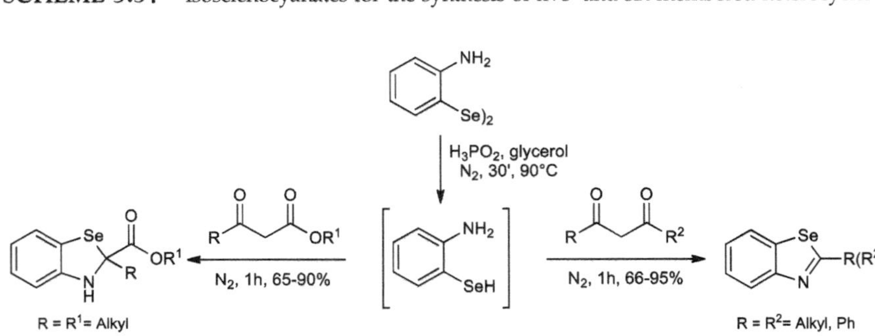

SCHEME 5.38 Synthesis of benzoselenoazoles and benzoselenazolines.

5.6 Synthesis of chalcogenoesters

Chalcogenoesters have great importance as intermediates in several organic transformations with applications in the synthesis of natural products, biologically active compounds, and new materials. Most of the synthetic methodologies employed acyl chlorides as starting material in ILs, PEG, water, or solvent-free conditions (Scheme 5.39).

The use of metallic nanostructures as catalysts has recently garnered massive interest. Eco-friendly syntheses of chalcogenoesters from acyl chlorides and stable diaryl diselenides or disulfides instead of malodorous and hazardous selenols or thiols have been developed in IL (BMIM-PF_6) with recyclable CuO nanopowder in the presence of Cs_2CO_3[120] or under microwave irradiation.[121] The catalytic systems were easily recovered and reused for further reactions without any loss of efficiency. Alternatively, indium metal was used as the reducing agent.[122] Moreover, hypophosphorous acid (H_3PO_2) gave selenoesters from diaryl diselenides in PEG-400.[123] Investigations on the use of zinc dust for the cleavage of dichalcogenides have received special attention because of its easier manipulation and better stability in air in respect with other metals. Braga and coworkers reported an efficient, quick, and eco-friendly method for the synthesis of selenol esters and selenocarbonates

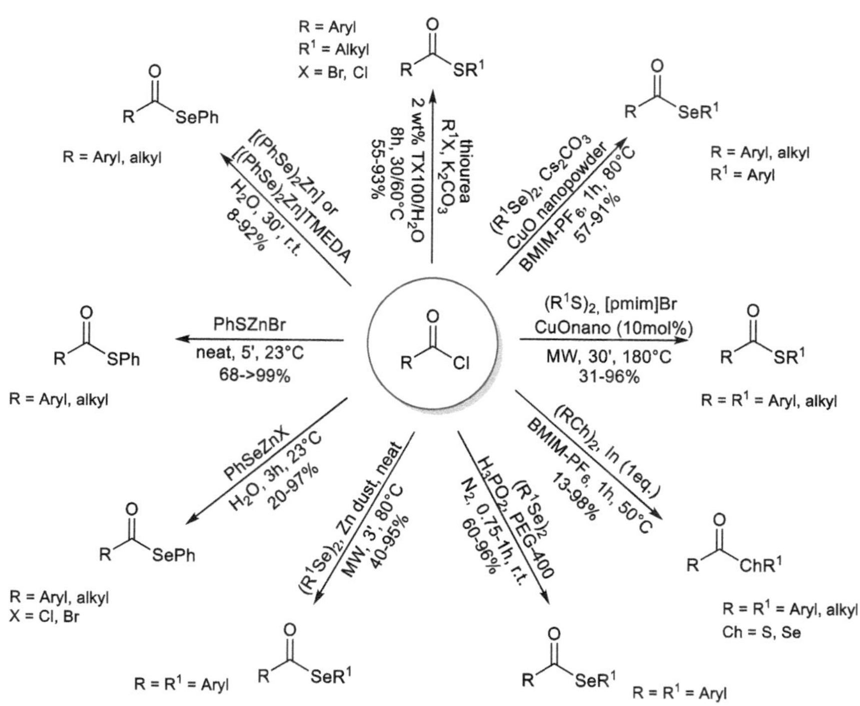

SCHEME 5.39 Synthesis of chalcogenoesters from acyl chlorides.

employing diorganyl diselenides, acyl chlorides, and zinc dust in solvent-free conditions under microwave irradiation, in very short reaction times.[124] As described by Santi, PhSe-Zn-halides reacted with aromatic and aliphatic acid chlorides to afford a variety of selenoesters in high yields under "on water" condition.[125] A considerable rate acceleration was observed. This bench-stable phenylchalcogenolate zinc(II) halide were also applied in a solvent-free procedure to prepare thioesters in excellent yields by simply grinding a mixture of PhSZnBr and acyl chloride for 5 minutes.[126] Discrete Fourier transform calculations showed that the coordination of solvent molecules to the metal center of PhSZnBr reduces the reactivity toward the acyl chloride in comparison with solvent-free conditions. Other atom-efficient preparations of zinc selenates such as [$(PhSe)_2Zn$] or [$(PhSe)_2Zn$]TMEDA were also successfully investigated.[127] Thioesters were prepared in water through one-pot three component reactions of organic halides, thiourea, and benzoyl chlorides. The S-alkylisothiouronium salts, generated in situ by reaction of organic halides and thiourea, were hydrolyzed in aqueous Triton X-100 to produce thiolate moiety and urea.[128] The presence of a nonionic surfactant that self-assembles in micelles may prevent the hydrolysis of benzoyl chlorides. New methods for the synthesis of chalcogenoesters from aldehydes, anhydrides, or carboxylic acid derivatives were reported (Scheme 5.40).

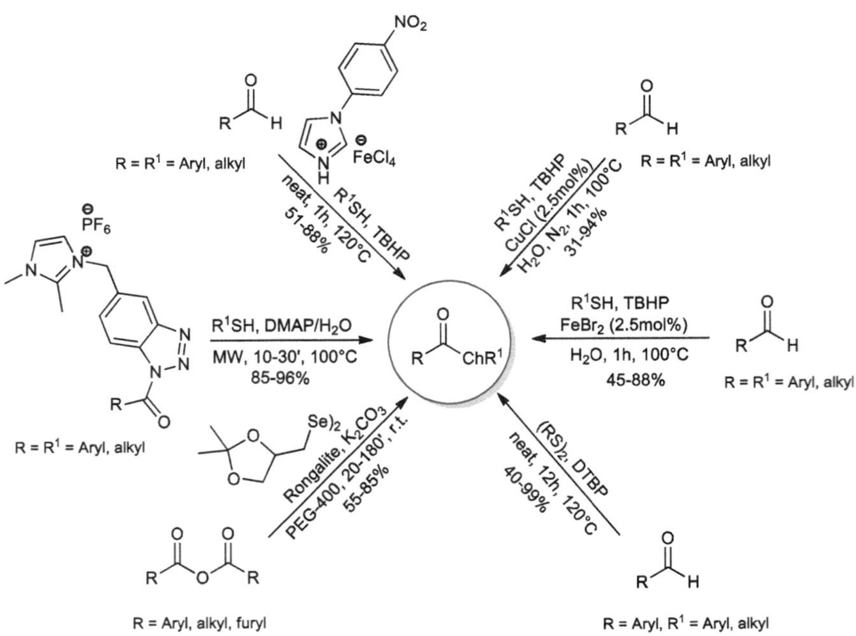

SCHEME 5.40 Synthesis of chalcogenoesters from aldehydes, anhydrides, and carboxylic acid derivatives.

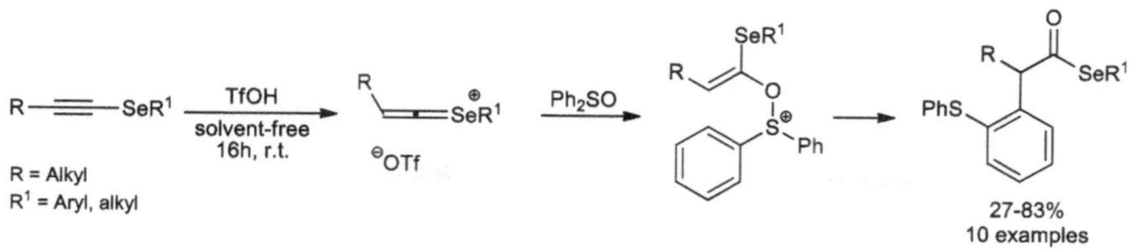

SCHEME 5.41 Acid-catalyzed redox-neutral oxyarylation of selenoalkynes.

Aryl imidazolium ILs successfully catalyzed thioesterification in sealed tube using aromatic and alkyl aldehydes with a variety of alkyl and aromatic thiols in the presence of TBHP as oxidants.[129] Transition metals efficiently catalyzed the formation of carbon-heteroatom bonds starting from aldehydes. Thioesters were synthesized through copper- or iron-catalyzed cross-coupling reactions between aryl and alkyl aldehydes with aryl and alkyl thiols in the presence of TBHP as an oxidant.[130] The metal-catalyzed formation of acyl radicals was followed by reaction with iron(III) thiolate complexes to provide the thioesters.[131] Metal- and solvent-free reactions of aryl aldehydes with disulfides in the presence of di-*tert*-butyl peroxide (DTBP) gave also access to thioesters.[132] A green method for the synthesis of glycerol-based selenoesters was developed by reactions between aromatic, aliphatic, and heteroaromatic anhydrides and bis-(2,2,-dimethyl-1,3-dioxolanylmethyl)diselenide using PEG-400 and Rongalite.[133] Water-soluble selenoesters were produced by deprotection of the ketal-protecting group with an acidic cation-exchange resin. An efficient and recyclable imidazolium-supported benzotriazole reagent (Im-CH_2-BtH) was used to activate carboxylic acid in the preparation of thioesters in water under microwave irradiation.[134] Quantitative conversions were obtained.

An alternative redox neutral oxyarylation of selenoalkynes for the synthesis of selenoesters is reported (Scheme 5.41).[135] The reaction involved a Brønsted acid activation of the triple bond to generate selenium-stabilized vinyl cation, which was trapped by the aryl sulfoxide. A subsequently sigmatropic rearrangement gave the α-arylated selenoester products.

5.7 Synthesis of compounds containing chalcogens at higher oxidation states

In the last decade, great efforts have been devoted to the synthesis of sulfoxides and sulfones (Scheme 5.42). The strong electron-withdrawing effect of both the sulfoxide and the sulfone functional groups and the broad possibility of chemical manipulations make these compounds

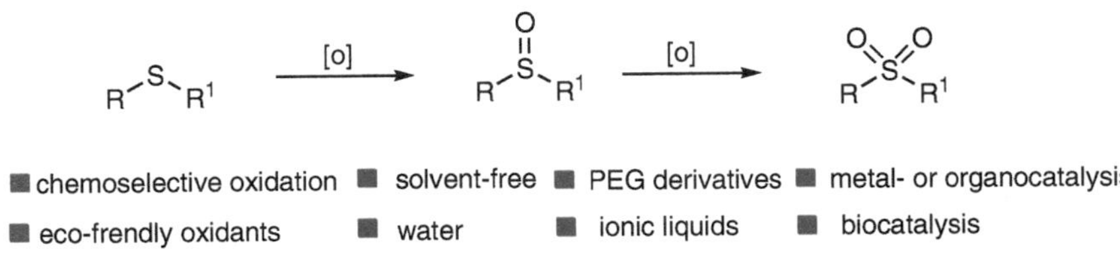

SCHEME 5.42 Oxidation of sulfides to sulfoxides and sulfones.

interesting building-blocks in organic synthesis. The chemoselective formation of both compounds by controlled oxidation of sulfides has been successfully investigated, and a plethora of new methods in alternative solvents have appeared in the literature. Some reactions were successfully performed with conventional organic oxidants such as iodine derivatives[136] and peroxyacids.[137] Moreover, in situ electrochemically generated hypochlorite[138] and inorganic peroxyacid salts[139–142] have been successfully investigated, avoiding the use of polluting heavy metals. For instance, solid sodium percarbonate (SPC) or perborate (SPB) were used for the green chemoselective oxidation of sulfides to sulfoxides or sulfones.[139] Sulfoxides were formed with SPC and Amberlyst under heating in solvent-free conditions, whereas sulfones were generated with SPB in water under microwave irradiations. The authors suggested that selectivity depends on the nature of the oxidant, the reaction medium, and the process activation technique. Controlled oxidations were also carried out with commercially available Oxone (triple salt: $2KHSO_5 \cdot KHSO_4 \cdot K_2SO_4$).[140] In this case, the choice of the solvent was crucial to switch the oxidation toward the selective formation of sulfoxides (in EtOH) or sulfones (in H_2O). Sulfones were also accessible in the absence of any solvent, using mechanical milling conditions.[141] Interestingly, Oxone in water was also employed in the oxidation of vinyl selenides into vinyl selenones,[142] a recently rediscovered class of selenium (VI) compounds with potential application in asymmetric catalysis and heterocyclic synthesis. It is interesting to point out that only few oxidizing agents are strong enough to convert selenides into the corresponding selenones, because of the stronger dipolar character of the Se = O in respect to sulfur analogs.

Hydrogen peroxide is a particularly attractive oxidant in consideration of the easy availability, low cost, safety, atom-efficiency, and formation of water as the only byproduct. Hydrogen peroxide–mediated oxidations of sulfides in water are often catalyzed by transition metal catalysts or Brønsted acids. For instance, iron oxide NPs on mesoporous silica[143] or molybdenum-containing surfactants[144] as easy recyclable catalysts have been recently used to catalyze the controlled oxidation of sulfides to sulfoxides in water. In the latter case, several sensitive functional groups on the sulfides were well tolerated. Heterogeneous nanocatalysts based on propylsulfamic acid[145] or sulfonate–PEG (PEG-SO_3H)[146] supported onto ferromagnetic magnetite (Fe_3O_4) particles were developed for an easy retrieval through an external magnetic tool. A hydrogen peroxide-promoted transition metal–free sulfoxidation employed camphor-10-sulfonic acid as catalyst.[147] Catalyst-free hydrogen peroxide oxidation protocols were carried out in poli(ethylene glycol) dimethy ether ($PEGDME_{500}$) under sonication[148] and in the IL $[Bmim]BF_4$.[149] Interesting enantioselective oxidation of sulfides to chiral sulfoxides were recently developed. Flavin–cyclodextrin conjugates were designed, prepared, and used as organocatalysts for enantioselective sulfoxidations with hydrogen peroxide and phosphate buffer at pH 7.5.[150] Low loading of the catalyst (0.5%–1%) converted alkyl methyl and bulkier sulfides in good yields, short reaction times, and moderate-to-good enantioselectivities (up to 91% *ee*). A chloroperoxidase was employed in the enzymatic enantioselective oxidation of prochiral methyl aryl sulfides with hydrogen peroxide.[151]

The enzyme was microencapsulated in a microporous silica gel cage to ensure a sufficient stability. Acceptable conversions and good-to-excellent *ee* were observed. The biocatalyst was recycled four times with only a moderate loss of activity and retained enantioselectivity. Alternatively, enantioselective enzymatic air oxidation of sulfides was performed with mutant cyclohexanone monooxygenases.[152] Artificial metalloenzymes, obtained through incorporation of abiotic metal cofactors within a protein scaffold, were also designed, synthesized, and tested for the enantioselective sulfoxidation with hydrogen peroxide. The noncovalently associated serum albumin-manganese (III) corrole system found application in the synthesis of *R*-modafinil (up to 75% *ee*), a drug approved for the treatment of narcolepsy and shift work sleep disorders.[153] Moreover, an artificial metalloenzyme based on the incorporation of a cobalt (II) Schiff base complex with BSA, gave variously substituted sulfoxides with excellent conversions and moderate-to-good enantioselectivity.[154] In the above-mentioned protocols, overoxidations to sulfones have been usually prevented. However, in some cases, sulfones were formed as single products when reaction conditions were modified, or an excess of the oxidant employed.[147,149] As an example, the above-mentioned Fe_3O_4/PEG-SO_3H gave the selective oxidation of sulfides into sulfones by tuning the amounts of hydrogen peroxide and catalyst. Moreover, a high atom-efficient oxidation of sulfides into sulfones was developed using 10 mol% excess of 30% aqueous hydrogen peroxide under heating in catalyst- and organic solvent-free conditions.[155] Both hydrophilic and hydrophobic sulfides were employed. Another simple procedure for the oxidation of sulfides into sulfones used hydrogen peroxide and V_2O_5 as catalyst in an IL.[156] The products were recovered by extractions with organic solvents, and the catalytic system was recycled and reused without any loss of catalytic activity. In the last years, mild oxidations of sulfides with molecular oxygen as a cheap, safe, and clean oxidant have attracted great interest. However, the use of molecular oxygen is challenging since reactions have a high-activation barrier and usually require catalysts or radical initiators. Sulfoxides were prepared in PEG,[157] ILs,[158] or water[159] using metal- or organocatalysts. Quite recently, eco-friendly high boiling point ethers, such as PEGDME and bis(2-butoxyethyl)ether, acted as both solvents and catalysts for chemoselective oxidations with molecular oxygen. Sulfoxides or sulfones were formed without addition of any metal catalyst or additive, respectively. Finally, an aerobic photoxidation of sulfides to sulfoxides under visible light irradiation in water at room temperature was successfully investigated.[160] New synthetic strategies for vinyl sulfones were recently established starting from alkenes or alkynes and sodium sulfinates, sulfinic acids, or sulfonyl hydrazides in water (Scheme 5.43).

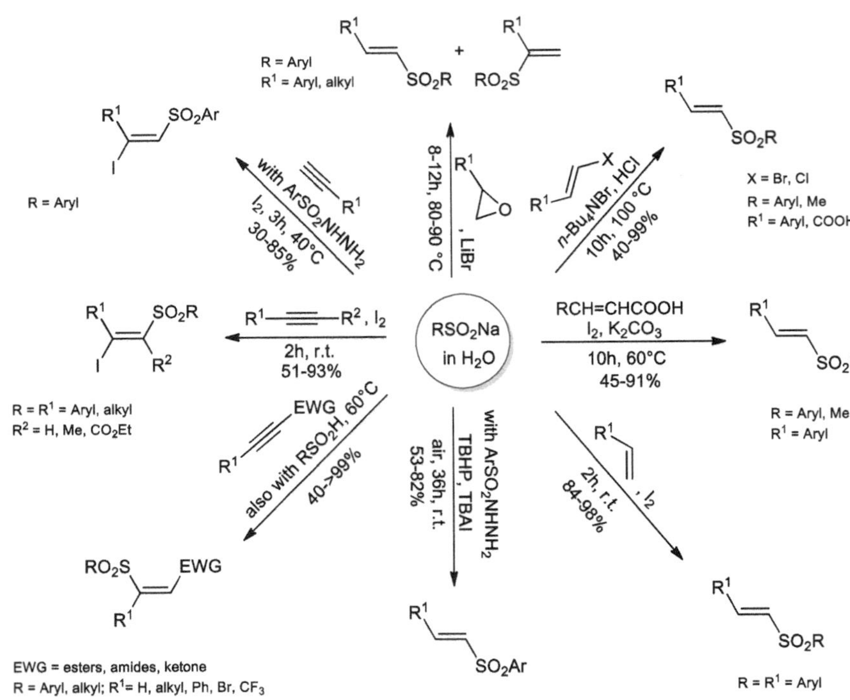

SCHEME 5.43 Synthesis of vinyl sulfones from sulfinates, sulfinic acids, and tosyl hydrazides.

From a mechanistic point of view, such interesting transformations can involve anionic or radical sulfonyl intermediates. Practical and efficient transition metal-free couplings with vinyl halides gave (*E*)-vinyl sulfones in good-to-excellent yields.[161] Iodine-mediated decarboxylative couplings of cinnamic acids[162] and sulfonylations of alkenes[163] were also successfully investigated. In both cases, the iodine-assisted addition of sulfonyl radicals to alkenes was followed by the elimination of CO_2 and/or HI. Good results in terms of yield and stereoselectivity were obtained in the direct C–S couplings of sulfinates or sulfonyl hydrazides to activated[164] or unactivated[165] alkynes. Alkynes containing ester, amide, acyl, or other activating groups reacted with sulfinate anions with formation of (*Z*)-sulfones. On the contrary, (*E*)-β-iodo vinyl sulfones were regio- and stereoselectively obtained from unactivated alkynes in the presence of molecular iodine, through a radical process. Finally, an LiBr-catalyzed ring opening of epoxides by aryl sulfinates followed by dehydration gave the one-pot formation of vinyl sulfones.[166] Alkyl epoxides gave products with complete regioselectivity due to the exclusive nucleophilic attack at the less-hindered carbon. Sodium sulfinates were competent reagents in the synthesis of β-ketosulfones via oxysulfonylation of alkenes or decarboxylative hydroxysulfonylation of arylpropiolic acids.[167]

5.8 Conclusions

Organochalcogen chemistry is a well-recognized tool in organic synthesis with interesting applications in pharmaceutical and bioorganic chemistry, biology, and material science. In the last two decades, a plethora of new synthetic strategies for organochalcogen compounds in environmentally benign solvents/reaction media have appeared in the literature. These valuable methods combine efficient and selective access to functionally and structurally diverse molecules with eco-compatibility and offer advantages such as high yields, good regio- and stereo-selectivities, short reaction times, safe handling, easy workup, and efficient recovery and recycling of the solvent. Although significant results have been already achieved, further advances in the field are expected and desirable.

References

1. (a) Cremlyn, R. J., Ed. *An Introduction to Organosulfur Chemistry;* John Wiley & Sons: Chichester, 1996.
   (b) Organosulfur Chemistry; In *Asymmetric Synthesis;* Toru, T.; Bolm, C., Eds. ; Wiley-VCH Verlag GmbH & Co. KGaA: Weinheim, 2008.
   (c) Wirth, T., Ed. *Organoselenium Chemistry: Synthesis and Reactions;* Wiley-VCH: Wenheim, 2012.
   (d) Santi, C., Ed. *Organoselenium Chemistry: Between Synthesis and Biochemistry;* Bentham e-books, 2014.

(e) Lenardão, E. J.; Santi, C.; Sancineto, L., Eds. *New Frontiers in Organoselenium Compounds;* Springer: Cham, 2018.
(f) Organoselenium Compounds in Biology and Medicine: Synthesis; In *Biological and Therapeutic Treatments;* Jain, V. K.; Priyadarsini, K. I., Eds. ; Royal Society of Chemistry: London, 2018.
(g) Laitinen, R.; Oilunkaniemi, R., Eds. *Selenium and Tellurium Reagents in Chemistry and Material Science;* De Gruyter: Germany, 2019.
(h) Kaiser, D.; Klose, I.; Oost, R.; Neuhaus, J.; Maulide, N. *Bond-Forming and -Breaking Reactions at Sulfur(IV): Sulfoxides, Sulfonium Salts, Sulfur Ylides, and Sulfinate. Chem. Rev.* **2019,** *119,* 8701–8780.

2. (a) Nogueira, W.; Rocha, J. B. T. *Organoselenium and Organotellurium Compounds: Toxicology and Pharmacology. PATAI's Chem. Funct. Groups;* John Wiley & Sons: New York, 2011, pp 1–81.
(b) Putnik, P.; Gabrić, D.; Roohinejad, S.; Barba, F. J.; Granato, D.; Mallikarjunan, K.; Lorenzo, J. M.; Kovačevića, D. B. An Overview of Organosulfur Compounds from *Allium* spp.: From Processing and Preservation to Evaluation of their Bioavailability, Antimicrobial, and Anti-Inflammatory Properties. *Food Chem.* **2019,** *276,* 680–691.

3. (a) Santoro, S.; Azeredo, J. B.; Nascimento, V.; Sancineto, L.; Braga, A. L.; Santi, C. The Green Side of the Moon: Ecofriendly Aspects of Organoselenium Chemistry. *RSC Adv.* **2014,** *4,* 31521–31535.
(b) Perin, G.; Alves, D.; Jacob, R. G.; Barcellos, A. M.; Soares, L. K.; Lenardão, E. J. Synthesis of Organochalcogen Compounds Using Non-Conventional Reaction Media. *ChemistrySelect* **2016,** *1,* 205–258.
(c) Santi, C.; Jacob, R. G.; Monti, B.; Bagnoli, L.; Sancineto, L.; Lenardão, E. J. Water and Aqueous Mixtures as Convenient Alternative Media for Organoselenium Chemistry. *Molecules* **2016,** *21,* 1482.
(d) Kostić, M. D.; Divac, V. M. Green Solvents in Organoselenium Chemistry. *Environ. Chem. Lett.* **2019,** *17,* 897–915.

4. (a) Leino, R.; Lonnqvist, J. E. A Very Simple Method for the Preparation of Symmetrical Disulfides. *Tetrahedron Lett.* **2004,** *45,* 8489–8491.
(b) Lenardão, E. J.; Lara, R. G.; Silva, M. S.; Jacob, R. G.; Perin, G. Clean and Fast Oxidative Transformation of Thiols to Disulfides under Solvent-Free Conditions. *Tetrahedron Lett.* **2007,** *48,* 7668–7670.
(c) Khazaei, A.; Zolfigol, M. A.; Rostami, A. 1,3-Dibromo-5,5-Dimethylhydantoin [DBDMH] as an Efficient and Selective Agent for the Oxidation of Thiols to Disulfides in Solution or under Solvent-Free Conditions. *Synthesis* **2004,** 2959–2961.

5. (a) Singh, D.; Galetto, F. Z.; Soares, L. C.; Rodrigues, O. E. D.; Braga, A. L. Metal-Free Air Oxidation of Thiols in Recyclable Ionic Liquid: A Simple and Efficient Method for the Synthesis of Disulfides. *Eur. J. Org. Chem.* **2010,** 2661–2665.
(b) Thurow, S.; Pereira, V. A.; Martinez, D. M.; Alves, D.; Perin, G.; Jacob, R. G.; Lenardão, E. J. Base-Free Oxidation of Thiols to Disulfides Using Selenium Ionic Liquid. *Tetrahedron Lett.* **2011,** *52,* 640–643.

6. Hudwekar, D. A.; Verma, P. K.; Kour, J.; Balgotra, S.; Sawant, S. D. Transition Metal-Free Oxidative Coupling of Primary Amines in Polyethylene Glycol at Room Temperature: Synthesis of Imines, Azobenzenes, Benzothiazoles, and Disulfides. *Eur. J. Org. Chem.* **2019,** 1242–1250.
7. Liu, Y.; Wang, H.; Wang, C.; Wan, J.-P.; Wen, C. Bio-Based Green Solvent Mediated Disulfide Synthesis *via* Thiol Couplings Free of Catalyst and Additive. *RSC Adv.* **2013,** *3,* 21369–21372.
8. Shard, A.; Kumar, R.; Saima.; Sharma, N.; Sinha, A. K. Amino Acid and Water-Driven Tunable Green Protocol to Access S–S/C–S Bonds *via* Aerobic Oxidative Coupling and Hydrothiolation. *RSC Adv.* **2014,** *4,* 33399–33407.

9. Leitemberger, A.; Böhs, L. M. C.; Rosa, C. H.; Da Silva, C.; Galetto, F. Z.; Godoi, M. Synthesis of Symmetrical Diorganyl Disulfides Employing WEB as an Eco-Friendly Oxidative System. *ChemistrySelect* **2019,** *4,* 7686–7690.
10. Firouzabadi, H.; Iranpoor, N.; Abbasi, M. A One-Pot, Efficient, and Odorless Synthesis of Symmetrical Disulfides using Organic Halides and Thiourea in the Presence of Manganese Dioxide and Wet Polyethylene Glycol (PEG-200). *Tetrahedron Lett.* **2010,** *51,* 508–509.
11. Abbasi, M.; Sabet, A. Europhtal (8020) Efficiently Catalyzes the Aerobic Oxidation of In Situ Generated Thiols to Symmetric Disulfides (disulfanes). *J. Organomet. Chem.* **2017,** *833,* 10–17.
12. Abbasi, M.; Nowrouzi, N.; Mousavi, S. Aerobic Oxidation of Thiols and In Situ Generated Thiols to Symmetrical Disulfides (disulfanes) Catalyzed by $Na_2S_4O_6$. *ChemistrySelect* **2019,** *4,* 12227–12231.
13. Lu, X.; Wang, H.; Gao, R.; Sun, D.; Bi, X. Microwave-Assisted Synthesis of Asymmetric Disulfides. *RSC Adv.* **2014,** *4,* 28794–28797.
14. Lu, X.; Fu, F.; Gao, R.; Liu, H.; Wang, H.; Xiao, J. Environmentally Friendly Synthesis of Unsymmetrical Dialkyl Disulfides by Reacting Organic Halides with Thiourea and Sodium Thiosulfate in an Aqueous Medium. *N. J. Chem.* **2019,** *43,* 16046–16049.
15. Li, Z.; Ke, F.; Deng, H.; Xu, H.; Xiang, H.; Zhou, X. Synthesis of Disulfides and Diselenides by Copper-Catalyzed Coupling Reactions in Water. *Org. Biomol. Chem.* **2013,** *11,* 2943–2946.
16. Soleiman-Beigi, M.; Yavari, I.; Sadeghizadeh, F. The Direct Synthesis of Symmetrical Disulfides and Diselenides by Metal–Organic Framework MOF-199 as an Efficient Heterogenous Catalyst. *RSC Adv.* **2015,** *5,* 87564–87570.
17. Tanini, D.; Ricci, L.; Capperucci, A. Rongalite-Promoted *on Water* Synthesis of Functionalised Tellurides and Ditellurides. *Adv. Synth. Catal.* **2020,** *362,* 1–11.
18. For the Synthesis of Biologically Active Diaryl Diselenides in a DMF/Water Mixture see also Pacuła, A. J.; Obieziurska, M.; Ścianowski, J.; Kaczor, K. B.; Antosiewicz, J. Water-Dependent Synthesis of Biologically Active Diaryl Diselenides. *Arkivoc* **2018,** *iii,* 153–164.
19. Krasowska, D.; Begini, F.; Santi, C.; Mangiavacchi, F.; Drabowicz, J.; Sancineto, L. Ultrasound-Assisted Synthesis of Alkali Metals Diselenides (M_2Se_2) and their Application for the Gram-Scale Preparation of 2,2′-diselenobis(benzoic acid). *Arkivoc* **2019,** *ii,* 24–37.
20. (a) Sancineto, L.; Mariotti, A.; Bagnoli, L.; Marini, F.; Desantis, J.; Iraci, N.; Santi, C.; Pannecouque, C.; Tabarrini, O. Design and Synthesis of Diselenobisbenzamides (DiSeBAs) as Nucleocapsid Protein 7 (NCp7) Inhibitors with Anti-HIV Activity. *J. Med. Chem.* **2015,** *58,* 9601–9614.
    (b) Nascimento, V.; Ferreira, N. L.; Schott, K. L.; Waczuk, E. P.; Sancineto, L.; Santi, C.; Braga, A. L. Synthesis and Biological Evaluation of New Nitrogen-Containing Diselenides. *Eur. J. Med. Chem* **2014,** *87,* 131–139.
    (c) Wang, L.; Yang, Z.; Fu, J.; Yin, H.; Xiong, K.; Tan, Q.; Jin, H.; Li, J.; Wang, T.; Tang, W., et al. Ethaselen: A Potent Mammalian Thioredoxin Reductase 1 Inhibitor and Novel Organoselenium Anticancer Agent. *Free Radic. Biol. Med.* **2012,** *52,* 898–908.
21. (a) Fringuelli, F.; Pizzo, F.; Tortoioli, S.; Vaccaro, L. Thiolysis of Alkyl- and Aryl-1,2-Epoxides in Water Catalyzed by $InCl_3$. *Adv. Synth. Catal.* **2002,** *344,* 379–384.
    (b) Fringuelli, F.; Pizzo, F.; Tortoioli, S.; Vaccaro, L. Easy and Environmentally Friendly Uncatalyzed Synthesis of β-Hydroxy Arylsulfides by Thiolysis of 1,2-Epoxides in Water. *Green Chem.* **2003,** *5,* 436–440.
22. (a) Chen, J.; Wu, H.; Jin, C.; Zhang, X.; Xie, Y.; Su, W. Highly Regioselective Ring-Opening of Epoxides with Thiophenols in Ionic Liquids Without the Use of Any Catalyst. *Green Chem.* **2006,** *8,* 330–332.

(b) Kamal, A.; Reddy, D. R. Rajendar Direct One-Pot Synthesis of β-Hydroxysulfides from Terminal Olefins in a Mixture of [bmim][BF_4] and Water in Presence of Molecular Oxygen. *J. Mol. Cat. A: Chem.* **2007,** *272,* 26–30.
(c) Yang, M.-H.; Yan, G.-B.; Zheng, Y.-F. Regioselective Ring-Opening Reactions of 1,2-Epoxides with Thiols and Arylselenols Directly Promoted by [Bmim]BF_4. *Tetrahedron Lett.* **2008,** *49,* 6471–6474.
23. (a) Fringuelli, F.; Pizzo, F.; Tortoioli, S.; Vaccaro, L. Thiolysis of 1,2-Epoxides by Thiophenol Catalyzed under Solvent-Free Conditions. *Tetrahedron Lett.* **2003,** *44,* 6785–6787.
(b) Fringuelli, F.; Pizzo, F.; Tortoioli, S.; Vaccaro, L. One-Pot Synthesis of Benzo[e]1,4-oxathiepin-5-Ones under Solvent-Free Condition via Self-Promoted Thiolysis of 1,2-Epoxides. *J. Org. Chem.* **2004,** *69,* 8780–8785.
(c) Lu, H.-F.; Zhou, J.-T.; Cheng, H.-L.; Sun, L.-L.; Yang, F.-F.; Wu, R.-Z.; Gao, Y.-H.; Luo, Z.-B. An Efficient Catalyst for Ring Opening of Epoxides with Phenol and Thiophenol under Solvent-Free Conditions. *Tetrahedron* **2013,** *69,* 11174–11184.
24. Boudou, M.; Ogawa, C.; Kobayashi, S. Chiral Scandium-Catalysed Enantioselective Ring-Opening of *Meso*-Epoxides with N-Heterocycle, Alcohol and Thiol Derivatives in Water. *Adv. Synth. Catal.* **2006,** *348,* 2585–2589.
25. Azizi, N.; Batebi, E. Highly Efficient Deep Eutectic Solvent catalyzed Ring Opening of Epoxides. *Catal. Sci. Technol.* **2012,** *2,* 2445–2448.
26. Azizi, N.; Edrisi, M. Biodegradable Choline Hydroxide Promoted Environmentally Benign Thiolysis of Epoxides. *Tetrahedron Lett.* **2016,** *57,* 525–528.
27. Azizi, N.; Yadollahy, Z.; Rahimzadeh-Oskooee, A. Odourless Strategy for Deep Eutectic Solvent-Mediated Ring Opening of Epoxides with In Situ Generated S-Alkylisothiouronium Salts. *Synlett* **2014,** *25,* 1085–1088.
28. Shridhar, R.; Srinivas, B.; Surendra, K.; Krishnaveni, N. S.; Rao, K. R. Synthesis of β-Hydroxy Selenides Using Benzeneselenol and Oxiranes under Supramolecular Catalysis in the Presence of β-Cyclodextrin in Water. *Tetrahedron Lett.* **2005,** *46,* 8837–8839.
29. Santi, C.; Santoro, S.; Battistelli, B.; Testaferri, L.; Tiecco, M. Preparation of the First Bench-Stable Phenyl Selenolate: An Interesting "on water" Nucleophilic Reagent. *Eur. J. Org. Chem.* **2008,** 5387–5390.
30. Silva, P. C.; Borges, E. L.; Lima, D. B.; Jacob, R. G.; Lenardão, E. J.; Perin, G.; Silva, M. S. A Simple and Non-Conventional Method for the Synthesis of Selected β-Arylalkylchalcogeno Substituted Alcohols, Amines and Carboxylic Acids. *Arkivoc* **2016,** *v,* 376–389.
31. Reddy, M. S.; Narender, M.; Rao, K. R. Regioselective Ring-Opening of Aziridines with Thiophenol in the Presence of β-Cyclodextrin in Water. *Synlett* **2005,** *3,* 489–490.
32. Reddy, M. S.; Narender, M.; Nageswar, Y. V. D.; Rao, K. R. Regioselective Ring-Opening of Aziridines with Potassium Thiocyanate in the Presence of β-Cyclodextrin in Water. *Tetrahedron Lett.* **2005,** *46,* 6437–6439.
33. Kamal, A.; Reddy, D. R.; Rajendar. Polyethylene glycol (PEG) as an Efficient Recyclable Medium for the Synthesis of β-Amino Sulfides. *Tetrahedron Lett.* **2006,** *47,* 2261–2264.
34. Ingebrigtsen, T.; Lejon, T. Ring Opening of Aziridines by Aromatic Thiols Followed by Amino-Substitution. *Tetrahedron Lett.* **2006,** *47,* 3949–3951.
35. Vale, J. R.; Siopa, F.; Branco, P. S.; Afonso, C. A. M. Ring Opening of 6-Azabicyclo [3.1.0]Hex-3-en-2-ols in Water under Mild Conditions. *Eur. J. Org. Chem.* **2016,** 2048–2053.
36. Lauder, K.; Masci, D.; Toscani, A.; Al Mekdad, A.; Black, G. W.; Brown, N. L.; Turner, N. J.; Luisi, R.; Castagnolo, D. A Facile and Regioselective Multicomponent Synthesis of Chiral Aryl-1,2-Mercaptoamines in Water Followed by Monoamine Oxidase (MAO-N) Enzymatic Resolution. *Org. Biomol. Chem.* **2019,** *17,* 8982–8986.

37. Salman, S. M.; Schwab, R. S.; Alberto, E. E.; Vargas, J.; Dornelles, L.; Rodrigues, O. E. D.; Braga, A. L. Efficient Ring Opening of Protected and Unprotected Aziridines Promoted by Stable Zinc Selenolate in Ionic Liquid. *Synlett* **2011,** *1,* 69–72.
38. J Vargas, J.; Narayanaperumal, S.; Gul, K.; Ravanello, B. B.; Dornelles, L.; Soares, L. C.; Alves, C. F. S.; Schneider, T.; de A. Vaucher, R.; Santos, R. C. V., et al. Synthesis of Chiral β-Chalcogen Amine Derivatives and Gram-Positive Bacteria Activity. *Tetrahedron* **2012,** *68,* 10444–10448.
39. Braga, A. L.; Schwab, R. S.; Alberto, E. E.; Salman, S. M.; Vargas, J.; Azeredo, J. B. Ring Opening of Unprotected Aziridines by Zinc Selenolates in a Biphasic System. *Tetrahedron Lett.* **2009,** *50,* 2309–2311.
40. (a) Ranu, B. C. ;; Dey, S. S. Catalysis by Ionic Liquid: A Simple, Green and Efficient Procedure for the Michael Addition of Thiols and Thiophosphate to Conjugated Alkenes in Ionic Liquid, [pmIm]Br. *Tetrahedron* **2004,** *60,* 4183–4188.
    (b) Meciarova, M.; Toma, S.; Kotrusz, P. Michael Addition of Thiols to α-Enones in Ionic Liquids With and Without Organocatalysts. *Org. Biomol. Chem.* **2006,** *4,* 1420–1424.
    (c) Ranu, B. C.; Dey, S. S.; Hajra, A. Catalysis by an Ionic Liquid: Efficient Conjugate Addition of Thiols to Electron Deficient Alkenes Catalyzed by Molten Tetrabutylammonium Bromide under Solvent-Free Conditions. *Tetrahedron* **2003,** *59,* 2417–2421.
    (d) Firouzabadi, H.; Iranpoor, N.; Jafarpour, M.; Ghaderi, A. Silica Gel Catalyzed Highly Selective C S Bond Formation via Michael Addition of Thiols to α,β-Unsaturated Ketones under Solvent-Free Conditions. *J. Mol. Catal A: Chem.* **2006,** *249,* 98–102.
    (e) Kamal, A.; Reddy, D. R.; Rajendar. A Simple and Green Procedure for the Conjugate Addition of Thiols to Conjugated Alkenes Employing Polyethylene Glycol (PEG) as an Efficient Recyclable Medium. *Tetrahedron Lett.* **2005,** *46,* 7951–7953.
    (f) Hussain, S.; Bharadwaj, S. K.; Chaudhuri, M. K.; Kalita, H. Borax as an Efficient Metal-Free Catalyst for Hetero-Michael Reactions in an Aqueous Medium. *Eur. J. Org. Chem.* **2007,** 374–378.
    (g) Payra, S.; Saha, A.; Banerjee, S. On-Water Magnetic $NiFe_2O_4$ Nanoparticle Catalyzed Michael Additions of Active Methylene Compounds, Aromatic/Aliphatic Amines, Alcohols and Thiols to Conjugated Alkenes. *RSC Adv.* **2016,** *6,* 95951–95956.
    (h) Krishnaveni, N. S.; Surendra, K.; Rao, K. R. Study of the Michael Addition of β-Cyclodextrin–Thiol Complexes to Conjugated Alkenes in Water. *Chem. Commun.* **2005,** 669–671.
41. Liu, Y.; Lai, Z.; Yang, P.; Xu, Y.; Zhang, W.; Liu, B.; Lu, M.; Chang, H.; Ding, T.; Xu, H. Thio-Michael Addition of α,β-Unsaturated Amides Catalyzed by Nmm-Based Ionic Liquids. *RSC Adv.* **2017,** *7,* 43104–43113.
42. Thopate, Y.; Singh, R.; Sinha, A. K.; Kumar, V.; Siddiqi, M. I. NMR and DFT Insight into the Synergistic Role of Bovine Serumalbumin–Ionic Liquid for Multicomponent Cascade Aldol/Knoevenagel–Thia-Michael/Michael Reactions in Onepot. *ChemCatChem* **2016,** *8,* 3050–3056.
43. Singh, R.; Thopate, Y.; Equbal, D.; Sinha, A. K. Synergistic Cooperative Effect of l-Arginine-[bmim]br in Cascade Decarboxylative Knoevenagel-Thia-Michael Addition Reactions: Green Approach Towards C – S Bond Formation with In Situ Generated Unactivated α,β-Unsaturated Ester. *Adv. Synth. Catal.* **2018,** *360,* 4412–4421.
44. Bonollo, S.; Lanari, D.; Pizzo, F.; Vaccaro, L. Sc(III)-Catalyzed Enantioselective Addition of Thiols to α,β-Unsaturated Ketones in Neutral Water. *Org. Lett.* **2011,** *13,* 2150–2152.

45. Ueno, M.; Kitanosono, T.; Sakai, M.; Kobayashi, S. Chiral Sc-Catalyzed Asymmetric Michael Reactions of Thiols with Enones in Water. *Org. Biomol. Chem.* **2011,** *9,* 3619–3621.
46. Kitanosono, T.; Sakai, M.; Ueno, M.; Kobayashi, S. Chiral-Sc Catalyzed Asymmetric Michael Addition/Protonation of Thiols with Enones in Water. *Org. Biomol. Chem.* **2012,** *10,* 7134–7147.
47. Suresh, P.; Pitchumani, K. Per-6-Amino-β-Cyclodextrin Catalyzed Asymmetric Michael Addition of Nitromethane and Thiols to Chalcones in Water. *Tetrahedron: Asymmetry* **2008,** *19,* 2037–2044.
48. Kalaria, P. N.; Avalani, J. R.; Raval, D. K. Highly Enantioselective Sulpha-Michael Addition to α,β-Unsaturated Carbonyl Scaffolds Catalysed by Allyloxy-*N*-(1-benzyl) Cinchonidinium Bromide in Water at Room Temperature. *Tetrahedron: Asymmetry* **2016,** *27,* 947–953.
49. Gaggero, N.; Albanese, D. C. M.; Celentano, G.; Banfi, S.; Aresi, A. Stereoselective Thio-Michael Addition to Chalcones in Water Catalyzed by Bovine Serum Albumin. *Tetrahedron: Asymmetry* **2011,** *22,* 1231–1233.
50. Gholinejad, M.; Firouzabadi, H. One-Pot Odorless Thia-Michael Reaction by Copper Ferrite Nanoparticle-Catalyzed Reaction of Elemental Sulfur, Aryl Halides and Electron-Deficient Alkenes. *N. J. Chem.* **2015,** *39,* 5953–5959.
51. (a) Zhao, Y.; Ge, Z.-M.; Cheng, T.-M.; Li, R.-T. Thia-Michael Addition Using Cheap and Odorless S-Alkylisothiouronium Salts as Thiol Equivalents in Water. *Synlett* **2007,** *10,* 1529–1532.
    (b) Firouzabadi, H.; Iranpoor, N.; Abbasia, M. Pronounced Catalytic Effect of a Micellar Solution of Sodium Dodecyl Sulfate (SDS) on the Efficient C-S Bond Formation via an Odorless Thia-Michael Addition Reaction through the In Situ Generation of S-Alkylisothiouronium Salts. *Adv. Synth. Catal.* **2009,** *351,* 755–766.
52. Azizi, N.; Yadollahy, Z.; Rahimzadeh-Oskooee, A. An Atom-Economic and Odorless Thia-Michael Addition in a Deep Eutectic Solvent. *Tetrahedron Lett.* **2014,** *55,* 1722–1725.
53. (a) Perin, G.; Borges, E. L.; Peglow, T. J.; Lenardão, E. J. Direct Michael Addition to Electron-Deficient Alkenes Using Diorganyl Dichalcogenides (Te/S) and $NaBH_4$/PEG-400. *Tetrahedron Lett.* **2014,** *55,* 5652–5655.
    (b) Perin, G.; Borges, E. L.; Rosa, P. C.; Carvalho, P. N.; Lenardão, E. J. Simple Cleavage of Diorganyl Diselenides with $NaBH_4$/PEG-400 and Direct Michael Addition to Electron-Deficient Alkenes. *Tetrahedron Lett.* **2013,** *54,* 1718–1721.
54. Banerjee, S.; Adak, L.; Ranu, B. C. Ionic Liquid/PPh_3 Promoted Cleavage of Diphenyl Disulfide and Diselenide: A Straightforward Metal-Free One-Pot Route to the Synthesis of Unsymmetrical Sulfides and Selenides. *Tetrahedron Lett.* **2012,** *53,* 2149–2152.
55. B Srinivas, B.; Kumar, V. P.; Sridhar, R.; Reddy, V. P.; Nageswar, Y. V. D.; Rao, K. R. β-Cyclodextrin-Promoted Addition of Benzeneselenol to Conjugated Alkenes in Water. *Helv. Chim. Acta* **2009,** *92,* 1080–1084.
56. Battistelli, B.; Testaferri, L.; Tiecco, M.; Santi, C. "On-water" Michael-Type Addition Reactions Promoted by PhSeZnCl. *Eur. J. Org. Chem.* **2011,** 1848–1851.
57. Nacca, F. G.; Monti, B.; Lenardão, E. J.; Evans, P.; Santi, C. A Simple Zinc-Mediated Method for Selenium Addition to Michael Acceptors. *Molecules* **2020,** *25,* 2018.
58. (a) Qureshi, Z. S.; Deshmukh, K. M.; Dhake, K. P.; Bhanage, B. M. Brønsted Acidic Ionic Liquid: A Simple, Efficient and Recyclable Catalyst for Regioselective Alkylation of Phenols and Anti-Markovnikov Addition of Thiols to Alkenes. *RSC Adv.* **2011,** *1,* 1106–1112.
    (b) Upadhyay, R.; Rana, R.; Sood, A.; Maurya, S. K. Formic Acid-Driven Rapid and Green Anti-Markovnikov Hydrothiolation of Styrenes. *ACS Omega* **2019,** *4,* 15101–15106.

(c) Rosa, C. H.; Peixoto, M. L. B.; Rosa, G. R.; Godoi, B.; Galetto, F. Z. Montes D'Oca, M. G.; Godoi, M. Sulfamic acid: An Efficient and Recyclable Catalyst for the Regioselective Hydrothiolation of Terminal Alkenes and Alkynes with Thiols. *Tetrahedron Lett.* **2017,** *58,* 3777–3781.

59. (a) Silveira, C. C.; Mendes, S. R.; Líbero, F. M. Solvent-free Anti-Markovnikov Addition of Thiols to Alkenes Using Anhydrous Cerium(III) Chloride as Catalyst Anti-Markovnikov Addition of Thiols to Alkenes. *Synlett* **2010,** *5,* 790–792.
(b) Kucinski, K.; Pawluc, P.; Hreczycho, G. Scandium(III)triflate-catalyzed anti-Markovnikov Hydrothiolation of Functionalized Olefins. *Adv. Synth. Catal.* **2015,** *357,* 3936–3942.
60. Banerjee, S.; Das, J.; Santra, S. Native Silica Nanoparticle Catalyzed Anti-Markovnikov Addition of Thiols to Inactivated Alkenes and Alkynes: A New Route to Linear and Vinyl Thioethers. *Tetrahedron Lett.* **2009,** *50,* 124–127.
61. (a) Lou, F.-W.; Xu, J.-M.; Liu, B.-K.; Wu, Q.; Pan, Q.; Lin, X.-F. Highly Selective Anti-Markovnikov Addition of Thiols to Vinyl Ethers under Solvent- and Catalyst-Free Conditions. *Tetrahedron Lett.* **2007,** *48,* 8815–8818.
(b) Choudhuri, K.; Mandal, A.; Mal, P. Aerial Dioxygen Activation *vs.* Thiol–Ene Click Reaction Within a System. *Chem. Commun.* **2018,** *54,* 3759–3762.
62. Kumar, R.; Saima.; Shard, A.; Andhare, N. H.; Richa.; Sinha, A. K. Thiol–ene "click" Reaction Triggered by Neutral Ionic Liquid: The "ambiphilic" Character of [hmim]br in the Regioselective Nucleophilic Hydrothiolation. *Angew. Chem. Int. (Ed.)* **2015,** *54,* 828–832.
63. E Lenardão, E. J.; Jacob, R. G.; Mesquita, K. D.; Lara, R. G.; Webber, R.; Martinez, D. M.; Savegnago, L.; Mendes, S. R.; Alves, D.; Perin, G. Glycerol as a Promoting and Recyclable Medium for Catalyst-Free Synthesis of Linear Thioethers: New Antioxidants from Eugenol. *Green. Chem. Lett. Rev.* **2013,** *6,* 269–276.
64. Tucker-Schwartz, A. K.; Farrell, R. A.; Garrell, R. L. Thiol_ene Click Reaction as a General Route to Functional Trialkoxysilanes for Surface Coating Applications. *J. Am. Chem. Soc.* **2011,** *133,* 11026–11029.
65. (a) Surendra, K.; Krishnaveni, N. S.; Sridhar, R.; Rao, K. R. Synthesis of β-Hydroxysulfides from Alkenes under Supramolecular Catalysis in the Presence of β-Cyclodextrin in Water. *J. Org. Chem.* **2006,** *71,* 5819–5821.
(b) Zhang, B.; Liu, T.; Bian, Y.; Lu, T.; Feng, J. Micelle Enhanced Auto-Oxidative Hydroxysulfenylation of Alkenes. *ACS Sustainable Chem. Eng.* **2018,** *6,* 2651–2655.
66. Kamal, A.; Reddy, D. R. Rajendar Direct One-Pot Synthesis of β-Hydroxysulfides from Terminal Olefins in a Mixture of [bmim][BF_4] and Water in Presence of Molecular Oxygen. *J. Mol. Cat. A: Chem.* **2007,** *272,* 26–30.
67. Vieira, A. A.; Azeredo, J. B.; Godoi, M.; Santi, C.; da Silva Júnior, E. N.; Braga, A. L. Catalytic Chalcogenylation under Greener Conditions: A Solvent-Free Sulfur- and Seleno-Functionalization of Olefins via I_2/DMSO Oxidant System. *J. Org. Chem.* **2015,** *80,* 2120–2127.
68. Silva, M. S.; Lara, R. G.; Marczewski, J. M.; Jacob, R. G.; Lenardão, E. J.; Perin, G. Synthesis of vinyl sulfides via hydrothiolation of alkynes using Al_2O_3/KF under solvent-free conditions. *Tetrahedron Lett.* **2008,** *49,* 1927–1930.
69. Thurow, S.; Ostosi, N. T.; Mendes, S. R.; Jacob, R. G.; Lenardão, E. J. Synthesis of Vinyl Sulfides under Base-Free Conditions Using Selenium Ionic Liquid. *Tetrahedron Lett.* **2012,** *53,* 2651–2653.
70. Lenardão, E. J.; Silva, M. S.; Lara, R. G.; Marczewski, J. M.; Sachini, M.; Jacob, R. G.; Alves, D.; Perin, G. Synthesis of Vinyl Sulfides using Glycerol as a Recyclable Solvent. *Arkivoc* **2011,** *ii,* 272–282.
71. Godoi, M.; Leitemberger, A.; Martins, L.; Böhs, C.; Silveira, M. V.; Rafque, J.; Gonçalves, M.; D'Oca, M. Rice Straw Ash Extract, an Efficient Solvent for Regioselective Hydrothiolation of Alkynes. *Environ. Chem. Lett.* **2019,** *17,* 1441–1446.

72. (a) Lenardão, E. J.; Dutra, L. G.; Saraiva, M. T.; Jacob, R. G.; Perin, G. Hydroselenation of Alkynes Using $NaBH_4$/$BMIMBF_4$: Easy Access to Vinyl Selenides. *Tetrahedron Lett.* **2007,** *48,* 8011–8013.
(b) Lenardão, E. J.; Gonçalves, L. C. C.; Mendes, S. R.; Saraiva, M. T.; Alves, D.; Jacob, R. G.; Perin, G. $NaBH_4$/[bmim]BF_4: A New Reducing System to Access Vinyl Selenides and Tellurides. *J. Braz. Chem. Soc.* **2010,** *21,* 2093–2099.
73. Tidei, C.; Sancineto, L.; Bagnoli, L.; Battistelli, B.; Marini, F.; Santi, C. A Recyclable Biphasic System for Stereoselective and Easily Handled Hydrochalcogenations. *Eur. J. Org. Chem.* **2014,** 5968–5975.
74. (a) Lin, Y.-M.; Lu, G.-P.; Cai, C.; Yi, W. B. Odorless, One-Pot Regio- and Stereoselective Iodothiolation of Alkynes with Sodium Arenesulfinates under Metal-Free Conditions in Water. *Org. Lett.* **2015,** *17,* 3310–3313.
(b) Lin, Y.-M.; Lu, G.-P.; Wang, G.-X.; Yi, W.-B. Acid/Phosphide-Induced Radical Route to Alkyl and Alkenyl Sulfides and Phosphonothioates from Sodium Arylsulfinates in Water. *J. Org. Chem.* **2017,** *82,* 382–389.
75. Perin, G.; Barcellos, A. M.; Luz, E. Q.; Borges, E. L.; Jacob, R. G.; Lenardão, E. J.; Sancineto, L.; Santi, C. Green Hydroselenation of Aryl Alkynes: Divinyl Selenides as a Precursor of Resveratrol. *Molecules* **2017,** *22,* 327.
76. Luz, E. Q.; Lopes, E. F.; Ricordi, V. G.; Santi, C.; Barcellos, T.; Lenardão, E. J.; Perin, G.; Alves, D. Water-Dependent Selective Synthesis of Mono- Or Bis-Selanyl Alkenes from Terminal Alkynes. *ChemistrySelect* **2016,** *1,* 4289–4294.
77. Lopes, E. F.; Gonçalves, L. C.; Vinueza, J. C. G.; Jacob, R. G.; Perin, G.; Santi, C.; Lenardão, E. J. DES as a Green Solvent to Prepare 1,2-Bis-Organylseleno Alkenes. Scope Limitations. *Tetrahedron Lett.* **2015,** *56,* 6890–6895.
78. Gonçalves, L. C. C.; Victória, F. N.; Lima, D. B.; Borba, P. M. Y.; Perin, G.; Savegnago, L.; Lenardão, E. J. CuI/Glycerol Mediated Stereoselective Synthesis of 1,2-Bis-Chalcogen Alkenes from Terminal Alkynes: Synthesis of New Antioxidants. *Tetrahedron Lett.* **2014,** *55,* 5275–5279.
79. Lara, R. G.; Rosa, P. C.; Soares, L. K.; Silva, M. S.; Jacob, R. G.; Perin, G. A Simple and Stereoselective Synthesis of (Z)-1,2-Bis-Arylselanyl Alkenes from Alkynes Using KF/Al_2O_3. *Tetrahedron* **2012,** *68,* 10414–10418.
80. Perin, G.; Borges, E. L.; Alves, D. Highly Stereoselective Method to Prepare Bis-Phenylchalcogen Alkenes via Addition of Chalcogenolate to Phenylseleno Alkynes. *Tetrahedron Lett.* **2012,** *53,* 2066–2069.
81. Sarrafi, Y.; Sadatshahabi, M.; Alimohammadi, K.; Tajbakhsha, M. A Green and Rapid Approach for the Stereoselective Vinylation of Phenol, Thiol and Amine Derivatives in Water. *Green. Chem.* **2011,** *13,* 2851–2858.
82. de Lima Borges, E.; Nobre, P. C.; da Silva, M. S.; Jacob, R. G.; Lenardão, E. J.; Perin, G. PEG-400 as a Recyclable Solvent in the Synthesis of β-Arylthio-α,β-Unsaturated Esters, Ketone and Aldehyde under Base and Catalyst-Free Conditions. *J. Environ. Chem. Eng.* **2016,** *4,* 2004–2007.
83. Pramanik, M.; Choudhuri, K.; Chakraborty, S.; Ghosh, A.; Mal, P. Z)-Selective Anti-Markovnikov or Markovnikov Thiol–Yne-Click Reactions of an Internal Alkyne by Amide Hydrogen Bond Control. *Chem. Commun.* **2020,** *56,* 2991–2994.
84. (a) Wang, Z.; Mo, H.; Bao, W. Mild, Efficient and Highly Stereoselective Synthesis of (Z)-Vinyl Chalcogenides from Vinyl Bromides Catalyzed by Copper(I) in Ionic Liquids Based on Amino Acids. *Synlett* **2007,** 91–94.
(b) Goncalves, L. C.; Fiss, G. F.; Perin, G.; Alves, D.; Jacob, R. G.; Lenardão, E. J. Glycerol as a Promoting Medium for Cross-Coupling Reactions of Diaryl Diselenides with Vinyl Bromides. *Tetrahedron Lett.* **2010,** *51,* 6772–6775.
(c) Goncalves, L. C. C.; Lima, D. B.; Borba, P. M. Y.; Perin, G.; Alves, D.; Jacob, R. G.; Lenardão, E. J. Glycerol/CuI/Zn as a Recyclable Catalytic System for Synthesis of Vinyl Sulfides and Tellurides. *Tetrahedron Lett.* **2013,** *54,* 3475–3480.

85. Saha, A.; Saha, D.; Ranu, B. C. Copper Nano-Catalyst: Sustainable Phenyl-Selenylation of Aryl Iodides and Vinyl Bromides in Water under Ligand Free Conditions. *Org. Biomol. Chem.* **2009,** *7,* 1652–1657.
86. Kundu, D.; Chatterjee, T.; Ranu, B. C. Magnetically Separable $CuFe_2O_4$ Nanoparticles Catalyzed Ligand-Free C-S Coupling in Water: Access to (*E*)- and (*Z*)-Styrenyl-, Heteroaryl and Sterically Hindered Aryl Sulfides. *Adv. Synth. Catal.* **2013,** *355,* 2285–2296.
87. Kundu, D.; Mukherjee, N.; Ranu, B. C. A General and Green Procedure for the Synthesis of Organochalcogenides by $CuFe_2O_4$ Nanoparticle Catalysed Coupling of Organoboronic Acids and Dichalcogenides in PEG-400. *RSC Adv.* **2013,** *3,* 117–125.
88. Chatterjee, T.; Ranu, B. C. Solvent-Controlled Halo-Selective Selenylation of Aryl Halides Catalyzed by Cu(II) Supported on Al_2O_3. a General Protocol for the Synthesis of Unsymmetrical Organo Mono- and Bis-Selenides. *J. Org. Chem.* **2013,** *78,* 7145–7153.
89. Webber, R.; Peglow, T. J.; Nobre, P. C.; Barcellos, A. M.; Roehrs, J. A.; Schumacher, R. F.; Perin, G. 1,1-Dibromoalkenes as Versatile Reagents to a Transition Metal-Free and Stereoselective Synthesis of (*E*)-1-Bromo-1-Selenoalkenes and Ketene Selenoacetals. *Tetrahedron Lett.* **2016,** *57,* 4128–4132.
90. (a) Wu, C.; Lu, L. H.; Peng, A.-Z.; Jia, G.-K.; Peng, C.; Cao, Z.; Tang, Z.; He, W.-M.; Xu, X. Ultrasound-Promoted Brønsted Acid Ionic Liquid-Catalyzed Hydrothiocyanation of Activated Alkynes under Minimal Solvent Conditions. *Green Chem.* **2018,** *20,* 3683–3688.
    (b) Sun, M.; Jiang, J.; Chen, J.; Yang, Q.; Yu, X. Deep Eutectic Solvent Promoted Hydrothiocyanation of Alkynoates Leading to Z-3-Thiocyanatoacrylates. *Tetrahedron* **2019,** *75,* 130456–130461.
    (c) Wu, C.; Xiao, H.-J.; Wang, S.-W.; Tang, M.-S.; Tang, Z.-L.; Xia, W.; Li, W.-F.; Cao, Z.; He, W.-M. Natural Deep Eutectic Solvent-Catalyzed Selenocyanation of Activated Alkynes via an Intermolecular h-Bonding Activation Process. *ACS Sustainable Chem. Eng.* **2019,** *7,* 2169–2175.
    (d) Wu, C.; Hong, L.; Shu, H.; Zhou, Q.-H.; Wang, Y.; Su, N.; Jiang, S.; Cao, Z.; He, W.-M. Practical Approach for Clean Preparation of z-β-Thiocyanate Alkenyl Esters. *ACS Sustainable Chem. Eng.* **2019,** *7,* 8798–8803.
91. Yang, Z.; Hu, L.; Cao, T.; An, L.; Li, L.; Yang, T.; Zhou, C. PIDA-Mediated α-C–H Functionalization of Enaminones: The Synthesis of Thiocyano Enaminones and Chromones in Water. *N. J. Chem.* **2019,** *43,* 16441–16444.
92. Lu, L.-H.; Zhou, S.-J.; Sun, M.; Chen, J.-L.; Xia, W.; Yu, X.; Xu, X.; He, W.-M. Metal- and Solvent-Free Ultrasonic Multicomponent Synthesis of (z)-β-Iodo Vinylthiocyanates. *ACS Sustain. Chem. Eng.* **2019,** *7,* 1574–1579.
93. Heredia, A. A.; Peñéñory, A. B. Transition-Metal-Free One-Pot Synthesis of Alkynyl Selenides from Terminal Alkynes under Aerobic and Sustainable Conditions. *Beilstein J. Org. Chem.* **2017,** *13,* 910–918.
94. Wu, G.; Min, L.; Li, H.; Gao, W.; Ding, J.; Huang, X.; Liu, M.; Wu, H. Metal-Free Synthesis of Alkynyl Alkyl Selenides via Three-Component Coupling of Terminal Alkynes, Se, and Epoxides. *Green Chem.* **2018,** *20,* 1560–1563.
95. Rafique, J.; Saba, S.; Franco, M. S.; Bettanin, L.; Schneider, A. R.; Silva, L. T.; Braga, A. L. Direct, Metal-Free C(sp2) – H Chalcogenation of Indoles and Imidazopyridines with Dichalcogenides Catalysed by KIO_3. *Chem. Eur. J.* **2018,** *24,* 4173–4180.
96. Dey, A.; Hajra, A. Iodine-Catalyzed Selenylation of 2*h*-Indazole. *J. Org. Chem.* **2019,** *84,* 14904–14910.
97. Liu, S.; Yang, H.; Jiao, L.-Y.; Zhang, J.-H.; Zhao, C.; Ma, Y.; Yang, X. Regioselective Deoxygenative Chalcogenation of 7-Azindole N-Oxides Promoted by I_2/PEG-200. *Org. Biomol. Chem.* **2019,** *17,* 10073–10087.

98. Fang, Y.; Wang, J.; Liu, Y.; Yan, J. An Environmentally Benign and Efficient Synthesis of 4-Selanylpyrazoles Catalyzed by Haloid Salts in Water. *Appl. Organomet. Chem.* **2019,** *33,* e4921.
99. Kong, D.; Wang, Q.; Huang, T.; Liang, M.; Lin, Q. Metal-Free Direct Regioselective Thiolation of Imidazoheterocycles in Water at Room Temperature. *Tetrahedron* **2019,** *75,* 1457–1463.
100. Kong, D.; Huang, T.; Liang, M.; Wu, M.; Lin, Q. KIO_3-Catalyzed Cross Dehydrogenative Coupling Reaction: Sulfenylation of Phenol and Arylamine Derivatives in Water at Room Temperature. *Org. Biomol. Chem.* **2019,** *17,* 830–834.
101. Xiao, F.; Tian, J.; Xing, Q.; Huang, H.; Deng, G.-J.; Liu, Y. Piperidine Promoted Direct Sulfenylation of 2-Naphthol with Aryl Thiols under Aqueous Conditions. *ChemistrySelect* **2017,** *2,* 428–431.
102. Xiao, F.; Chen, S.; Li, C.; Huang, H.; Deng, G.-J. Copper-Catalyzed Three-Component One-Pot Synthesis of Aryl Sulfides with Sulfur Powder under Aqueous Conditions. *Adv. Synth. Catal.* **2016,** *358,* 3881–3886.
103. Xiao, F.; Chen, S.; Tian, J.; Huang, H.; Liu, Y.; Deng, G.-J. Chemoselective Cross-Coupling Reaction of Sodium Sulfinates with Phenols under Aqueous Conditions. *Green. Chem.* **2016,** *18,* 1538–1546.
104. Yang, Y.; Zhang, S.; Tang, L.; Hu, Y.; Zha, Z.; Wang, Z. Catalyst-Free Thiolation of Indoles with Sulfonyl Hydrazides for the Synthesis of 3-Sulfenylindoles in Water. *Green. Chem.* **2016,** *18,* 2609–2613.
105. Yang, Z.; Yan, Y.; Li, A.; Liao, J.; Zhang, L.; Yang, T.; Zhou, C. Iodine-Mediated Sulfenylation of 4-Hydroxycoumarins with Sulfonyl Hydrazides under Aqueous Conditions. *N. J. Chem.* **2018,** *42,* 14738–14741.
106. Equbal, D.; Singh, R.; Saima.; Lavekar, A. G.; Sinha, A. K. Synergistic Dual Role of [hmim]Br-$ArSO_2Cl$ in Cascade Sulfenylation–Halogenation of Indole: Mechanistic Insight into Regioselective C–S and C–S/C–X (X = Cl and Br) Bond Formation in One Pot. *J. Org. Chem.* **2019,** *84,* 2660–2675.
107. Xu, Z.-B.; Lu, G.-P.; Cai, C. Acid-Induced Chemoselective Arylthiolations of Electron-Rich Arenes in Ionic Liquids from Sodium Arylsulfinates: The Reducibility of Halide Anions in [Hmim]Br. *Org. Biomol. Chem.* **2017,** *15,* 2804–2808.
108. Duarte, l F. B.; Barbosa, E. S.; Oliveira, R. L.; Pinz, M. P.; Godoi, B.; Schumacher, R. F. ;; Luchese, C.; Wilhelm, E. A.; Alves, D. A Simple Method for the Synthesis of 4-arylselanyl-7-Chloroquinolines Used as *In Vitro* Acetylcholinesterase Inhibitors and *In Vivo* Memory Improvement. *Tetrahedron Lett.* **2017,** *58,* 3319–3322.
109. Guo, T.; Wei, X.-N.; Liu, Y.; Zhang, P.-K.; Zhao, Y.-H. Oxidative Dual C–H Selenation of Imidazoheterocycles with Ethers or Alkanes Using Selenium Powder *via* a Radical Pathway. *Org. Chem. Front.* **2019,** *6,* 1414–1422.
110. Perin, G.; Araujo, D. R.; Nobre, P. C.; Lenardão, E. J.; Jacob, R. G.; Silva, M. S.; Roehrs, J. A. Ultrasound-Promoted Synthesis of 2-Organoselanyl-Naphthalenes Using Oxone® in Aqueous Medium as an Oxidizing Agent. *PeerJ* **2018,** e4706.
111. Fujita, K.-I.; Hashimoto, S.; Kanakubo, M.; Oishi, A.; Taguchi, Y. Solid-Phase Intramolecular Oxyselenenylation and Deselenenylation Reactions using Polymer-Supported Selenoreagents and their Application to Aqueous Media Organic Synthesis. *Green Chem.* **2003,** *5,* 549–553.
112. Fujita, K.-I.; Hashimoto, S.; Oishi, A.; Taguchi, Y. Intramolecular Oxyselenenylation and Deselenenylation Reactions in Water, Conducted by Employing Polymer-Supported Arylselenenyl Bromide. *Tetrahedron Lett.* **2003,** *44,* 3793–3795.
113. Ericsson, C.; Engman, L. Microwave-Assisted Group-Transfer Cyclization of Organotellurium Compounds. *J. Org. Chem.* **2004,** *69,* 5143–5146.
114. Taniguchi, T.; Fujii, T.; Idota, A.; Ishibashi, H. Reductive Addition of the Benzenethiyl Radical to Alkynes by Amine-Mediated Single Electron Transfer Reaction to Diphenyl Disulfide. *Org. Lett.* **2009,** *11,* 3298–3301.

115. Ruberte, A. C.; Sanmartin, C.; Aydillo, C.; Sharma, A. K.; Plano, D. Development and Therapeutic Potential of Selenazo Compounds. *J. Med. Chem.* **2020,** *63,* 1473–1489.
116. Sashida, H.; Pan, C.; Kaname, M.; Minoura, M. A Facile and Practical Solvent-Free One-Pot Synthesis of (Z)-4-Methylene-3selenaquinoline Derivatives from *o*-Ethynylanilines and Isoselenocyanates1 Synthesis of (Z)-4-Methylene-3-Selenaquinoline Derivatives. *Synthesis* **2010,** *18,* 3091–3096.
117. Sengoden, M.; Punniyamurthy, T. "On water": Efficient Iron-Catalyzed Cycloaddition of Aziridines with Heterocumulenes. *Angew. Chem. Int. (Ed.)* **2013,** *52,* 572–575.
118. Sengoden, M.; Vijay, M.; Balakumar, E.; Punniyamurthy, T. Efficient Pyrrolidine Catalyzed Cycloaddition of Aziridines with Isothiocyanates, Isoselenocyanates and Carbon Disulfide "on water.". *RSC Adv.* **2014,** *4,* 54149–54157.
119. Balaguez, R. A.; Krüger, R.; Radatz, C. S.; Rampon, D. S.; Lenardão, E. J.; Schneider, P. H.; Alves, D. Synthesis of Benzoselenazoles and Benzoselenazolines by Cyclization of 2-Amino-Benzeneselenol with β-Dicarbonyl Compounds. *Tetrahedron Lett.* **2015,** *56,* 2735–2740.
120. Singh, D.; Narayanaperumal, S.; Gul, K.; Godoi, M.; Rodrigues, O. E. D.; Braga, A. L. Efficient Synthesis of Selenoesters from Acyl Chlorides Mediated by CuO Nanopowder in Ionic Liquid. *Green Chem.* **2010,** *12,* 957–960.
121. Azeredo, J. B.; Godoi, M.; Schwab, R. S.; Botteselle, G. V.; Braga, A. L. Synthesis of Thiol Esters Using Nano CuO/Ionic Liquid as an Eco-Friendly Reductive System under Microwave Irradiation. *Eur. J. Org. Chem.* **2013,** 5188–5194.
122. Tabarelli, G.; Alberto, E. E.; Deobald, A. M.; Marin, G.; Rodrigues, O. E. D.; Dornelles, L.; Braga, A. L. Ionic Liquid: An Efficient and Reusable Media for Seleno- and Thioester Synthesis Promoted by Indium. *Tetrahedron Lett.* **2010,** *51,* 5728–5731.
123. Perin, G.; Silveira, M. B.; Barcellos, A. M.; Jacob, R. G.; Alves, D. Polyethylene Glycol-400/H_3PO_2: An Eco-Friendly Reductive System for the Synthesis of Selanylesters. *Org. Chem. Front.* **2015,** *2,* 1531–1535.
124. Godoi, M.; Ricardo, E. W.; Botteselle, G. V.; Galetto, F. Z.; Azeredo, J. B.; Braga, A. L. Efficient Synthesis of Selenoesters from Acyl Chlorides Mediated by CuO Nanopowder in Ionic Liquid. *Green Chem.* **2012,** *14,* 456–460.
125. Santi, C.; Battistelli, B.; Testaferri, L.; Tiecco, M. On Water Preparation of Phenylselenoesters. *Green Chem.* **2012,** *14,* 1277–1280.
126. Sancineto, L.; Tidei, C.; Bagnoli, L.; Marini, F.; Lippolis, V.; Arca, M.; Lenardão, E. J.; Santi, C. Synthesis of Thiol Esters Using PhSZnBr as Sulfenylating Agent: A Dft-Guided Optimization of Reaction Conditions. *Eur. J. Org. Chem.* **2016,** 2999–3005.
127. Sancineto, L.; Vargas, J. P.; Monti, B.; Arca, M.; Lippolis, V.; Perin, G.; Lenardão, E. J.; Santi, C. Atom Efficient Preparation of Zinc Selenates for the Synthesis of Selenol Esters under "on water" Conditions. *Molecules* **2017,** *22,* 953.
128. Lu, G.-P.; Cai, C. An Odorless, One-Pot Synthesis of Thioesters from Organic Halides, Thiourea and Benzoyl Chlorides in Water. *Adv. Synth. Catal.* **2013,** *355,* 1271–1276.
129. Lin, Y. –J.; Wu, Y.-P.; Thul, M.; Hung, M.-W.; Chou, S.-H.; Chen, W.-T.; Lin, W.; Lin, M.; Reddy, D. M.; Wu, H.-R., et al. Tunable Aryl Imidazolium Recyclable Ionic Liquid with Dual Brønsted–Lewis Acid as Green Catalyst for Friedel-Crafts Acylation and Thioesterification. *Molecules* **2020,** *25,* 352.
130. Yi, C.-L.; Huang, Y.-T.; Lee, C.-F. Synthesis of Thioesters through Copper-Catalyzed Coupling of Aldehydes with Thiols in Water. *Green Chem.* **2013,** *15,* 2476–2484.
131. Huang, Y.-T.; Lu, S.-Y.; Yi, C.-L.; Lee, C.-F. Iron-Catalyzed Synthesis of Thioesters from Thiols and Aldehydes in Water. *J. Org. Chem.* **2014,** *79,* 4561–4568.
132. Zeng, J.-W.; Liu, Y.-C.; Hsiegh, P.-A.; Huang, Y.-T.; Yi, C.-L.; Badsara, S. S.; Lee, C.-F. Metal-Free Cross-Coupling Reaction of Aldehydes with Disulfides by Using DTBP as an Oxidant under Solvent-Free Conditions. *Green Chem.* **2014,** *16,* 2644–2652.

133. Perin, G.; Silveira, M. B.; Barcellos, A. M.; Araujo, D. R.; Jacob, R. G.; Barcellos, T.; Lenardão, E. J. Rongalite®/PEG-400 as Reducing System in the Synthesis of New Glycerol-Derived Selenol Esters Using Anhydrides and Bis-(2,2-dimethyl-1,3-dioxolanylmethyl)diselenide as Substrates. *Arkivoc* **2017,** *ii*, 138–148.
134. Abdul Shakoor, S. M.; Choudhary, S.; Bajaj, K.; Muthyala, M. K.; Kumar, A.; Sakhuja, R. Imidazolium-Supported Benzotriazole: An Efficient and Recoverable Activating Reagent for Amide, Ester and Thioester Bond Formation in Water. *RSC Adv.* **2015,** *5*, 82199–82207.
135. L. Baldassari, L. L.; Mantovani, A. C.; Senoner, S.; Maryasin, B.; Maulide, N.; Ludtke, D. S. Redox-Neutral Synthesis of Selenoesters by Oxyarylation of Selenoalkynes under Mild Conditions. *Org. Lett.* **2018,** *20*, 5881–5885.
136. (a) Yu, B.; Guo, C.-X.; Zhong, C.-L.; Diao, Z.-F.; He, L.-N. Metal-Free Chemoselective Oxidation of Sulfides by In Situ Generated Koser's Reagent in Aqueous Media. *Tetrahedron Lett.* **2014,** *55*, 1818–1821.
(b) Ahammed, S.; Kundu, D.; Siddiqui, M. N.; Ranu, B. C. Metal and Solvent Free Selective Oxidation of Sulfides to Sulfone Using Bifunctional Ionic Liquid [pmim] IO_4. *Tetrahedron Lett.* **2015,** *56*, 335–337.
137. Gayakwad, E. M.; Patil, V. V.; Shankarling, G. S. Nonanebis(Peroxoic Acid) Mediated Efficient and Selective Oxidation of Sulfide. *N. J. Chem.* **2016,** *40*, 223–230.
138. Buckley, B. R.; Chan, Y.; Dreyfus, N.; Elliott, C.; Marken, F.; Page, P. C. B. Harnessing Applied Potential to Oxidation in Water. *Green Chem.* **2012,** *14*, 2221–2225.
139. Gómez, M. V.; Caballero, R.; Vázquez, E.; Moreno, A.; de la Hoz, A.; Díaz-Ortiza, A. Green and Chemoselective Oxidation of Sulfides with Sodium Perborate and Sodium Percarbonate: Nucleophilic and Electrophilic Character of the Oxidation System. *Green Chem.* **2007,** *9*, 331–336.
140. Yu, B.; Liu, A.-H.; He, L.-N.; Li, B.; Diao, Z.-F.; Li, Y.-N. Catalyst-Free Approach for Solvent-Dependent Selective Oxidation of Organic Sulfides with Oxone. *Green Chem.* **2012,** *14*, 957–962.
141. Cravotto, G.; Garella, D.; Carnaroglio, D.; Calcio Gaudino, E.; Rosati, O. Solvent-Free Chemoselective Oxidation of Thioethers and Thiophenes by Mechanical Milling. *Chem. Commun.* **2012,** *48*, 11632–11634.
142. Palomba, M.; Trappetti, F.; Bagnoli, L.; Santi, C.; Marini, F. Oxone-Mediated Oxidation of Vinyl Selenides in Water. *Eur. J. Org. Chem.* **2018,** 3914–3919.
143. Rajabi, F.; Naserian, S.; Primo, A.; Luque, R. Efficient and Highly Selective Aqueous Oxidation of Sulfides to Sulfoxides at Room Temperature Catalysed by Supported Iron Oxide Nanoparticles on SBA-15. *Adv. Synth. Catal.* **2011,** *353*, 2060–2066.
144. Chakravarthy, R. D.; Ramkumar, V.; Chand, D. K. A Molybdenum Based Metallomicellar Catalyst for Controlled and Selective Sulfoxidation Reactions in Aqueous Medium. *Green Chem.* **2014,** *16*, 2190–2196.
145. Rostami, A.; Tahmasbi, B.; Abedi, F.; Shokri, Z. Magnetic Nanoparticle Immobilized *N*-Propylsulfamic Acid: The Chemoselective, Efficient, Green and reusable Nanocatalyst for Oxidation of Sulfides to Sulfoxides using H_2O_2 under Solvent-Free Conditions. *J. Mol. Cat. A: Chem.* **2013,** *378*, 200–205.
146. Mirfakhraei, S.; Hekmati, M.; Eshbala, F. H.; Veisi, H. Fe_3O_4/PEG-SO_3H as a Heterogeneous and Magnetically-Recyclable Nanocatalyst for the Oxidation of Sulfides to Sulfones or Sulfoxides. *N. J. Chem.* **2018,** *42*, 1757–1761.
147. Shen, H.-M.; Zhou, W.-J.; Ma, X.; Wu, H.-K.; Yu, W.-B.; Ai, N.; Ji, H.-B.; Shi, H.-X.; She, Y.-B. pH-Dependence of the Aqueous Phase Room Temperature Brønsted Acid-Catalyzed Chemoselective Oxidation of Sulfides with H_2O_2. *Molecules* **2015,** *20*, 16709.
148. Khatri, P. K.; Jain, S. L.; Sain, B. Ultrasound-Promoted Oxidation of Sulfides with Hydrogen Peroxide under Catalyst-Free Conditions. *Ind. Eng. Chem. Res.* **2011,** *50*, 701–704.

149. Zhang, B.; Zhou, M.-D.; Cokoja, M.; Mink, J.; Zang, S.-L.; Kuhn, F. E. Oxidation of Sulfides to Sulfoxides Mediated by Ionic Liquids. *RSC Adv.* **2012,** *2,* 8416–8420.
150. (a) Mojr, V.; Budesınsky, M.; Cibulka, R.; Kraus, T. Alloxazine–Cyclodextrin Conjugates for Organocatalytic Enantioselective Sulfoxidations. *Org. Biomol. Chem.* **2011,** *9,* 7318–7326.
(b) Tomanová, P.; Šturala, J.; Budešínský, M.; Cibulka, R. A Click Chemistry Approach towards Flavin-Cyclodextrin Conjugates—Bioinspired Sulfoxidation Catalysts. *Molecules* **2015,** *20,* 19837.
151. Trevisan, V.; Signoretto, M.; Colonna, S.; Pironti, V.; Strukul, G. Microencapsulated Chloroperoxidase as a Recyclable Catalyst for the Enantioselective Oxidation of Sulfides with Hydrogen Peroxide. *Angew. Chem. Int. (Ed.)* **2004,** *43,* 4097–4099.
152. Reetz, M. T.; Daligault, F.; Brunner, B.; Hinrichs, H.; Deege, A. Directed Evolution of Cyclohexanone Monooxygenases: Enantioselective Biocatalysts for the Oxidation of Prochiral Thioethers. *Angew. Chem. Int. (Ed.)* **2004,** *43,* 4078–4081.
153. Nigel-Etinger, I.; Mahammed, A.; Gross, Z. Covalent vs Non-Covalent (Biocatalytic) Approaches for Enantioselective Sulfoxidation Catalyzed by Corrole Metal Complexes. *Catal. Sci. Technol.* **2011,** *1,* 578–581.
154. Tang, J.; Huang, F.; Wei, Y.; Bian, H.; Zhang, W.; Liang, H. Bovine Serum Albumin–Cobalt(II) Schiff Base Complex Hybrid: An Efficient Artificial Metalloenzyme for Enantioselective Sulfoxidation Using Hydrogen Peroxide. *Dalton Trans.* **2016,** *45,* 8061–8072.
155. Jereb, M. Highly Atom-Economic, Catalyst- and Solvent-Free Oxidation of Sulfides into Sulfones using 30% Aqueous H_2O_2. *Green Chem.* **2012,** *14,* 3047–3052.
156. Hu, Y.-L.; Liu, X.-B.; Fang, D. Efficient and Convenient Oxidation of Sulfides to Sulfones Using H_2O_2 Catalyzed by V_2O_5 in Ionic Liquid [C12mim][HSO_4]. *Catal. Sci. Technol.* **2014,** *4,* 38–42.
157. Li, B.; Liu, A.-H.; He, L.-N.; Yang, Z.-Z.; Gao, J.; Chen, K.-H. Iron-Catalyzed Selective Oxidation of Sulfides to Sulfoxides with the Polyethylene Glycol/O_2 System. *Green Chem.* **2012,** *14,* 130–135.
158. Hu, Y.-L.; Fang, D.; Xing, R. Efficient and Convenient Oxidation of Sulfides to Sulfoxides with Molecular Oxygen Catalyzed by $Mn(OAc)_2$ in Ionic Liquid [C12mim][NO_3]. *RSC Adv.* **2014,** *4,* 51140–51145.
159. (a) Imada, Y.; Kitagawa, T.; Wang, H.-K.; Komiya, N.; Naota, T. Flavin-Catalyzed Aerobic Oxidation of Sulfides in Aqueous Media. *Tetrahedron Lett.* **2013,** *54,* 621–624.
(b) Liu, K.-J.; Deng, J.-H.; Yang, J.; Gong, S.-F.; Lin, Y.-W.; He, J.-Y.; Cao, Z.; He, W.-M. Selective Oxidation of (Hetero)Sulfides with Molecular Oxygen under Clean Conditions. *Green Chem.* **2020,** *22,* 433–438.
160. L. Wei, L.-Q.; Ye, B.-H. Cyclometalated Ir – Zr metal – Organic Frameworks as Recyclable Visible-Light Photocatalysts for Sulfide Oxidation into Sulfoxide in Water. *ACS Appl. Mater. Interfaces* **2019,** *11,* 41448–41457.
161. Liang, S.; Zhang, R.-Y.; Wang, G.; Chen, S.-Y.; Yu, X.-Q. Transition-Metal-Free Synthesis of (*E*)-Vinyl Sulfones from Vinyl Halides in Water. *Eur. J. Org. Chem.* **2013,** 7050–7053.
162. Gao, J.; Lai, J.; Yuan, G. Iodine-Mediated Synthesis of (*E*)-Vinyl Sulfones from Sodium Sulfinates and Cinnamic Acids in Aqueous Medium. *RSC Adv.* **2015,** *5,* 66723–66726.
163. (a) Zhang, N.; Yang, D.; Wei, W.; Yuan, L.; Cao, Y.; Wang, H. Metal-Free Iodine-Mediated Synthesis of Vinyl Sulfones at Room Temperature Using Water as Solvent. *RSC Adv.* **2015,** *5,* 37013–37017.
(b) Bao, W.-H.; Ying, W.-W.; Xu, X.-D.; Zhou, G.-D.; Meng, X.-X.; Wei, W.-T.; Liu, Y.-Y.; Li, Q. Radical Heck-Type Reaction of Styrenes with Sulfonyl Hydrazides on Water at Room Temperature. *Tetrahedron Lett.* **2019,** *60,* 55–58.

164. (a) Li, W.; Yin, G.; Huang, L.; Xiao, Y.; Fu, Z.; Xin, X.; Liu, F.; Li, Z.; He, W. Regioselective and Stereoselective Sulfonylation of Alkynylcarbonyl Compounds in Water. *Green Chem.* **2016,** *18,* 4879–4883.
(b) Wu, C.; Yang, P.; Fu, Z.; Peng, Y.; Wang, X.; Zhang, Z.; Liu, F.; Li, W.; Li, Z.; He, W. Regio- and Stereoselective Hydrosulfonation of Alkynylcarbonyl Compounds with Sulfinic Acid in Water. *J. Org. Chem.* **2016,** *81,* 10664–10671.
165. (a) Sun, Y.; Abdukader, A.; Lu, D.; Zhang, H.; Liu, C. Synthesis of (*E*)-β-Iodo Vinylsulfones via Iodine Promoted Iodosulfonylation of Alkynes with Sodium Sulfinates in an Aqueous Medium at Room Temperature. *Green Chem.* **2017,** *19,* 1255–1258.
(b) Hou, Y.; Zhu, L.; Hu, H.; Chen, S.; Li, Z.; Liu, Y.; Gong, P. Synthesis of (*E*)-β-Iodo Vinylsulfones via Iodinepromoted Iodosulfonylation of Alkynes with Sodium Sulfinates in an Aqueous Medium at Room Temperature. *New J. Chem.* **2018,** *42,* 8752–8755.
166. Chawla, R.; Kapoor, R.; Singh, A. K.; Yadav, L. D. S. A One-Pot Regioselective Synthetic Route to Vinyl Sulfones from Terminal Epoxides in Aqueous Media. *Green Chem.* **2012,** *14,* 1308–1313.
167. (a) Xiong, Y.-S.; Weng, J.; Lu, G. Manganese(III)-Mediated and -Catalyzed Decarboxylative Hydroxysulfonylation of Arylpropiolic Acids with Sodium Sulfinates in Water. *Adv. Synth. Catal.* **2018,** *360,* 1611–1616.
(b) Chawla, R.; Singh, A. K.; Yadav, L. D. S. $K_2S_2O_8$-Mediated Aerobic Oxysulfonylation of Olefins into β-Keto Sulfones in Aqueous Media. *Eur. J. Org. Chem.* **2014,** 2032–2036.

PART II

Organochalcogens Promoting Green Chemistry

CHAPTER

6

Synthesis and catalytic functions of selenopeptides

Michio Iwaoka and Shingo Shimodaira

Department of Chemistry, School of Science, Tokai University, Hiratsuka-shi, Kanagawa, Japan

6.1 Introduction

Selenocysteine (Sec, U) (Fig. 6.1A), the 21st proteinogenic amino acid,[1] is a selenium analog of natural amino acids serine (Ser, S) and cysteine (Cys, C). In the last few decades, the research on this enigmatic amino acid has developed rapidly in various fields of chemistry and biology. As a result, the outline of how this natural amino acid is synthesized and introduced to proteins in cells has been elucidated.[2] Various methodologies for the chemical synthesis of Sec and its derivatives are also available now.[3–5]

Selenoproteins, that is, Sec-containing proteins, are commonly found in most domains of life, except for plants and fungi.[2] Those domains equip similar machineries to express the selenoproteins according to the genetic information encoded in the DNA.[6,7] In humans, there are 25 selenoproteins,[8] most of which are the enzymes related to the antioxidative defense systems against the oxidative stress. These selenoenzymes usually contain one Sec residue at the catalytic active site. Glutathione peroxidase (GPx),[9] thioredoxin reductase (TrxR),[10–12] and iodothyronine deiodinase (ID)[13,14] are well-known representative selenoenzymes. Furthermore, selenoprotein P is a unique selenoprotein as it contains 10 Sec residues in the molecule.[15]

Selenoproteins are now available from artificial sources as well. Several synthetic methods for selenoproteins have been developed to date, which include the chemical modification of natural proteins with

Organochalcogen Compounds
DOI: https://doi.org/10.1016/B978-0-12-819449-2.00004-5

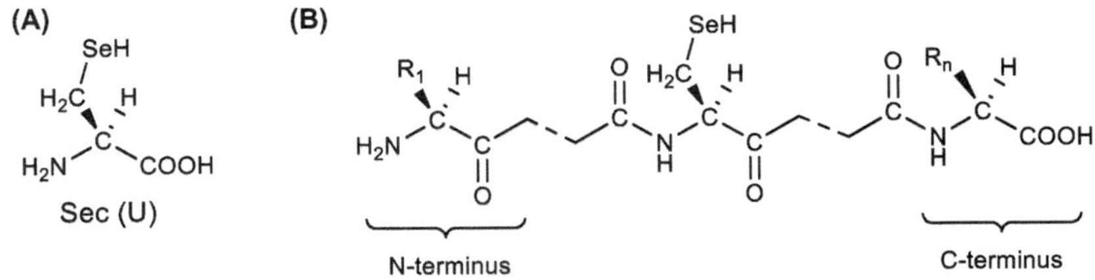

FIGURE 6.1 Molecular structures of (A) selenocysteine and (B) a selenopeptide.

a selenating reagent,[16,17] the genetic expression of the recombinant DNA modified with a selenocysteine insertion sequence,[18,19] and the total chemical synthesis via ligation of the peptide segments.[20–22] The research on these experimental methodologies, that is, the artificial selenoprotein syntheses, has witnessed a rapid progress. A variety of selenoproteins are now available for a wide range of purposes.

Compared to selenoproteins, Sec-containing short peptides (Fig. 6.1B) are rather easily obtained by organic synthesis thanks to the development of peptide synthesis methodologies.[23,24] Selenopeptides prepared are frequently used for the chemical synthesis of long polypeptides and proteins. On the other hand, the chemical and biological assessments of the selenopeptides as mimics and mechanistic probes of selenoenzymes and novel peptide-based antioxidants and therapeutic drugs have been insufficiently explored. In this chapter, the current status of experimental methodologies available for the chemical synthesis of selenopeptides and the catalytic functions of the selenopeptides as selenoenzyme mimics are summarized. Perspectives for the other applications, such as selenopeptide-based drug discovery, are also outlined briefly.

6.2 Synthesis of selenopeptides

6.2.1 General strategies

Two methods, namely, the liquid-phase peptide synthesis (LPPS) and the solid-phase peptide synthesis (SPPS), are available for the chemical synthesis of selenopeptides (Scheme 6.1). Like the cases of general peptides, these methods employ the common scheme that a short peptide chain is grown step by step by attaching one amino acid unit from the C- to the N-termini. If the peptide chain is grown toward the C-terminus like the biosynthesis of proteins in cells, racemization of the α-carbon atom of the amino acid, which is activated for condensation to the resin, can take place via the azlactone intermediate.[25] This racemization might also be possible even in the normal peptide synthesis from the C- to N-terminal direction through the α-proton abstraction under the basic

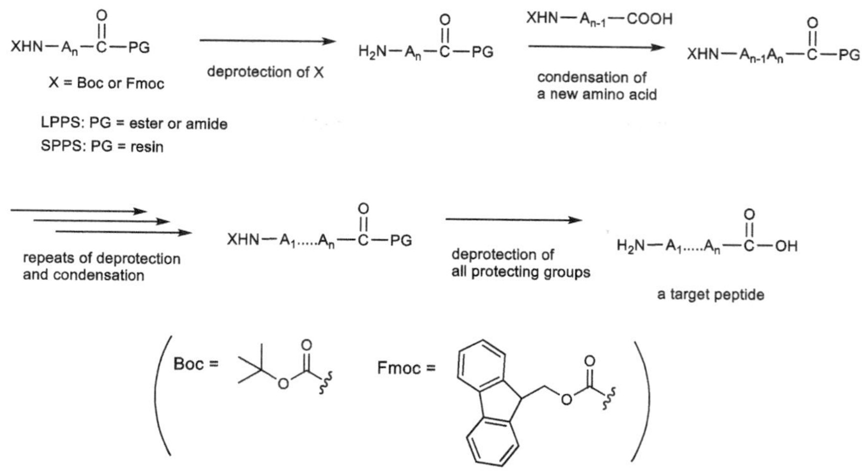

SCHEME 6.1 A common scheme of the peptide syntheses by LPPS and SPPS.

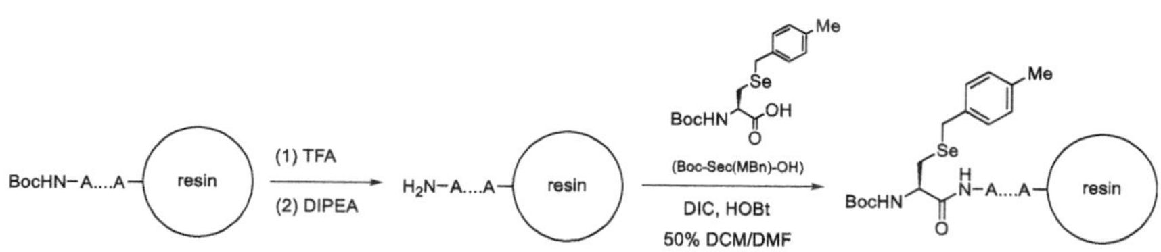

SCHEME 6.2 The reaction of Boc-Sec(MBn)-OH in the Boc strategy.

condition. However, this unfavorable process can be suppressed by using 1-hydroxybenzotriazole (HOBt) or 1-hydroxy-7-azabenzotriazole (HOAt) as a condensation reagent.[26] As a protecting group for the amino group, a *tert*-butoxycarbonyl (Boc) or 9-fluorenylmethyloxycarbonyl (Fmoc) group is widely applied. These peptide syntheses are called the Boc and Fmoc strategies, respectively. Both the strategies have been applied to the synthesis of selenopeptides, although different deprotection and condensation conditions are necessarily applied in each strategy.[3] In SPPS, an excess amount of the Sec derivative (>3 molar equivalents) is usually used to complete the condensation reaction.

In the Boc strategy, the Sec derivative, which has a *p*-methylbenzyl (MBn) protection on the selanyl (SeH) group, is usually applied.[27] The standard protocol for the synthesis of selenopeptides using the Boc strategy is outlined in Scheme 6.2. The N-terminal Boc group of the peptide chain on the resin is deprotected by trifluoroacetic acid (TFA), and then the resin is neutralized with *N,N*-diisopropylethylamine (DIPEA). Boc-Sec(MBn)-OH is activated with *N,N*′-diisopropylcarbodiimide (DIC) and HOBt in 50% dichloromethane (DCM)/*N,N*-dimethylformamide

(DMF) at 0°C for 5 minutes and reacted with the deprotected resin for 1 hour to afford the Sec-attached peptide on the resin. The peptide elongated is cleaved from the resin by applying strong acidic conditions, such as in hydrofluoric acid (HF) or trifluoromethanesulfonic acid (TfOH). The MBn group on the Sec residue can also be deprotected by HF.

On the other hand, in the Fmoc strategy, the SeH group of Sec is usually protected with a *p*-methoxybenzyl (PMB, MPM, or Mob) group, which can be deprotected with iodine (I_2)[28] or 2,2′-dithiobis(5-nitropyridine) (DTNP)/TFA.[29] The frequently applied protocol for the synthesis of selenopeptides by the Fmoc strategy is outlined in Scheme 6.3. The N-terminal Fmoc group of the peptide chain on the resin is deprotected by 20% piperidine in *N*-methylpyrrolidone (NMP) or DMF, and the resin is thoroughly washed with DCM. Fmoc-Sec(PMB)-OH is activated with DIC and HOBt in DCM at room temperature for 20 minutes and then reacted with the resin for 30 minutes to afford the Sec-attached peptide on the resin. The peptide elongated is cleaved from the resin by applying TFA.

Flemer reported the synthesis of Sec derivatives with a trityl (Trt) or xanthyl (Xan) protection on the Se atom (Fig. 6.2).[30] These protecting groups could be deprotected more easily than a PMB group in the mixture of TFA, triisopropylsilane, and water (96:2:2) at room temperature for 1.5 hours. These variations of Sec derivatives enabled a selective deprotection of the specific Sec residue when more than two Sec residues existed in a peptide chain. The benzhydryl (Bzh) protection, which can be deprotected under similar mild conditions, was also applied by Shimodaira and Iwaoka.[31]

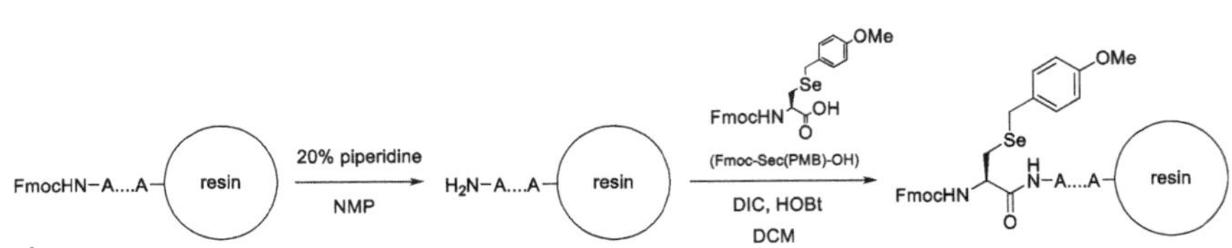

SCHEME 6.3 The reaction of Fmoc-Sec(PMB)-OH in the Fmoc strategy.

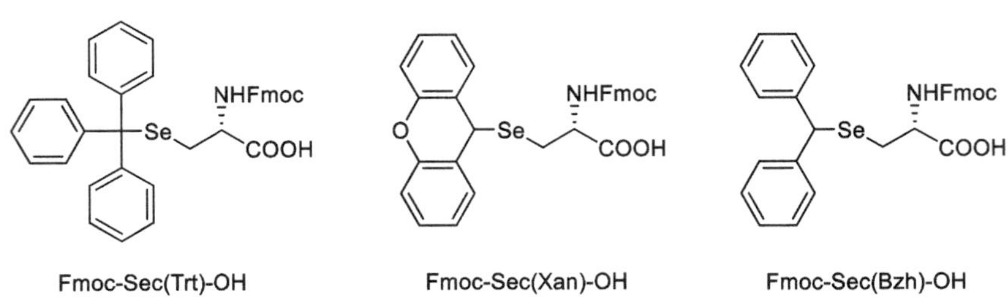

FIGURE 6.2 New Sec derivatives useful for the Fmoc strategy.

6.2.2 Liquid-phase peptide synthesis

LPPS is a conventional method for the peptide synthesis based on the protocols of basic organic synthesis. In LPPS, the synthetic intermediates should be purified every time after the condensation of a new amino acid unit to the growing peptide chain. The longer peptide tends to become less soluble in organic solvents. Owing to this problem, the length of the peptide chain that can be accessed by LPPS is usually up to about a 10-mer, that is, a peptide with 10 amino acid residues.

Short selenopeptides are usually synthesized by LPPS. For example, selenoglutathione (GSeH), a selenium analog of glutathione (GSH), was first synthesized by Soda and coworkers by applying LPPS.[32] In Soda's method, *N*-(*p*-methoxybenzyloxycarbonyl)-Se-(*p*-methoxybenzyl)-L-selenocysteine (Z(OMe)-Sec(PMB)-OH) was connected to glycine (Gly) and γ-glutamic acid (γGlu) step by step to obtain the oxidized dimer of GSeH (GSeSeG) in 9% yield. Recently, the method was significantly improved by Shimodaira et al. as shown in Scheme 6.4.[33] Fmoc-Sec (PMB)-OH was condensed to glycine *tert*-butyl ester in the presence of 1-ethyl-3-(3-dimethylaminopropyl)carbodiimide (EDCI) and HOBt in DMF. The resulting Fmoc-Sec(PMB)-Gly-O*t*Bu was reacted with diethylamine and then Boc-Glu-O*t*Bu to give the protected tripeptide. Finally, all the protecting groups were deprotected with 20% thioanisole (TA)/TFA. GSeSeG was obtained up to 98% in three steps. Synthesis of GSeSeG by SPPS was also reported.[34,35]

6.2.3 Solid-phase peptide synthesis

SPPS is a more effective method for the synthesis of long peptide chains.[36] In SPPS, a peptide chain is grown on an insoluble resin. Therefore, the synthetic intermediates can be easily purified by filtering the reaction mixture, which contains excess reagents and the byproducts. Thus, the target peptide can be obtained with much simple experimental

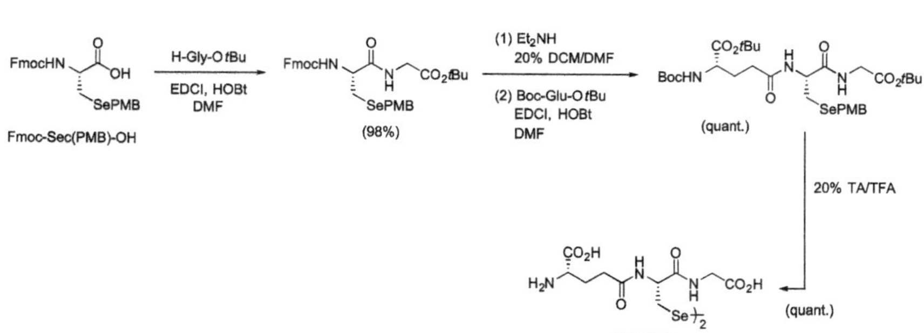

SCHEME 6.4 High-yield synthesis of selenoglutathione by LPPS.

procedures compared to LPPS. Although by applying SPPS, long peptides up to about a 50-mer can be synthesized, the longer peptides are usually not obtained in a reasonable yield because of self-aggregation of the peptide chains on the resin, which significantly reduces the effectiveness of the amino acid condensation reaction. It is also known that the yields of long peptides by SPPS can be low because of the formation of significant amounts of byproducts lacking some of the amino acid residues because of the incomplete condensation reactions.

Scheme 6.5 shows an example of SPPS of long selenopeptides. Metanis and coworkers synthesized a human insulin A-chain analog, which has two Sec residues instead of the Cys6 and Cys20.[37] The double-mutated 21-mer selenopeptide was synthesized by SPPS using the Fmoc strategy with a yield of 8%. Bovine insulin A-chain (21-mer) and B-chain (30-mer) analogs, which have one Sec residue along the chain, were also synthesized by Arai et al. by the similar Fmoc-SPPS method.[38] These selenopeptide chains were utilized for the preparation of selenoinsulins, selenium analogs of a pancreatic hormone insulin.[37,38]

6.2.4 Selenocysteine-mediated native chemical ligation

Development of SPPS enabled us to obtain long peptide chains relatively easily. However, as mentioned above, chemical synthesis of the peptide chains with more than 50 amino acid residues by SPPS is still a difficult challenge unless another useful synthetic technique, namely, peptide ligation, is applied. In the peptide ligation, two peptide segments synthesized by SPPS are linked together under the optimal conditions. Among these ligation methods, native chemical ligation (NCL) is most useful and most widely applied currently.[39]

In NCL, one segment having a thioester group at the C-terminus is condensed with the other segment having a Cys residue at the N-terminus (Scheme 6.6). This coupling reaction proceeds in two steps under a neutral pH condition. In the first step, the C-terminal thioester reacts with the sulfanyl group (SH) of the N-terminal Cys residue of the other peptide segment to exchange the thioester, yielding a long-chain intermediate, in which the two peptide segments are linked together with a thioester bond. In the second step, the thioester bond undergoes

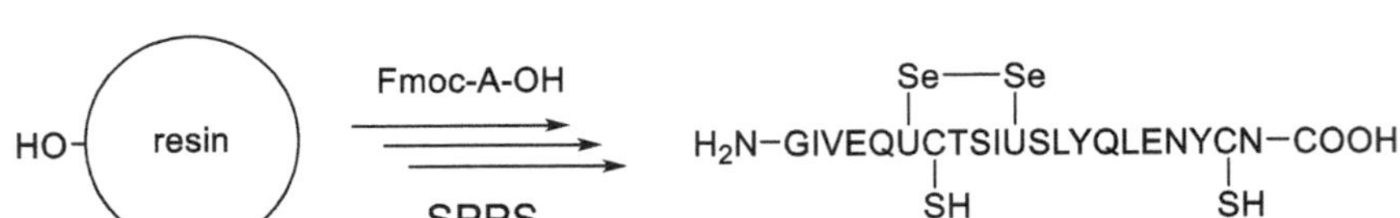

SCHEME 6.5 Synthesis of human insulin A-chain analog, [C6U, C11U], by SPPS.

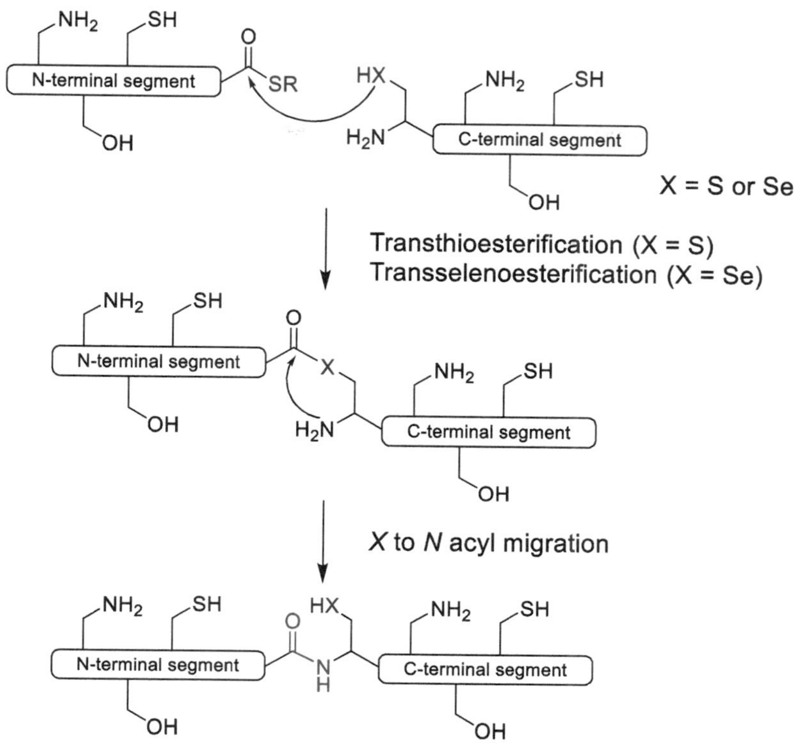

SCHEME 6.6 Reaction mechanisms for native chemical ligation (NCL) (X = S) and selenocysteine-mediated native chemical ligation (Sec-NCL) (X = Se).

S to *N* acyl migration to transform the thioester bond to a more stable amide bond, thus affording a long peptide. The NCL method has an advantageous feature in that a long peptide can be obtained in a high yield without applying protections of any functional groups on both the peptide segments, because of the highly chemo- and regioselective reactions. However, natural proteins contain Cys only in 3% on the average. Therefore, the possible ligation point of NCL is limited. Moreover, for the protein, which does not have a Cys residue, NCL cannot be applied for the chemical synthesis.

To overcome the above limitations of NCL, different variations of the original NCL technique have been devised to date, in which unique amino acid units are sometimes applied at the pivotal ligation site. One useful variation is the desulfurization of the Cys residue at the ligation point to alanine (Ala) after ligation.[40] The transformation can be achieved by H_2/metal reagents[40] or tris(2-carboxyethyl)phosphine (TCEP).[41] A similar desulfurization was already applied for SH-containing nonnatural amino acids, which were introduced at the ligation site instead of Cys and then desulfurized to the corresponding natural amino acid residues, such as Phe[42] and Pro.[43] However, this protocol requires protection of the SH groups of the other Cys residues present in the peptide segments because the desulfurization is not site-specific.

Another useful variation is a selenocysteine-mediated native chemical ligation (Sec-NCL), in which Sec is applied at the ligation site of NCL (Scheme 6.6).[20,21,44] As the SeH group of Sec is more acidic than the SH group of Cys, the ligation proceeds at pH 5.0 chemoselectively for the peptide segments having Cys residues[20] through the reaction mechanism similar to NCL via the formation of a selenoester intermediate.

By applying the Sec-NCL, various Cys- to Sec-substituted proteins, such as bovine pancreatic ribonuclease A (RNase A),[20] bovine pancreatic trypsin inhibitor (BPTI),[21,45] and glutaredoxin 1 (Grx1),[46] were chemically synthesized. These protein variants possessed almost the same folded structures to the native proteins.

Payne developed a selenocystine–selenoester peptide ligation method, which allowed rapid and efficient formation of long selenopeptide chains at pH 6.2 without using any additional reagents.[47] The Sec residue at the ligation point could be easily converted to Ala through deselenization using TCEP.[48,49] This method was applied for the synthesis of long peptides having a Pro-Pro sequence at the ligation site,[50] hairpin molecular architectures of peptide nucleic acids (PNAs),[51] and lipoproteins under highly diluted conditions.[52] Deselenization of β-selenophenylalanine and β-selenoleucine at the ligation site to Phe and Leu, respectively, was also feasible.[53]

6.2.5 Cyclic selenopeptides

Cyclic peptides are frequently found in natural products.[54–56] Many types of cyclic peptides are known. For example, peptides can be cyclized with an amide or ester bond using the terminal or sidechain carboxy group and the intramolecular amino or hydroxy group. The cysteinyl SH groups are also usable for cyclization through formation of a disulfide (SS) bridge. These cyclic peptides frequently possess cell membrane permeability and tolerance against the hydrolysis by protease enzymes, compared to linear peptides. Moreover, their affinity toward the target proteins is enhanced with the limited conformational flexibility. Accordingly, cyclic peptides have garnered increasing attention of researchers in their applications to medicinal chemistry areas.[57]

To obtain cyclic peptides, which have a closed peptide backbone, the NCL method is quite useful. Indeed, the peptide that has an N-terminal Cys and a C-terminal thioester efficiently cyclizes under the NCL conditions.[58–60] The similar intramolecular NCL method can be applied for the synthesis of cyclic selenopeptides by introducing a Sec residue at the N-terminus of the linear precursor peptide instead of the Cys residue.

Quaderer and Hilvert reported the first synthesis of a cyclic selenopeptide as shown in Scheme 6.7.[61] A linear peptide having an N-terminal Sec residue was grown on the resin by SPPS. The cleavage of the peptide from the resin using an aluminum-thiolate reagent[62,63] and the subsequent removal of the protecting groups on the sidechains using a cleavage cocktail liberated a precursor selenopeptide thioester, which spontaneously cyclized under the Sec-NCL condition in the presence of thiophenol to obtain a cyclic 16-mer selenopeptide.

Recently, Shimodaira et al. reported the synthesis of a cyclic selenopeptide by applying a new cyclization protocol, wherein the peptide having an N-terminal Sec residue and a C-terminal *N*-ethyl-cysteine residue was employed as the potential precursor (Scheme 6.8).[64] The selenopeptide obtained by SPPS was converted into a cyclic selenopeptide in one pot via *N* to *S* acyl migration, followed by intramolecular Sec-NCL. The cyclic 8-mer peptide, c-KQGUGWGN-, was isolated as a diselenide form and showed an enhanced GPx-like activity (see Section 6.3.2).

6.2.6 Selenization of dehydroalanine

Another interesting approach to selenopeptides was reported by Reddy and Mugesh.[65] They used dehydroalanine (Dha) as a building block for the Sec residue (Scheme 6.9). The Dha residue of Boc-Leu-Ala-Dha-OMe was converted to Sec by the reaction with *p*-methoxybenzyl selenol (PMBSeH) and the subsequent deprotection of the PMB group on the Se atom with iodine. Although this method gave a mixture of the epimers because of the low diastereoselectivity of the selenization

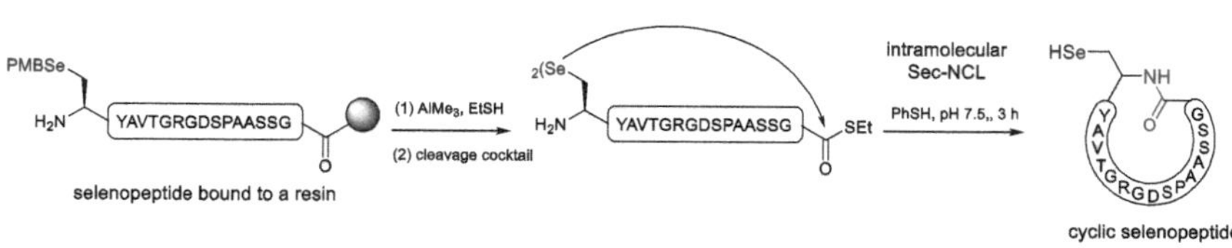

SCHEME 6.7 Synthesis of a cyclic selenopeptide by Quaderer and Hilvert.

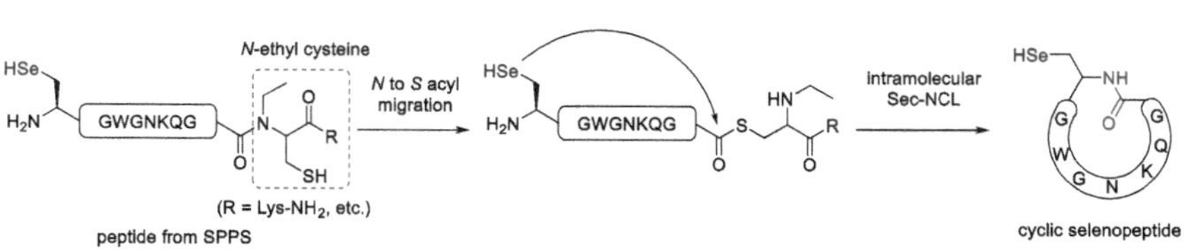

SCHEME 6.8 Synthesis of a cyclic selenopeptide by Shimodaira et al.

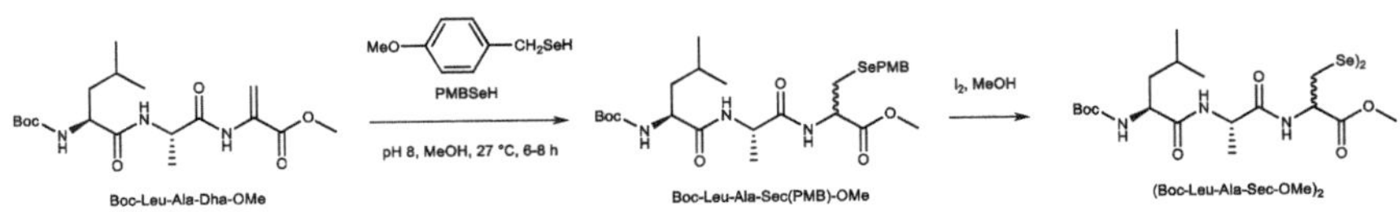

SCHEME 6.9 Synthesis of (Boc-Leu-Ala-Sec-OMe)$_2$ via selenization of dehydroalanine (Dha).

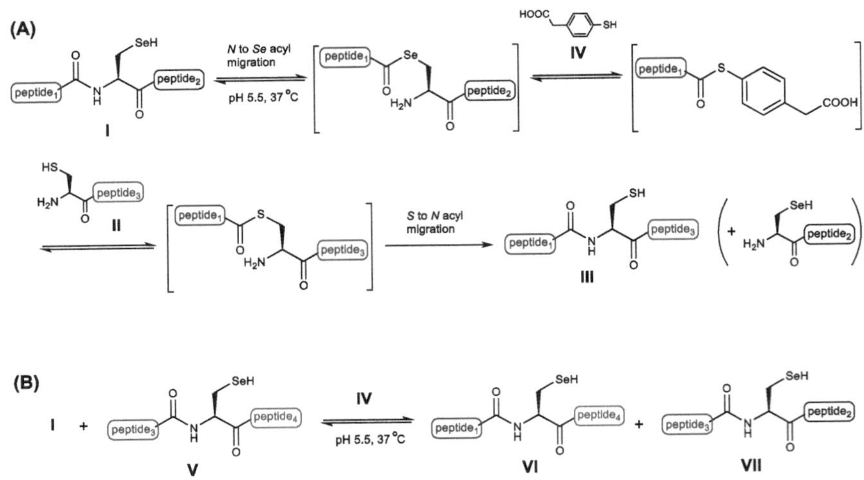

SCHEME 6.10 (A) Transamidation and (B) metathesis of selenopeptides.

reaction, it could be applied to the conversion of GSH to a GSeH derivative through the two-step transformation of the Cys to Sec via Dha.

6.2.7 Other methods

Transamidation of a selenopeptide (I) with an N-terminal cysteinyl peptide (II) was reported by Ollivier et al. (Scheme 6.10A).[66] The reaction proceeded under mild conditions (pH 5.5, 37°C) to give selenopeptide III probably through a reverse NCL in the presence of arylthiol catalyst IV. Under similar conditions, two selenopeptide molecules (I and V) underwent the metathesis reaction to give new selenopeptides (VI and VII), which have their C-terminal peptide segments exchanged with each other (Scheme 6.10B).

6.3 Catalytic functions of selenopeptides

6.3.1 Selenoglutathione

Glutathione (GSH), a water-soluble tripeptide comprising γ-Glu, Cys, and Gly, is abundant in our body and exhibits versatile biological functions as a specific substrate or cofactor to various enzymes,[67] an effective antioxidant against reactive oxygen species (ROS), such as free radicals and hydroperoxides,[68,69] a redox reagent to assist oxidative folding of proteins by accelerating the SS-bond formation and reshuffling,[70] and so on.

Selenoglutathione (GSeH) is the selenium analog of GSH. This small nonnatural seleno-tripeptide possesses unique biochemical properties. Owing to the lower redox potential of GSeSeG/GSeH (−407 mV) than that of GSSG/GSH (−256 mV),[34] the reducing ability of GSeH is significantly higher than that of GSH. Indeed, selenoglutathione is typically isolated as an oxidized dimeric form, that is, GSeSeG, under aerobic conditions. Since the molecular structure of GSeSeG is almost identical to that of GSSG, the former can be reduced to GSeH by the function of glutathione reductase (GR) in the presence of β-nicotinamide adenine dinucleotide 2′-phosphate (NADPH) as a reducing cofactor (Scheme 6.11),[34,35] although the substrate specificity decreases about 10 times from that of GSSG.[34] This suggests that active GSeH can be repeatedly produced in the presence of NADPH and GR. These unique properties of GSeH, that is, the high reducing ability and the possible regeneration in vivo, make GSeH an interesting target of a wide range of biological applications. The examples are shown below.

Hilvert and coworkers applied GSeSeG as a redox folding reagent for disulfide-containing proteins (Scheme 6.12).[34,71–73] Although the SeSe

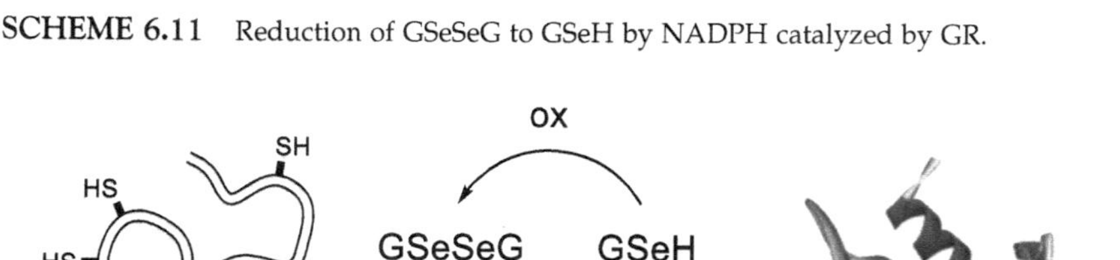

SCHEME 6.11 Reduction of GSeSeG to GSeH by NADPH catalyzed by GR.

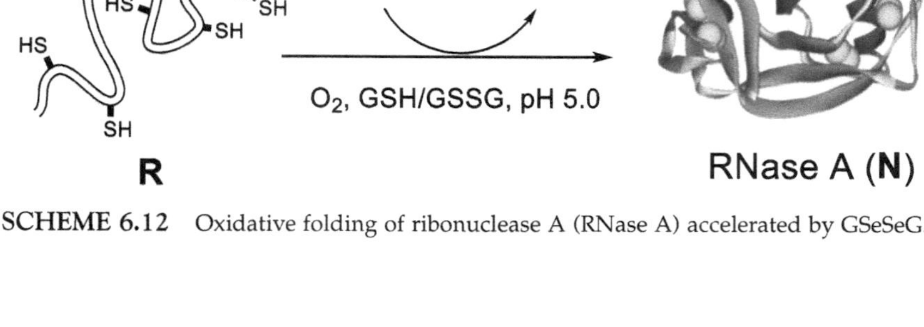

SCHEME 6.12 Oxidative folding of ribonuclease A (RNase A) accelerated by GSeSeG.

bond of GSeSeG is 7 kcal/mol more stable than the SS bond of GSSG, GSeSeG could not only oxidize the Cys residues in a reduced and unfolded protein (R) to guide it to the native folded structure (N) faster than GSSG but also catalyze the oxidative folding reactions in the presence of O_2 and GSH/GSSG even at pH 5.0, where GSH/GSSG alone was inactive.

GSeSeG is also usable as a catalyst for refolding of misfolded protein species, which have scrambled SS bonds, to the native protein. Shimodaira et al. reported that by addition of a catalytic amount of GSeSeG with GR and NADPH into the buffer solution containing misfolded bovine pancreatic RNase A having four scrambled SS bonds (4S), native RNase A (N) with four native SS bonds regenerated gradually (Scheme 6.13).[33] The yield of N was much higher than that was achieved when GSSG was employed. The superior activity of GSeSeG to repair the misfolded protein would be due to the high reducing ability of GSeH, which is catalytically reproduced by the function of GR.

Koppenol and coworkers measured the rate constants for the reactions of GSeH with tyrosyl radicals and found that GSeH reacts with the radicals 10^2 to 10^5 times faster than GSH.[74,75] A similar rapid reaction of GSeSeG to hypochlorous acid (HOCl) was observed by Davies and coworkers.[76] The results suggested that selenoglutathione is a much better radical scavenger or antioxidant than glutathione.

Wang synthesized the methylmercury complex of GSeH (GSeHgMe) by reacting GSeH with MeHgOH and characterized the molecular structure through X-ray analysis.[77,78] It was found that GSeHgMe adopted a similar structure to the sulfur analog (i.e., GSHgMe), while GSeH showed a stronger binding affinity to Hg than GSH. When incubated at 37°C, the product degraded in 150 minutes into HgSe solid via $(MeHg)_2Se$. The results suggested that GSeH is also a good scavenger of toxic methylmercury ions.

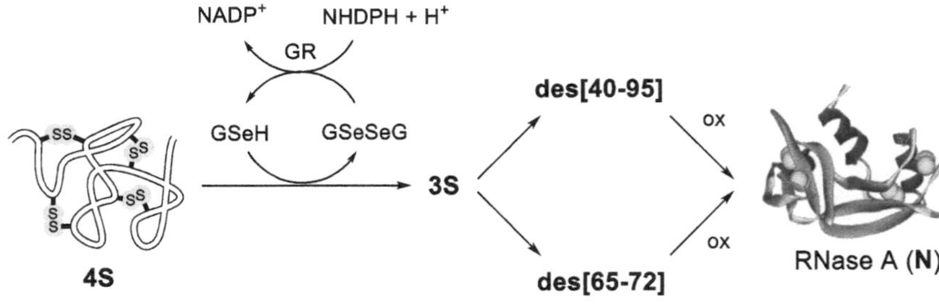

SCHEME 6.13 Refolding of native RNase A (N) from the misfolded 4S catalyzed by GSeH. 3S represents an ensemble of the intermediates with three random SS bonds. The des[40-95] and des[65-72] represent the discrete intermediates with three native SS bonds but lacking the C40–C95 or C65–C72 SS bond, respectively.

Iwaoka and coworkers applied GSeSeG as a potential GPx model.[33,35] The details are presented in the next section.

6.3.2 Glutathione peroxidase mimics

GPx is a representative selenoprotein.[9] Its enzymatic function is to reduce harmful ROS, such as H_2O_2 and ROOH, to water and/or alcohol using GSH as a reducing cofactor. GPx family proteins (GPx1–8) share similar peroxidase functions and protect biological molecules against the oxidative damage. These GPx variants are expressed in different cells depending on the targets. Among these, GPx1–4 and GPx6 contain Sec at the active site, while in the other GPxs, Cys is located at the active site instead of Sec.

In 1984, Seis and Wendel first reported that 2-phenyl-1,2-benzoselenazol-3-one, called ebselen, exhibits the GPx-like antioxidant activity.[79,80] Since then, a number of small organoselenium compounds have been tested as GPx mimics.[81] Most of those compounds were aromatic selenium compounds, like ebselen, having a relatively stable Se–C(Ar) bond. On the other hand, less attention has been paid to aliphatic selenium compounds as GPx mimics.[82–84] This was because the aliphatic selenium compounds are usually unstable upon oxidation of the Se atom. It is, therefore, of significant interest to note that life uses Sec with an aliphatic Se–C bond as a redox active center of various enzymes.

The attempts to use selenopeptides as GPx mimics have embarked rather recently. Singh et al. pursued the reaction of selenocystine, a stable diselenide dimer of selenocysteine (CysSeSeCys), by ^{77}Se NMR spectroscopy and elucidated the GPx-like catalytic cycle of Sec (Fig. 6.3).[85] Since the reduction of the SeSe bond of CysSeSeCys was difficult by GSH, the initiation step was assumed to be the oxidation to a labile selenenic acid intermediate (CysSeOH), which would be further

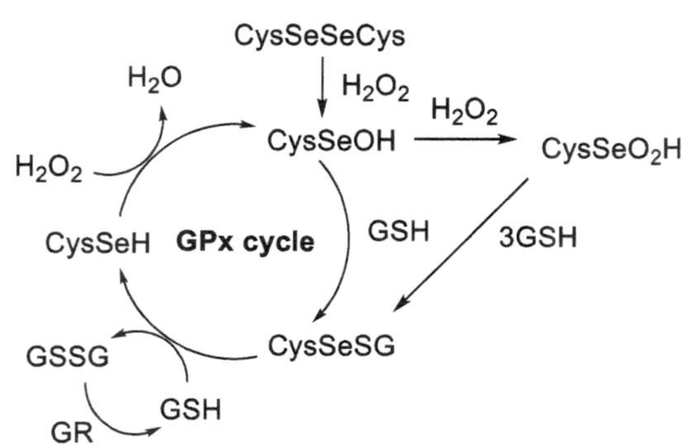

FIGURE 6.3 A GPx-like catalytic cycle of CysSeSeCys.

oxidized to a seleninic acid species ($CysSeO_2H$) or reduced by two molecules of GSH to a selenol intermediate (CysSeH) via a selenenyl sulfide intermediate (CysSeSG). The CysSeH would react with H_2O_2 rapidly to reproduce CysSeOH, thus completing the GPx-like catalytic cycle.

Yoshida et al. synthesized GSeSeG by SPPS and measured the GPx-like catalytic activities under various conditions.[35] They found that GSeSeG exhibits unusually high GPx-like activity in the presence of GR and NADPH. Based on the analysis of the reaction mechanism by reverse-phase HPLC, the high activity of GSeSeG was ascribed mainly to the rapid formation of the active selenol intermediate (GSeH) from GSeSeG and GSeSG by the function of GR. The formation of the overoxidized intermediate $GSeO_2H$ was also characterized by ESI-MS analysis of the reaction mixture of GSeH or GSeSeG and H_2O_2.

The more distinct reaction mechanism for the GPx-like catalytic function of GSeSeG was analyzed by Shimodaira et al. by means of ^{77}Se NMR spectroscopy.[33] This analysis could be conducted since GSeSeG was synthesized in a large amount by applying LPPS (Scheme 6.4). The GPx-like catalytic cycle revealed is shown in Fig. 6.4. The reactive selenolate intermediate (GSe^-), which was in situ−generated by the reaction of GSeSeG with dithiothreitol (DTT^{red}) at pH 10 in D_2O, showed the ^{77}Se NMR signal at −210 ppm as a broad singlet peak. On the other hand, when GSeSeG was reacted with 1 eq H_2O_2 in D_2O, one-third of GSeSeG was oxidized to $GSeO_2H$, which showed a sharp signal at 1213 ppm in the ^{77}Se NMR spectrum. During the oxidation reaction, formation of a labile selenenic acid intermediate (GSeOH) was proposed as a small signal corresponding to the dimedone adduct (50 ppm) was observed in the presence of excess dimedone. However, GSeOH could not be directly observed under any applied conditions. Shimodaira further observed clear formation of the selenenyl sulfide intermediate (GSeSG) (367 ppm) when $GSeO_2H$ was reacted with 3 eq GSH. The

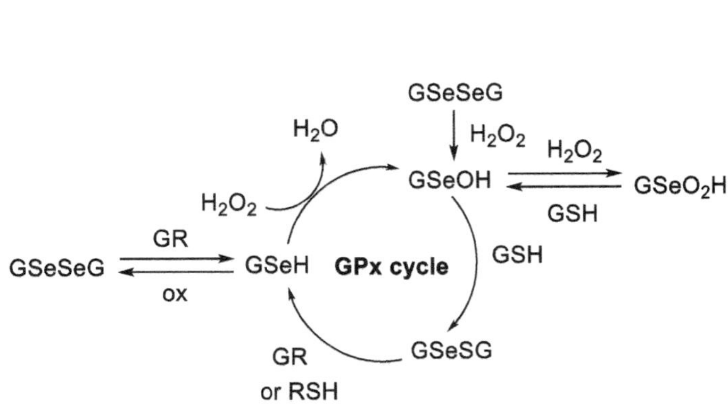

FIGURE 6.4 A GPx-like catalytic cycle of GSeSeG revealed by ^{77}Se NMR spectroscopy.

produced GSeSG was significantly stable, but it could be reduced to the selenol intermediate (GSeH) by the reaction with DTT^{red}.

The unsymmetrical selenenyl sulfides (GSeSR), which could be obtained by reacting $GSeO_2H$ with various RSH, such as CysSH, $H_2NCH_2CH_2SH$, $HOCH_2CH_2SH$, and $PhCH_2SH$, exhibited interesting properties.[33] When GSeSR was incubated in D_2O at room temperature, the disproportionation to GSeSeG and RSSR took place in a different reaction rate depending on R (Scheme 6.14). In the disproportionation, importance of the weak intramolecular interaction between the Se atom and the functional groups of RSH, such as NH_2 and C_6H_5, was suggested. These functional groups exist at the active site of GPx in the proximate Gln and Trp residues (vide infra), and thus, it was proposed that the similar interactions would accelerate the enzymatic reaction.

The selenopeptides, which have the same amino acid sequence as that around the active center Sec residue of GPx, were synthesized, and their GPx-like antioxidant activity was already investigated. Yoshida et al. prepared 3-mer (LUG) and 5-mer (SLUGT) selenopeptides by SPPS as an oxidized diselenide form and tested them as GPx mimics. However, these short selenopeptides exhibited the much less activity than CysSeSeCys.[35] Wehrle et al. synthesized a 10-mer selenopeptide with an amino acid sequence of UGTTVRDYTQ, which corresponds to the amino acids 49–58 of the GPx active site.[86] Interestingly, although the 10-mer selenopeptide exhibited a lower activity than ebselen, the activity could be enhanced by mutating the Sec (U) to α-methyl selenocysteine, (αMe)Sec. It was also found that the selenopeptide containing (αMe)Sec instead of Sec has superior stability because the β-syn elimination of the selenium moiety is prohibited for (αMe)Sec, which is attributable to the absence of an α-hydrogen atom. Nevertheless, the observed low activities indicated that the primary amino acid sequence around the Sec residue at the active site of GPx is not primarily important in increasing the antioxidant activity of Sec. Therefore, the

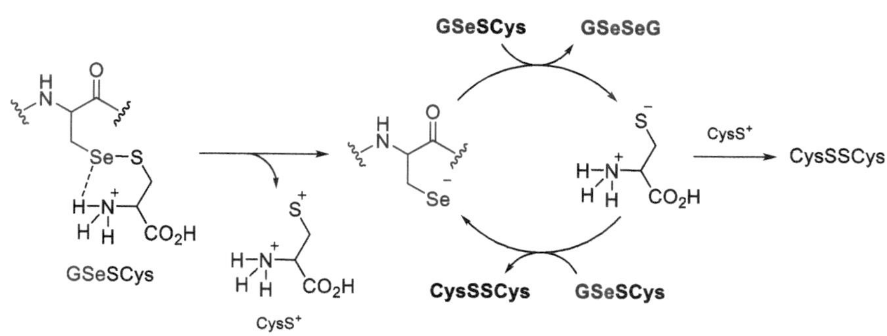

SCHEME 6.14 Disproportionation of GSeSCys to GSeSeG and CysSSCys in D_2O.

interactions between the Se atom and the proximate amino acid side-chains should be worth being explored further.

At the active site of GPx, the presence of a catalytic triad comprising Sec (U), Gln (Q), and Trp (W) is well established (Fig. 6.5A).[87,88] This catalytic triad was modeled by Takei et al. using a selenopeptide with an amino acid sequence of GQAUAWG (Fig. 6.5B).[89] The 7-mer selenopeptide, as well as its mutants lacking either Q or W, were synthesized as an oxidized diselenide form by SPPS using the Fmoc strategy. GPx-like activities of these model peptides clearly demonstrated that the antioxidant activity of GQAUAWG is cooperatively enhanced by the presence of Q and W proximate to U, as well as that the effect of Q on the activity is more important than that of W, supporting the presence of the Q· · ·U· · ·W catalytic triad at the active site of GPx.

Meanwhile, Tosatto et al.[90] proposed that at the active site of selenium-dependent GPx4 a catalytic tetrad, comprising Sec, Gln, Trp, and Asn (N), is present, based on the X-ray structure of the U46C mutant of human GPx4 determined by Scheerer et al. (Fig. 6.6A)[91] as

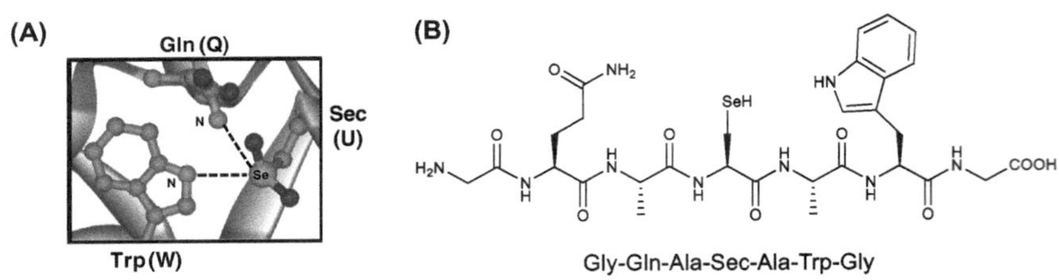

FIGURE 6.5 Mimicking the active site of GPx with a selenopeptide. (A) The catalytic triad of GPx. (B) The molecular formulae of GQAUAWG.

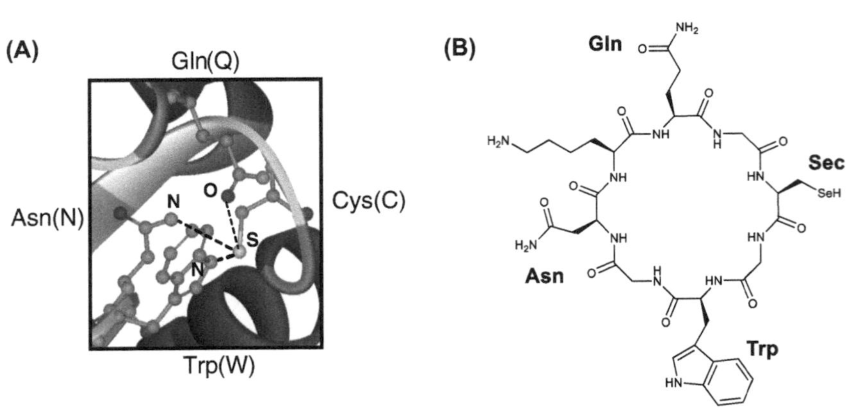

FIGURE 6.6 Mimicking the active site of human GPx4 with a selenopeptide. (A) The active site of the U46C mutant. (B) The molecular formulae of cyclic selenopeptide c-KQGUGWGN-.

well as the amino acid sequence analysis. While the catalytic triad, comprising Sec, Gln, and Trp, has been well established for GPx, the tetrad is still controversial owing to the lack of further corroboration.

To get information about the plausible tetrad, Shimodaira et al. applied a cyclic selenopeptide (Fig. 6.6B), which was designed so that the sidechains of the four amino acid residues, namely, Gln (Q), Sec (U), Trp (W), and Asn (N), came close to each other by placing them alternately with an amino acid spacer.[64] A series of such cyclic selenopeptides were successfully synthesized as a diselenide form according to the tandem *N* to *S* acyl migration and intramolecular Sec-mediated NCL as shown in Scheme 6.8. The GPx-like activity assay revealed that the cyclic peptides exhibit higher activities than the linear peptides, which have the same amino acid sequence, probably because of a decreased chain flexibility by cyclization, enabling stronger interactions among the tetrad amino acid residues. In addition, the mutation of either amino acid of Gln (Q), Trp (W), or Asn (N) resulted in a decrease in the activity. The results supported the presence of the catalytic tetrad at the activity site of GPx4 as proposed by Tosatto et al.[90]

6.3.3 Thioredoxin reductase mimics

TrxR is another representative antioxidant enzyme having a Sec residue at the active site.[10–12] TrxR reduces the SS bond of oxidized thioredoxin using NADPH (Fig. 6.7). The active site has an amino acid sequence of GCUG, which is located at the C-terminus. The CU dyad

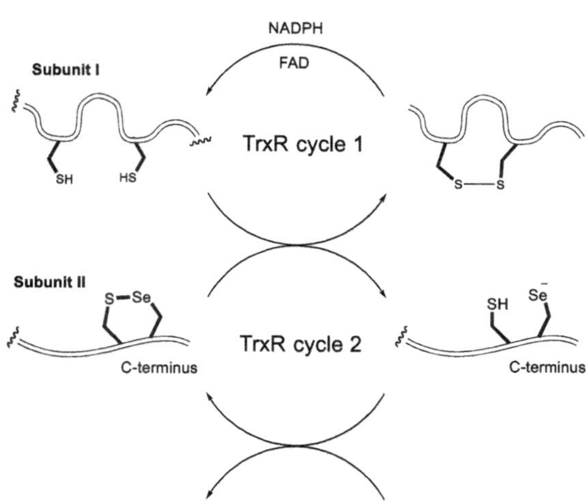

FIGURE 6.7 The catalytic cycles of TrxR. TrxR consists of two subunits. "Sub" represents the substrates of TrxR, such as thioredoxin (Trx), H_2O_2, etc.

makes an eight-membered ring with a Se–S bond when oxidized. This resting intermediate is converted by the reaction with the internal thioredoxin domain into the active selenolate intermediate (ESe^-), which rapidly reacts the oxidized thioredoxin substrate to reduce the SS bond.[92,93]

Metanis et al. demonstrated that glutaredoxin 3 (Grx3), which has a $^{11}CXX^{14}C$ motif at the redox active site, can be transformed to TrxR-like enzymes by introducing Sec (U) in the place of Cys (C).[27] The artificial selenoenzymes were obtained by NCL of the two peptide segments, Grx3(1–37) and Grx3(38–82), both of which were synthesized by SPPS using the Boc strategy.

On the other hand, to obtain the mechanistic insights into the enzymatic function of TrxR, Schneider et al. synthesized a 4-mer selenopeptide GCUG, which has the same amino acid sequence as the C-terminus of human TrxR, by SPPS using the Fmoc strategy.[94] The redox potential of the obtained 4-mer selenopeptide, which has an eight-membered cyclic structure with an internal Se–S bridge, was found to be significantly influenced by introducing Ser at the flanking positions of the catalytic dyad, CU, depending on the pH.

The effects of the amino acid components of the dyad as well as the order of Cys and Sec on the TrxR-like activity were studied by Shimodaira and Iwaoka.[31] Four dipeptides, namely., CU, UC, CC, and UU, were synthesized by LPPS, and their TrxR-like activity was evaluated by using H_2O_2 as a model substrate of TrxR in the presence of GR, GSSG, and NADPH. UC, which has a different order of U and C from the active site of TrxR, exhibited the greater catalytic activity than CU, and CC and UU did not exhibit any significant activity. The results suggested that the N-terminal amino group of UC interacts the Se atom to form NH· · ·Se hydrogen bond, which would facilitate the reduction of the Se–S bond of UC by GSH to produce an active selenol intermediate (Fig. 6.8). At the active site of TrxR, there are proximate His and Glu residues making a putative catalytic triad.[95] Thus, the importance of the NH· · ·Se hydrogen bond between the Sec and His residues was supported. Recently, Arai et al. reported a similar activation process of the

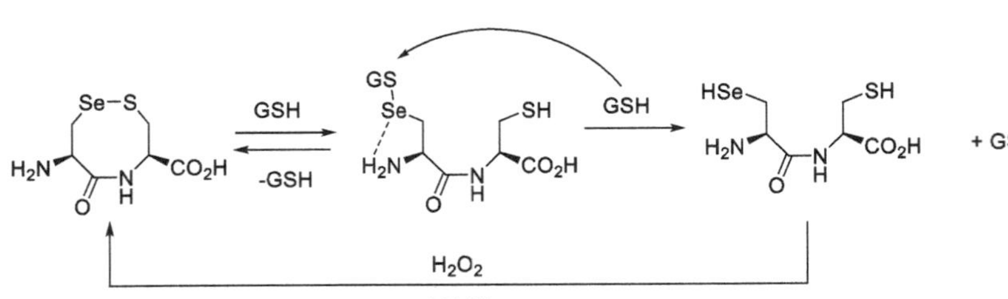

FIGURE 6.8 The TrxR-like catalytic cycle of the UC dipeptide.

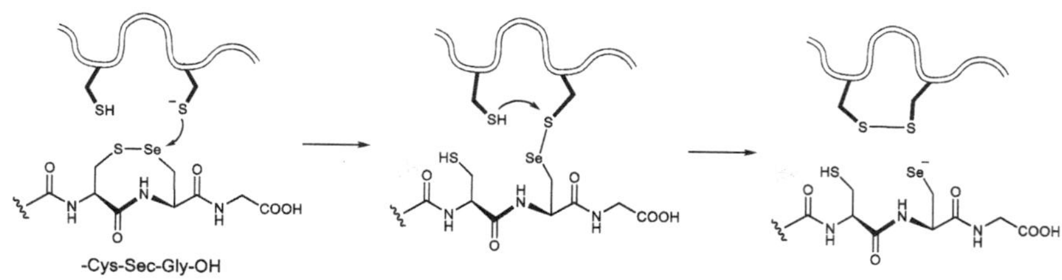

FIGURE 6.9 The activation mechanism of the Se–S state of TrxR.

Se–S bond using small cyclic selenenyl sulfide compounds with an amino substituent.[96]

The ring conformation of the catalytic dyad of TrxR was studied by Ruggles, et al. by ^{1}H NMR spectroscopy for the oxidized cyclic form of Ac-GCUG.[97] The 4-mer selenopeptide was synthesized by SPPS using the Fmoc strategy. The eight-membered ring with an Se–S linkage existed dominantly in the trans conformation (trans:cis ~91:9). The ring strain was reduced from that of the disulfide analog. The longer model selenopeptide, PTVTGCUG, was also synthesized by Lothrop et al. by SPPS using the Fmoc strategy, along with several analogs, and their TrxR-like activities were investigated.[98] It was suggested that the Se atom of the CU dyad in the C-terminal domain is attacked by the SH of the Cys residue present in the N-terminal domain to exchange the Cys SH group and the produced mixed Se–S intermediate is subsequently reduced by another Cys SH group to generate a highly reducible selenolate intermediate (Fig. 6.9).

6.3.4 Other functions of selenopeptides

Recently, new functions of selenopeptides other than the selenoenzyme mimics have been explored by a couple of research groups. For example, Gokula et al. reported that a 5-mer peptide consisting of five benzyl-protected Sec residues, which was synthesized by SPPS, is enriched with β-sheet structures in solution and can self-assemble to form amyloid fibrils (Fig. 6.10A).[99] Zhang et al. prepared a gold nanozyme by conjugating a Sec-containing 5-mer peptide, URGDC, with gold nanoparticles (AuNPs) (Fig. 6.10B).[100] The nanozyme exhibited a higher GPx-like activity than the free selenopeptide. The activity could be further increased by coconjugation of cysteamine on the surface of the AuNPs.

6.4 Perspectives

Selenopeptides can be obtained by LPPS and SPPS rather easily now, and thus, we can utilize the chemically synthesized selenopeptides in

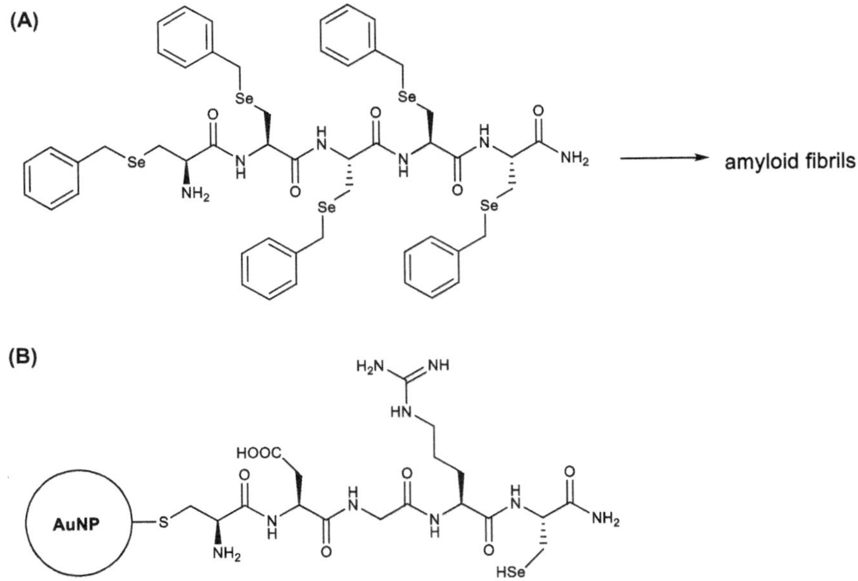

FIGURE 6.10 Applications of short selenopeptides to (A) the formation of amyloid fibrils and (B) gold-nanoparticle conjugates with a GPx-like catalytic activity.

various fields of chemistry and biology. In this chapter, one of these applications, that is, the design of peptide models for selenoenzymes, such as GPx and TrxR, was exemplified. Modeling the active site of selenoenzymes not only for GPx and TrxR but also for the other selenoenzymes, for which the reaction mechanism is not yet well known, will be of importance to understand the reaction mechanisms. Moreover, the study on the selenoenzyme mimics and antioxidants using selenopeptides would open a path for selenopeptide-based drug discovery. In such applications, cyclic selenopeptides, such as those described in Section 6.2.5, are especially intriguing as promising candidates because cyclic peptides possess good stability and cell permeability.

To expedite the applications of selenopeptides, molecular simulation methods, which can predict molecular structures of selenopeptides with satisfactory accuracy within a reasonable computation time, will be helpful. However, such simulation methods are not readily available currently. Therefore, the amino acid sequence of a selenopeptide has to be repeatedly designed until the desired activity is experimentally obtained. This trial-and-error process would be environmentally unfriendly and time-consuming. Iwaoka and coworkers have recently developed a new computation method, called the replica-exchange Monte Carlo simulation using the single amino acid potential force field (REMC/SAAP),[101] and applied the same for the design of some selenopeptides that mimic the active site of GPx.[64,89] Bickelhaupt and Orian developed the AMBER ff14SB force field parameters for Sec and used it for the molecular dynamic simulation for GPx4.[102] Such theoretical methods will become indispensable in future as a basic research tool to support the molecular design of selenopeptides.

Toward the selenopeptide drug discovery, the designed selenopeptides should be first tested for in vitro cell assays to evaluate the biological activities. The pharmacokinetic analysis and the toxicity assay of the selected selenopeptides should be subsequently undertaken using animal models. Selenopeptides would be hydrolyzed in vivo to produce Sec and its degraded products. As we intake a small amount of selenium from foods in our daily life, the degraded products of the selenopeptides will be confluent with the normal metabolites of selenium and would be excreted or reused for the synthesis of Sec and selenoenzymes.[103,104]

Since the selenopeptides can be routinely synthesized in the laboratory now, exploration of versatile applications of the selenopeptides is an interesting challenge in various fields of chemistry and biology.

References

1. Böck, A.; Forchhammer, K.; Heider, J.; Leinfelder, W.; Sawers, G.; Veprek, B.; Zinoni, F. *Mol. Microbiol.* **1991,** *5,* 515–520.
2. Labunskyy, V. M.; Hatfield, D. L.; Gladyshev, V. N. *Physiol. Rev.* **2014,** *94,* 739–777.
3. Flemer, S. *Molecules* **2011,** *16,* 3232–3251.
4. Mousa, R.; Notis Dardashti, R.; Metanis, N. *Angew. Chem. Int. (Ed.) Engl.* **2017,** *56,* 15818–15827.
5. Moroder, L.; Musiol, H. J. *J. Pept. Sci.* **2020,** *26,* e3232.
6. Böck, A.; Forchhammer, K.; Heider, J.; Baron, C. *Trends Biochem. Sci.* **1991,** *16,* 463–467.
7. Hatfield, D. L.; Gladyshev, V. N. *Mol. Cell. Biol.* **2002,** *22,* 3565–3576.
8. Kryukov, G. V.; Castellano, S.; Novoselov, S. V.; Lobanov, A. V.; Zehtab, O.; Guigó, R.; Gladyshev, V. N. *Science* **2003,** *300,* 1439–1443.
9. Brigelius-Flohé, R.; Maiorino, M. *Biochim. Biophys. Acta—Gen. Subj.* **2013,** *1830,* 3289–3303.
10. Mustacich, D.; Powis, G. *Biochem. J.* **2000,** *346,* 1–8.
11. Nordberg, J.; Arnér, E. S. J. *Free. Radic. Biol. Med.* **2001,** *31,* 1287–1312.
12. Arnér, E. S. J. *Biochim. Biophys. Acta—Gen. Subj.* **2009,** *1790,* 495–526.
13. St Germain, D. L.; Galton, V. A.; Hernandez, A. *Endocrinology* **2009,** *150,* 1097–1107.
14. Darras, V. M.; Van Herck, S. L. J. *J. Endocrinol.* **2012,** *215,* 189–206.
15. Burk, R. F.; Hill, K. E. *J. Nutr.* **1994,** *124,* 1891–1897.
16. Wu, Z. P.; Hilvert, D. *J. Am. Chem. Soc.* **1989,** *111,* 4513–4514.
17. Ren, X.; Jemth, P.; Board, P. G.; Luo, G.; Mannervik, B.; Liu, J.; Zhang, K.; Shen, J. *Chem. Biol.* **2002,** *9,* 789–794.
18. Arnér, E. S.; Sarioglu, H.; Lottspeich, F.; Holmgren, A.; Böck, A. *J. Mol. Biol.* **1999,** *292,* 1003–1016.
19. Vandemeulebroucke, A.; Aldag, C.; Stiebritz, M. T.; Reiher, M.; Hilvert, D. *Biochemistry* **2015,** *54,* 6692–6703.
20. Hondal, R. J.; Nilsson, B. L.; Raines, R. T. *J. Am. Chem. Soc.* **2001,** *123,* 5140–5141.

21. Quaderer, R.; Sewing, A.; Hilvert, D. *Helv. Chim. Acta* **2001,** *84*, 1197–1206.
22. Kulkarni, S. S.; Watson, E. E.; Premdjee, B.; Conde-Frieboes, K. W.; Payne, R. J. *Nat. Protoc.* **2019,** *14*, 2229–2257.
23. Muttenthaler, M.; Alewood, P. F. *J. Pept. Sci.* **2008,** *14*, 1223–1239.
24. Liu, J.; Cheng, R.; Rozovsky, S. *Curr. Opin. Chem. Biol.* **2018,** *46*, 41–47.
25. de Castro, P. P.; Batista, G. M. F.; Dos Santos, H. F.; Amarante, G. W. *ACS Omega* **2018,** *3*, 3507–3512.
26. Valeur, E.; Bradley, M. *Chem. Soc. Rev.* **2009,** *38*, 606–631.
27. Metanis, N.; Keinan, E.; Dawson, P. E. *J. Am. Chem. Soc.* **2006,** *128*, 16684–16691.
28. Koide, T.; Itoh, H.; Otaka, A.; Yasui, H.; Kuroda, M.; Fujii, N.; Esaki, N.; Soda, K. *Chem. Pharm. Bull.* **1993,** *41*, 502–506.
29. Schroll, A. L.; Hondal, R. J.; Flemer, S. *J. Pept. Sci.* **2012,** *18*, 155–162.
30. Flemer, S. *J. Pept. Sci.* **2015,** *21*, 53–59.
31. Shimodaira, S.; Iwaoka, M. *Phosphorus. Sulfur. Silicon Relat. Elem.* **2019,** *194*, 750–752.
32. Tamura, T.; Oikawa, T.; Ohtaka, A.; Fujii, N.; Esaki, N.; Soda, K. *Anal. Biochem.* **1993,** *208*, 151–154.
33. Shimodaira, S.; Asano, Y.; Arai, K.; Iwaoka, M. *Biochemistry* **2017,** *56*, 5644–5653.
34. Beld, J.; Woycechowsky, K. J.; Hilvert, D. *Biochemistry* **2007,** *46*, 5382–5390.
35. Yoshida, S.; Kumakura, F.; Komatsu, I.; Arai, K.; Onuma, Y.; Hojo, H.; Singh, B. G.; Priyadarsini, K. I.; Iwaoka, M. *Angew. Chem.—Int. (Ed.)* **2011,** *50*, 2125–2128.
36. Merrifield, R. B. *J. Am. Chem. Soc.* **1963,** *85*, 2149–2154.
37. Weil-Ktorza, O.; Rege, N.; Lansky, S.; Shalev, D. E.; Shoham, G.; Weiss, M. A.; Metanis, N. *Chem. – A Eur. J.* **2019,** *25*, 8513–8521.
38. Arai, K.; Takei, T.; Okumura, M.; Watanabe, S.; Amagai, Y.; Asahina, Y.; Moroder, L.; Hojo, H.; Inaba, K.; Iwaoka, M. *Angew. Chem.—Int. (Ed.)* **2017,** *56*, 5522–5526.
39. Dawson, P. E.; Muir, T. W.; Clark-Lewis, I.; Kent, S. B. H. *Science* **1994,** *266*, 776–779.
40. Yan, L. Z.; Dawson, P. E. *J. Am. Chem. Soc.* **2001,** *123*, 526–533.
41. Wan, Q.; Danishefsky, S. J. *Angew. Chem.—Int. (Ed.)* **2007,** *46*, 9248–9252.
42. Crich, D.; Banerjee, A. *J. Am. Chem. Soc.* **2007,** *129*, 10064–10065.
43. Gui, Y.; Qiu, L.; Li, Y.; Li, H.; Dong, S. *J. Am. Chem. Soc.* **2016,** *138*, 4890–4899.
44. Gieselman, M. D.; Xie, L.; van Der Donk, W. A. *Org. Lett.* **2001,** *3*, 1331–1334.
45. Metanis, N.; Hilvert, D. *Angew. Chem. Int. (Ed.) Engl.* **2012,** *51*, 5585–5588.
46. Casi, G.; Roelfes, G.; Hilvert, D. *ChemBioChem* **2008,** *9*, 1623–1631.
47. Mitchell, N. J.; Malins, L. R.; Liu, X.; Thompson, R. E.; Chan, B.; Radom, L.; Payne, R. J. *J. Am. Chem. Soc.* **2015,** *137*, 14011–14014.
48. Malins, L. R.; Mitchell, N. J.; Payne, R. J. *J. Pept. Sci.* **2014,** *20*, 64–77.
49. Metanis, N.; Keinan, E.; Dawson, P. E. *Angew. Chem.—Int. (Ed.)* **2010,** *49*, 7049–7053.
50. Sayers, J.; Karpati, P. M. T.; Mitchell, N. J.; Goldys, A. M.; Kwong, S. M.; Firth, N.; Chan, B.; Payne, R. J. *J. Am. Chem. Soc.* **2018,** *140*, 13327–13334.
51. Sayers, J.; Payne, R. J.; Winssinger, N. *Chem. Sci.* **2018,** *9*, 896–903.
52. Chisholm, T. S.; Kulkarni, S. S.; Hossain, K. R.; Cornelius, F.; Clarke, R. J.; Payne, R. J. *J. Am. Chem. Soc.* **2020,** *142*, 1090–1100.
53. Wang, X.; Corcilius, L.; Premdjee, B.; Payne, R. J. *J. Org. Chem.* **2020,** *85*, 1567–1578.
54. Rezai, T.; Yu, B.; Millhauser, G. L.; Jacobson, M. P.; Lokey, R. S. *J. Am. Chem. Soc.* **2006,** *128*, 2510–2511.
55. Rezai, T.; Bock, J. E.; Zhou, M. V.; Kalyanaraman, C.; Lokey, R. S.; Jacobson, M. P. *J. Am. Chem. Soc.* **2006,** *128*, 14073–14080.
56. Shibata, K.; Suzawa, T.; Soga, S.; Mizukami, T.; Yamada, K.; Hanai, N.; Yamasaki, M. *Bioorganic Med. Chem. Lett.* **2003,** *13*, 2583–2586.
57. Driggers, E. M.; Hale, S. P.; Lee, J.; Terrett, N. K. *Nat. Rev. Drug Discov.* **2008,** *7*, 608–624.
58. Boll, E.; Ebran, J. P.; Drobecq, H.; El-Mahdi, O.; Raibaut, L.; Ollivier, N.; Melnyk, O. *Org. Lett.* **2015,** *17*, 130–133.

59. Raibaut, L.; Drobecq, H.; Melnyk, O. *Org. Lett.* **2015,** *17*, 3636–3639.
60. Terrier, V. P.; Delmas, A. F.; Aucagne, V. *Org. Biomol. Chem.* **2017,** *15*, 316–319.
61. Quaderer, R.; Hilvert, D. *Chem. Commun.* **2002,** *22*, 2620–2621.
62. Swinnen, D.; Hilvert, D. *Org. Lett.* **2000,** *2*, 2439–2442.
63. Sewing, A.; Hilvert, D. *Angew. Chem. Int. (Ed.) Engl.* **2001,** *40*, 3395–3396.
64. Shimodaira, S.; Takei, T.; Hojo, H.; Iwaoka, M. *Chem. Commun.* **2018,** *54*, 11737–11740.
65. Reddy, K. M.; Mugesh, G. *RSC Adv.* **2019,** *9*, 34–43.
66. Ollivier, N.; Blanpain, A.; Boll, E.; Raibaut, L.; Drobecq, H.; Melnyk, O. *Org. Lett.* **2014,** *16*, 4032–4035.
67. Lu, S. C. *Biochim. Biophys. Acta—Gen. Subj.* **2013,** *1830*, 3143–3153.
68. Sies, H. *Free. Radic. Biol. Med.* **1999,** *27*, 916–921.
69. Pompella, A.; Visvikis, A.; Paolicchi, A.; De Tata, V.; Casini, A. F. *Biochem. Pharmacol.* **2003,** *66*, 1499–1503.
70. Chakravarthi, S.; Jessop, C. E.; Bulleid, N. J. *EMBO Rep.* **2006,** *7*, 271–275.
71. Beld, J.; Woycechowsky, K. J.; Hilvert, D. *Biochemistry* **2008,** *47*, 6985–6987.
72. Beld, J.; Woycechowsky, K. J.; Hilvert, D. *J. Biotechnol.* **2010,** *150*, 481–489.
73. Metanis, N.; Foletti, C.; Beld, J.; Hilvert, D. *Isr. J. Chem.* **2011,** *51*, 953–959.
74. Steinmann, D.; Nauser, T.; Beld, J.; Tanner, M.; Günther, D.; Bounds, P. L.; Koppenol, W. H. *Biochemistry* **2008,** *47*, 9602–9607.
75. Gebicki, J. M.; Nauser, T.; Domazou, A.; Steinmann, D.; Bounds, P. L.; Koppenol, W. H. *Amino Acids* **2010,** *39*, 1131–1137.
76. Carroll, L. *Revis. Requested* **2020.**
77. Khan, M. A. K.; Asaduzzaman, A. M.; Schreckenbach, G.; Wang, F. *Dalt. Trans.* **2009,** 5766–5772.
78. Khan, M. A. K.; Wang, F. *Chem. Res. Toxicol.* **2010,** *23*, 1202–1206.
79. Müller, A.; Cadenas, E.; Graf, P.; Sies, H. *Biochem. Pharmacol.* **1984,** *33*, 3235–3239.
80. Wendel, A.; Fausel, M.; Safayhi, H.; Tiegs, G.; Otter, R. *Biochem. Pharmacol.* **1984,** *33*, 3241–3245.
81. Jain, V. K.; Priyadarsini, K. I., Eds. *Organoselenium Compounds in Biology and Medicine: Synthesis, Biological and Therapeutic Treatments;* Royal Society Of Chemistry, 2017.
82. Reich, H. J.; Jasperse, C. P. *J. Am. Chem. Soc.* **1987,** *109*, 5549–5551.
83. Back, T. G.; Moussa, Z.; Parvez, M. *Angew. Chem.—Int. (Ed.)* **2004,** *43*, 1268–1270.
84. Kumakura, F.; Mishra, B.; Priyadarsini, K. I.; Iwaoka, M. *Eur. J. Org. Chem.* **2010,** 440–445.
85. Singh, B. G.; Bag, P. P.; Kumakura, F.; Iwaoka, M.; Priyadarsini, K. I. *Bull. Chem. Soc. Jpn.* **2010,** *83*, 703–708.
86. Wehrle, R. J.; Ste Marie, E. J.; Hondal, R. J.; Masterson, D. S. *J. Pept. Sci.* **2019,** *25*, e3173.
87. Epp, O.; Ladenstein, R.; Wendel, A. *Eur. J. Biochem.* **1983,** *133*, 51–69.
88. Ursini, F.; Maiorino, M.; Brigelius-Flohé, R.; Aumann, K. D.; Roveri, A.; Schomburg, D.; Flohé, L. *Methods Enzymol.* **1995,** *252*, 38–53.
89. Takei, T.; Urabe, Y.; Asahina, Y.; Hojo, H.; Nomura, T.; Dedachi, K.; Arai, K.; Iwaoka, M. *J. Phys. Chem. B* **2014,** *118*, 492–500.
90. Tosatto, S. C. E.; Bosello, V.; Fogolari, F.; Mauri, P.; Roveri, A.; Toppo, S.; Flohé, L.; Ursini, F.; Maiorino, M. *Antioxid. Redox Signal.* **2008,** *10*, 1515–1526.
91. Scheerer, P.; Borchert, A.; Krauss, N.; Wessner, H.; Gerth, C.; Höhne, W.; Kuhn, H. *Biochemistry* **2007,** *46*, 9041–9049.
92. Zhong, L.; Arner, E. S. J.; Holmgren, A. *Proc. Natl. Acad. Sci.* **2000,** *97*, 5854–5859.
93. Sandalova, T.; Zhong, L.; Lindqvist, Y.; Holmgren, A.; Schneider, G. *Proc. Natl. Acad. Sci.* **2001,** *98*, 9533–9538.
94. Schneider, A.; Brandt, W.; Wessjohann, L. A. *Biol. Chem.* **2007,** *388*, 1099–1101.
95. Gromer, S.; Wessjohann, L. A.; Eubel, J.; Brandt, W. *ChemBioChem* **2006,** *7*, 1649–1652.

96. Arai, K.; Matsunaga, T.; Ueno, H.; Akahoshi, N.; Sato, Y.; Chakrabarty, G.; Mugesh, G.; Iwaoka, M. *Chem. – A Eur. J.* **2019,** *25,* 12751–12760.
97. Ruggles, E. L.; Deker, P. B.; Hondal, R. J. *J. Pept. Sci.* **2014,** *20,* 349–360.
98. Lothrop, A. P.; Snider, G. W.; Ruggles, E. L.; Patel, A. S.; Lees, W. J.; Hondal, R. J. *Biochemistry* **2014,** *53,* 654–663.
99. Gokula, R. P.; Mahato, J.; Singh, H. B.; Chowdhury, A. *Chem. Commun.* **2018,** *54,* 11697–11700.
100. Zhang, D.; Shen, N.; Zhang, J.; Zhu, J.; Guo, Y.; Xu, L. *RSC Adv.* **2020,** *10,* 8685–8691.
101. Iwaoka, M.; Suzuki, T.; Shoji, Y.; Dedachi, K.; Shimosato, T.; Minezaki, T.; Hojo, H.; Onuki, H.; Hirota, H. *J. Comput. Aided. Mol. Des.* **2017,** *31,* 1039–1052.
102. Bortoli, M.; Torsello, M.; Bickelhaupt, F. M.; Orian, L. *ChemPhysChem* **2017,** *18,* 2990–2998.
103. Suzuki, K. T. *J. Heal. Sci.* **2005,** *51,* 107–114.
104. Takahashi, K.; Ogra, Y. *Metallomics* **2020,** *12,* 241–248.

CHAPTER

7

Oxygen-transfer reactions catalyzed by organoselenium compounds

Agata J. Pacuła-Miszewska[1] *and Luca Sancineto*[2]

[1]Department of Organic Chemistry, Faculty of Chemistry, Nicolaus Copernicus University, Torun, Poland, [2]Group of Catalysis, Synthesis and Organic Green Chemistry, Department of Pharmaceutical Sciences, University of Perugia, Perugia, Italy

7.1 Introduction

Oxidation reactions are one of the basic tools in organic syntheses. The number of oxygen-containing molecules, both of natural and synthetic origin, is huge and continuously increasing. The need for efficient protocols enabling the oxidative transformation of the substrate, by a stoichiometric oxygen donor or a specific oxygen-transfer agent, has triggered the development in the field. In this branch of redox chemistry, a relevant place is occupied by organic and inorganic selenium compounds. Numerous reviews and book chapters regarding the application of Se derivatives as reagents and catalysts in oxidation reaction have been published, proving their synthetic versatility.[1–4]

Since the early 1930s, when the potent oxidizing properties of selenium dioxide were identified with the well-known Riley allylic oxidation,[5] Se compounds started to be recognized as promising reagents for the oxofunctionalization of organic molecules. As stoichiometric oxidants, they were successfully applied in various oxidative organic transformations including the conversion of alcohols and phenols to carbonyl compounds with benzeneseleninic acid anhydride or a combination of diaryl diselenides and

Organochalcogen Compounds
DOI: https://doi.org/10.1016/B978-0-12-819449-2.00009-4

tert-buthyl hydroperoxide.[6–11] Oxidation of different nitrogen-containing compounds was also efficiently performed using Se reagents: hydrazines to different aza derivatives,[12–18] synthesis of nitriles and imines from amines,[19–22] and dehydrogenative transformation of indolines to indoles with benzeneseleninic acid anhydride.[22–24] Additionally, an efficient oxygen transfer from peroxyseleninic and selenonic acids to sulfides and phosphines could be conducted.[25] Benzeneseleninic acid anhydride was also efficiently adopted for the oxidative removal of protecting groups from ketones and aldehydes,[12,26–30] the transformation of thiocarbonyls to carbonyls,[31,32] the dehydrogenation of ketones, alcohols,[33] lactams,[34,35] and lactones,[36] the regioselective hydroxylation of polycyclic ketones[37] and carbonyl group introduction in the side chains of arenes.[38–40] Additionally, benzeneselenenic and benzeneseleninic acids were applied as oxygen donors in epoxidation of olefins[41–44] and Baeyer–Villiger oxidations.[45–47] Recently, an interesting synthesis of pyrazoles by electrophilic selenium reagent, another important class of Se-containing compounds,[48] generated by the reaction of diphenyl diselenide with Oxone was also conducted.[49]

In recent years, oxidation reactions as well as other basic synthetical transformations have evolved to more modern and "greener" procedures. In line with this trend, the use of organoselenium reagents as catalyst also started to develop. In 1978, Hori and Sharpless reported the application of areneseleninic acids as catalysts in the synthesis of allylic alcohols, through an oxidation/elimination sequence,[42] which initiated the rapid growth of new synthetic protocols utilizing organoselenium compounds as catalysts. In the same decade, the important role of the antioxidant selenoenzyme, namely, glutathione peroxidase (GPx), in maintaining redox homeostasis in living organisms was also well established. The direct correlation between the presence of the amino acid selenocysteine (Sec), at the enzymes active site, and the peroxide-scavenging activity of GPx was postulated.[50] The unique ability of Sec to catalyze peroxide reduction revealed the bio-catalytic potential of organoselenium molecules and their possible application as redox modulators in biological processes.[51,52] Several selenorganic compounds were developed since then, ebselen,[53] for example, was reported to have several biological activities; similarly, diphenyl diselenide[54] and diselenides, in general.[55,56] Besides, diselenobisbenzamides, also known as DiSeBas,[57] were reported as anti-HIV agents owing to their ability to inhibit the key viral protein NCp7.[58,59]

The transfer of oxygen atom from one molecule to another is a fundamental action widespread both in nature and synthetic organic chemistry. Over the years, the selenium moiety has been proven to be highly reactive toward oxygen donors and capable of forming an efficient oxygen-transfer system in various natural and artificial catalytic cycles. This chapter summarizes the role of organoselenium compounds as catalysts in oxidation reactions reported in the last few years and briefly as redox enzymes mimics.

7.2 Oxygen transfer in Se-catalyzed oxidation reactions

To perform a catalytic oxidation of an organic molecule by an organoselenium reagent, the applied catalyst, a selenide, diselenide, or seleninic acid, has to be first activated by a primary oxidant, for example, hydrogen peroxide (H_2O_2), *tert*-butyl hydroperoxide (TBHP), or iodoxybenzene ($PhIO_2$), to a catalytically active species, which can further oxidize the substrate. The catalyst is regenerated by the primary oxidant, usually a commercially available and environment-friendly oxygen donor. Other advantages of these metal-free procedures are selectivity and high reactivity toward different substrates including alkenes, alcohols, thiols, aldehydes, ketones, and oximes. This section will be divided according to the type of the Se-catalyzed reaction.

7.2.1 Oxidation of alkenes

Oxidations of carbon–carbon double bonds is a thoroughly evaluated field of Se catalysis, enabling to obtain variously functionalized molecules including epoxides, diols, and carbonyl compounds. First reports, presented in the 1970s,[42,60] proved that seleninic acids can be efficiently applied for oxidative purposes in catalytic quantities with additional milder terminal oxidants like hydrogen peroxide. Starting from the work of Betzemeier et al.,[61] several papers presented the utility of fluoroalkylated organoselenium derivatives as useful oxidation catalysts in the epoxidation of alkenes (Scheme 7.1). 2,4-Bisperfluorooctylphenyl buthylselenide (**1**) efficiently catalyzed oxidation of olefins, with moderate-to-high yields, in a fluorus biphasic system. The presence of fluoroalkyl substituents enabled the solubilization of the compounds in fluorinated solvents, which further allowed a facile recovery of the catalyst.[61] Next, Detty and coworkers applied selenoxides as oxygen-transfer agents with hydrogen peroxide as the primary oxidant. The best result for the epoxidation of cyclooctene was obtained for benzyl 3,5-bis(trifluoromethyl)phenyl selenoxide (**2**) (yield: 93%, catalyst loading 2.5 mol%).[62] Finally, the diselenide (**3**) was evaluated in a series of

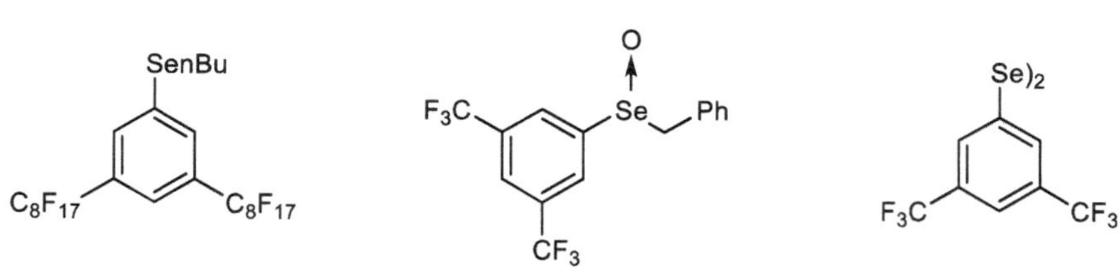

1 **2** **3**

SCHEME 7.1 The fluoroalkylated compounds (**1–3**) applied in epoxidation of alkenes.

investigations to efficiently convert various olefins to epoxides, including fatty esters. To accelerate the rate of the reaction and eliminate side-product formation, a considerable part of the research was focused on choosing the appropriate solvent and base. For the epoxidation of cyclohexene, the best results were obtained when the reaction was carried out in 2,2,2-trifluoroethanol in the presence of sodium acetate.[63]

To design more sustainable and environment-friendly epoxidation procedures, a series of glycerol-based solvents was also evaluated using the same catalytic system diselenide **3**-H_2O_2. It was concluded that the solvent should possess high Lewis acidity to enable the oxidation reaction.[64,65]

In 2016, Yu and coworkers presented the first selective and switchable Se-catalyzed oxidation. β-ionone, a natural compound with several possible reaction sites, was selectively oxidized depending on the used organoselenium catalyst. Compound (**3**) enabled a regioselective epoxidation to derivative (**5**). On the contrary, when the benzyl diselenide (**6**) was used as the precatalyst, a mixture of compounds was obtained but with the vinyl acetate (**7**) as the major product (Bayer–Villiger oxidation) (Scheme 7.2).[66] Additionally, an interesting heterogeneous catalyst Se/C fabricated using carbohydrates as the source of carbon, was applied and enabled to regiospecifically transform β-ionone to epoxide (**5**).[67]

After the synthesis of initial epoxide, the Se-catalyzed oxidation of alkene can further lead to epoxide ring opening and the formation of a dihydroxylated compounds. This way diols can be obtained as undesired side products or, if the procedure is intended, as the wanted final compounds. The first diphenyl diselenide catalyzed dihydroxylation was presented by Santi et al. Using hydrogen peroxide as oxidant and 10 mol% of the catalyst the reaction was performed with excellent yield

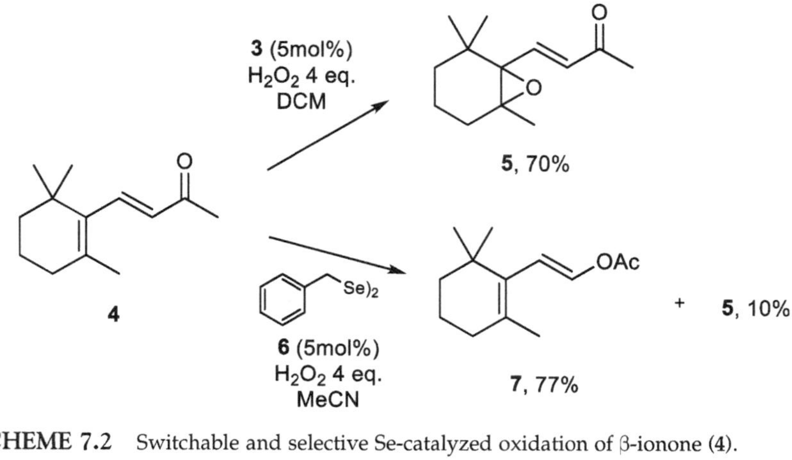

SCHEME 7.2 Switchable and selective Se-catalyzed oxidation of β-ionone (**4**).

and good diastereoselectivity (*Dr* 11:89).[68] Next, three chiral diselenides (**8**)–(**10**) were tested in the same dihydroxylation procedure and additionally in hydroxymethoxylation reaction. Exemplary results are presented for 1-methylcyclohexene (Scheme 7.3). In case of all precatalysts (**8**)–(**10**) the *trans*-diols were synthetized with good enantiomeric excess. However, the hydroxymethoxylation was conducted only with selenocystine (**8**) as the Se catalyst. Surprisingly, a significant decrease of enantioselectivity was observed.[69]

Recently, the same group presented the $(PhSe)_2$-catalyzed oxacyclization of alkenoic acids and alkenols. The reaction proceeded through the oxidation of the double bond with peroxyseleninic acid and further epoxide ring-opening by the internal nucleophilic group. The best result was obtained for *trans*-3-hexenoic acid (**11**) with *Dr* >99:1. The reaction was also tested with chiral terpene catalysts formed from dimenthyl (**12**) and dineomenthyl diselenide (**13**). Compound (**12**) afforded the target lactone with the same yield and diasteroselectivity but with moderate facial selectivity leading to a mixture of product with 28% of enantiomeric excess (*ee*). The reaction carried out with precatalyst (**13**) gave the same results in term of yields and diasteroselectivity, but no facial selectivity was observed and the major diasteroisomer was obtained in racemic mixture (Scheme 7.4).[70] Recently, this method was implemented with the aid of flow chemistry.[71]

Yu and coworkers also presented several procedures for the dihydroxylation of alkenes. The oxidation of cyclohexene with diphenyl diselenide as the precatalyst helped obtain a very high yield; however, the procedure required a long reaction time.[72] The improvement of the

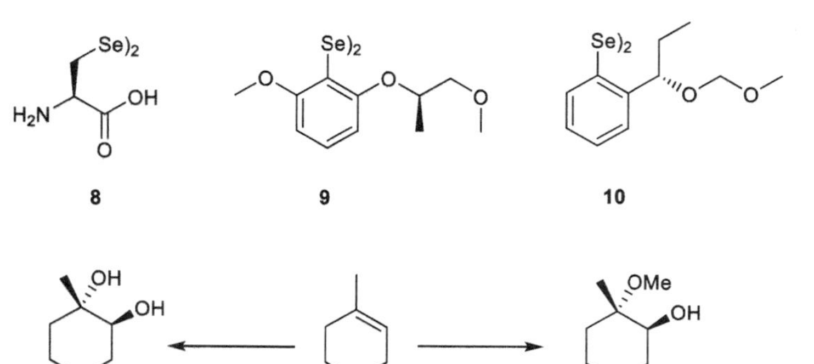

Precatalyst	Yield %	ee%
8 (1mol%, H_2O_2 4 eq.)	80	87
9 (10mol%, H_2O_2 40 eq.)	98	80
10 (10mol%, H_2O_2 40 eq.)	90	75

Precatalyst	Yield %	ee%
8 (1mol%, H_2O_2 4 eq.)	93	37

SCHEME 7.3 Dihydroxylation and hydroxymethoxylation catalyzed by diselenides (**8**)–(**10**).

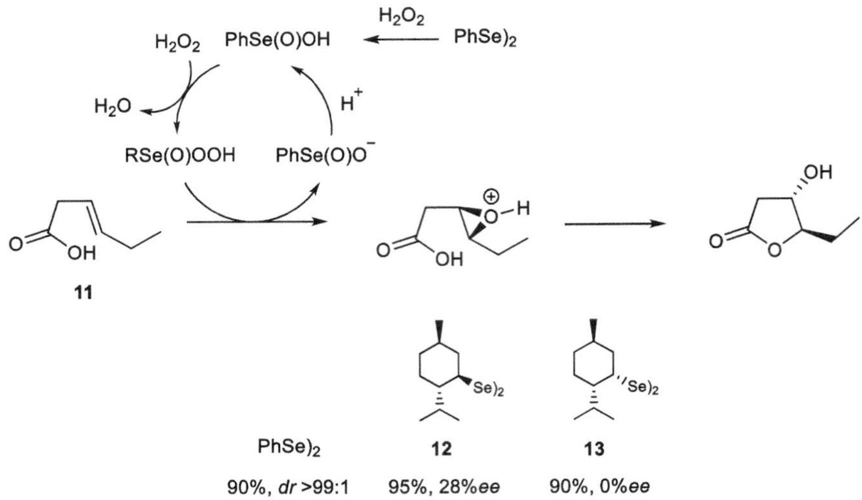

SCHEME 7.4 Oxacyclization of *trans*-3-hexenoic acid catalyzed by $(PhSe)_2$ and diselenides (**12**), (**13**).

Se catalyst by incorporating two $-CF_3$ groups on both *meta* positions of the phenyl ring (compound (**3**)) helped shorten the process to 3–5 hours.[73] The methodology was further modified to more sustainable by reducing the amount of hydrogen peroxide, using water as solvent and a polymer-supported Se catalyst (**14**), which was easily recovered by filtration (Scheme 7.5).[74] Organoselenium chemistry, especially the catalytic approaches, suits perfectly with water as a nonconventional reaction medium and uncommon energy sources, several examples were reported.[75–78]

Finally, the latest reports present the utility of organoselenium catalyst in the transformation of alkenes to carbonyl compounds through an oxidative cleavage of the double bond. This procedure was initially performed using selenium dioxide in 2007 by Konwar and coworkers.[79] Yet, the first successful application of organoselenium reagents was presented by Wang et al. in 2016.[80] The reported green protocol involved the use of dialkyl and diaryl diselenides as precatalysts, hydrogen peroxide as the oxidant, and ethanol as the solvent. The highest yield of the final carbonyl compound (**20**) was obtained in the presence of cyclohexyl diselenide (**15**), contrary to diphenyl diselenide and diselenide (**3**), which exhibited the best catalytic potential in the epoxidation and dihydroxylation reaction. The mechanism of the process, investigated by gas chromatography–mass spectrometry analysis, included several steps: epoxide (**17**) formation, hydrolysis to diol (**18**), oxidation to intermediate (**19**), and final ketone (**20**) formation (Scheme 7.6).

Precatalyst	Time	Yield %
PhSe)₂ (1mol%, H_2O_2 1,5 eq., MeCN)	42h	96
3 (1mol%, H_2O_2 1,5 eq., MeCN)	3-5h	80
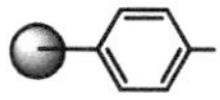 **14** (1mol%, H_2O_2 1.0 eq., H_2O)	20h	98

SCHEME 7.5 Dihydroxylation catalyzed by different Se catalysts.

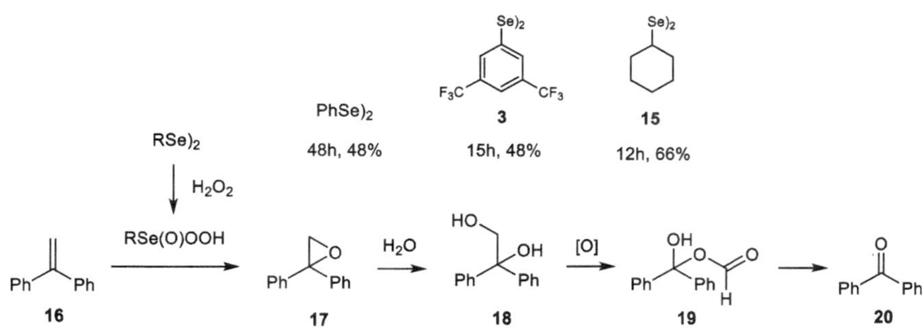

SCHEME 7.6 Oxidative cleavage of double bonds catalyzed by $(PhSe)_2$ and diselenides (**3**) and (**15**).

Recently, the same research group also tested the applicability of polymeric catalyst, formed from 1,4-bis(chloromethyl)benzene and NaHSe, in analogous transformation of olefins to carbonyls. However, the yield of the reaction was lower than for the cyclohexyl diselenide (**15**) (after 48 hours of reaction time, (**3**): 74%; polymeric catalyst: 47%).[81]

In 2015, Yu et al. presented an interesting ring-expansion reaction to obtain substituted cyclobutanones from corresponding methylenecyclopropanes. The reactions were performed using a diselenide–hydrogen peroxide catalytic system. The efficiency of the transformation of 1,1′-(cyclopropylidenemethylene)dibenzene (**21**) to cyclopropane derivative (**22**) was evaluated using different diselenides as precatalysts. When a *para*-methoxy substituent was attached to the phenyl ring, the yield of the process (47%) was lower than that in the presence of $(PhSe)_2$ (60%). The best result was observed for the electron-deficient diselenide $(2\text{-}FC_6H_4Se)_2$ with 71% yield (Scheme 7.7).[82]

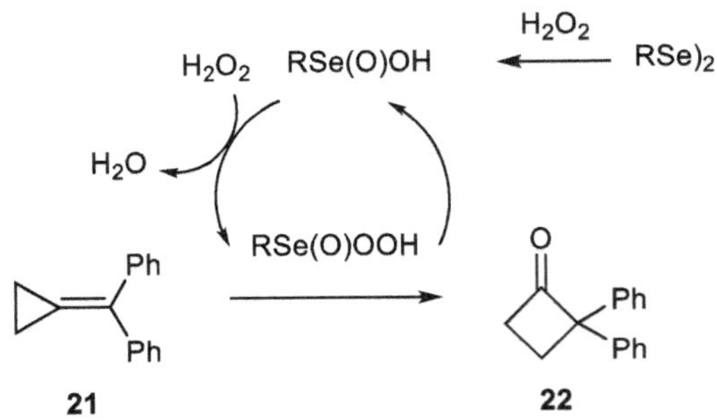

Precatalyst	Yield
$PhSe)_2$	60%
$(4\text{-}MeOC_6H_4Se)_2$	47%
$(2\text{-}FC_6H_4Se)_2$	71%

SCHEME 7.7 Oxidative ring expansion of methylenecyclopropanes (**21**).

7.2.2 Oxidation of compounds bearing an azomethine linkage

Oxidation of azomethine compounds, including azines, hydrazones, and oximes, was also efficiently performed using organoselenium catalysis. In a series of papers, starting from 2001, Młochowski and coworkers investigated the H_2O_2 oxidation of various C=N–R derivatives in the presence of Se polymers (**23**) and (**24**). Oximes, tosylhydrazones, and azines were converted to arenecarboxylic acids using catalyst (**23**). The procedure was also efficient for additionally substituted substrates (2,4-chloro- and 4-*tert*-buthyl derivatives). On the contrary, when a methoxy group was attached to the aromatic ring, the yield decreased significantly.[83] *N,N*-dimethylhydrazones, in the presence of (**24**), were oxidized to nitriles; however, in some cases, the formation of several side-products was observed[84] (Scheme 7.8).

A different polymeric catalyst was applied in 2003 by Sheng et al. A polystyrene with bounded seleninic acid moiety (**25**) was efficiently used in an H_2O_2-mediated ester synthesis from methanol and corresponding oximes[85] (Scheme 7.9).

Besides the oxidation of dimethylhydrazones, organonitrile was also obtained by the oxidative dehydration of aldoximes. Yu and coworkers used different aromatic diselenides (**26**) as a precatalyst, which were further converted with hydrogen peroxide to selenenic acids (**27**) and underwent H_2O-elimination to form the areneselenenic anhydrides (**28**). The active intermediate (**28**) reacted with the –OH group of the oxime and further rearranged to selenoxide (**29**). *Syn*-elimination of selenenic acid (**27**) furnished the final organonitrile (**30**). The best result was obtained for fluorinated derivative (**31**) (yield: 91%), slightly better than for $(PhSe)_2$ (85%)[86] (Scheme 7.10).

In the same reaction, the authors also applied benzenseleninic acid $(PhSe)_2$ as the precatalyst. In the presence of air, the compound was transformed to the active selenenic acid PhSeOH, which catalyzed the dehydration reaction.[87] Additionally, the procedure was modified to obtain

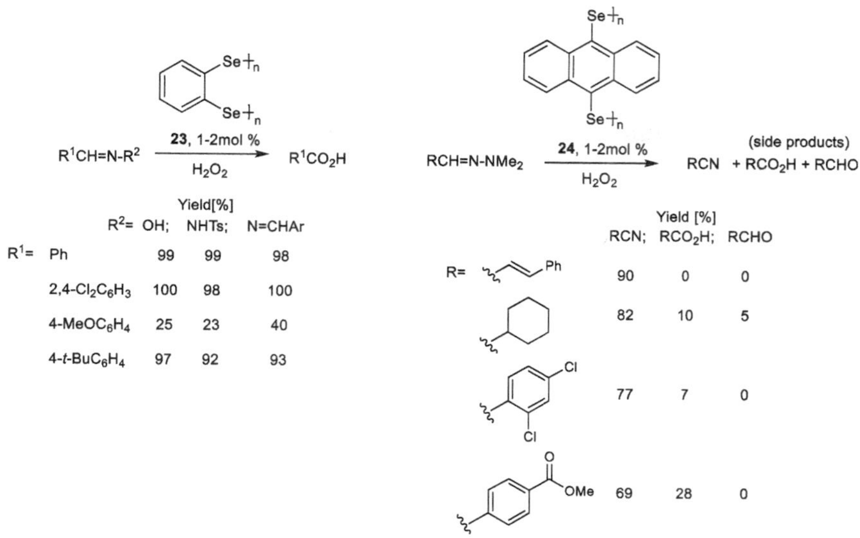

SCHEME 7.8 Oxidation of various azomethine derivatives using polymeric catalysts (23) and (24).

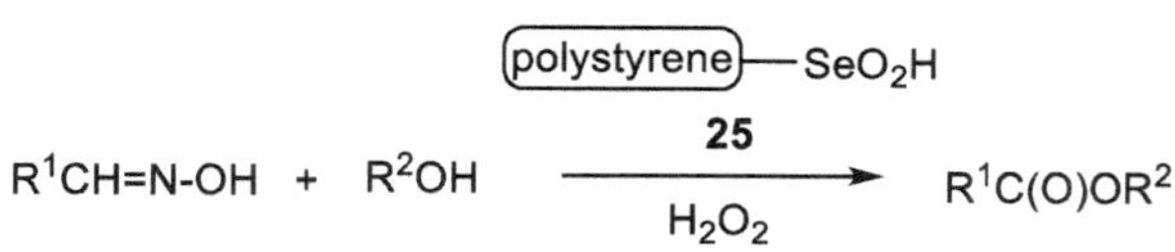

R^1	R^2	Yield
Ph	Et	81%
4-$NO_2C_6H_4$	Me	87%
-CH_2-Ph	Me	78%

SCHEME 7.9 Ester synthesis catalyzed by polystyrene-bounded seleninic acid (25).

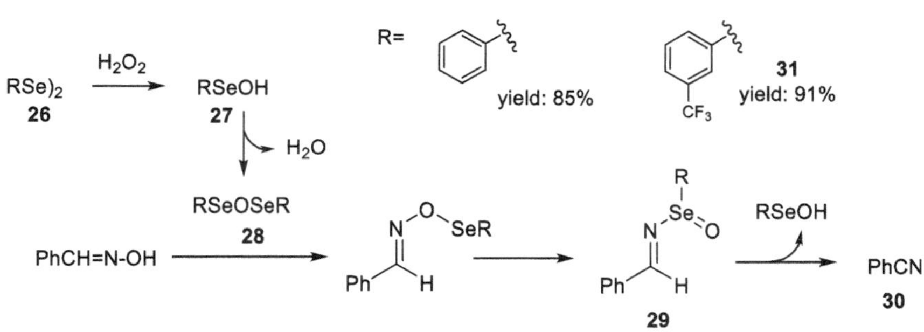

SCHEME 7.10 Dehydration of oximes to organonitriles (30).

anisonitrile (**33**) from anisaldoxime (**32**) in solvent-free conditions[88] (Scheme 7.11).

Recently, dibenzyl diselenide–hydrogen peroxide system was utilized in the deoximation of oximes to carbonyl derivatives. The transformation was initiated by the reaction of the benzylseleninic acid (**35**), formed in situ from the diselenide (**34**), with the oxime (**36**). The addition product (**37**) rearranged to the selenoxide (**38**) with the elimination of the final ketone (**39**). The catalyst was regenerated in the presence of air or excess of H_2O_2[89] (Scheme 7.12).

Xu and coworkers further modified this procedure, and by adding 1.25 mol% of iron(II) salt, they were able to efficiently perform the reaction using air as the primary oxidant instead of hydrogen peroxide (Method A, Scheme 7.13).[90] Recently, a poly-selenide was also applied as

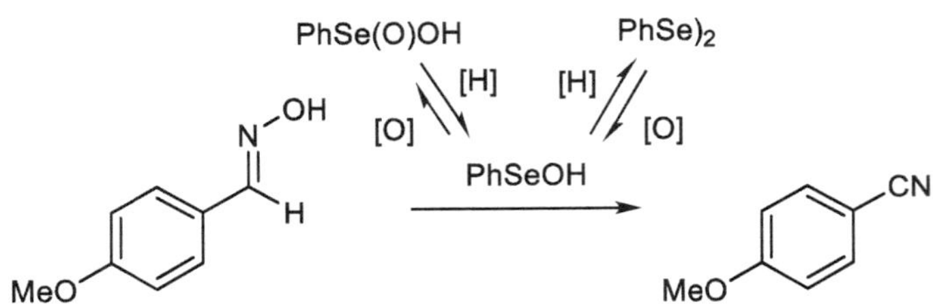

32 **33**

SCHEME 7.11 PhSeOH-catalyzed synthesis of anisonitrile (**33**) from anisaldoxime (**32**).

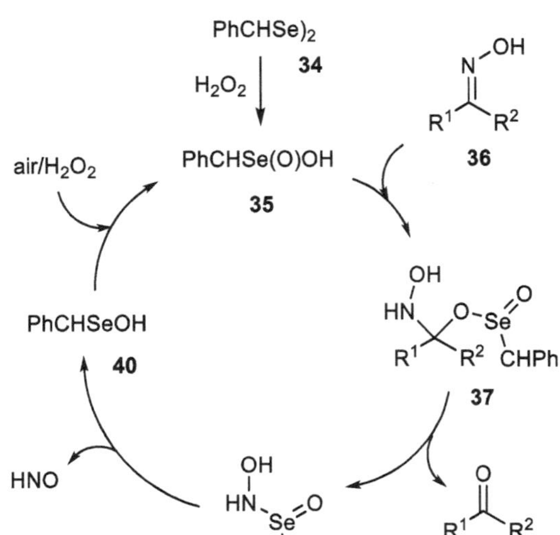

34 **35** **36** **37** **38** **39** **40**

SCHEME 7.12 Catalytic cycle of the deoximination of oximes with diselenide (**34**).

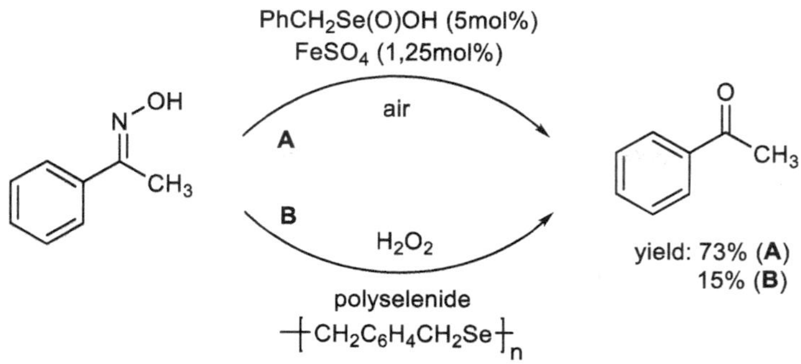

SCHEME 7.13 The deoximination of oximes in different conditions **A** and **B**.

catalyst in the H_2O_2-mediated deoximation of ketoximes. However, very low yield of the final carbonyl compound was obtained (Method B, Scheme 7.13).[80]

7.2.3 Oxidation of alcohols and phenols

Primary and secondary alcohols can be oxidatively converted into the carbonyl-containing compounds by using selenium reagents as first reported by Back and Shimizu using stoichiometric amounts of benzenseleninic anhydride[6,34,35] and diaryl diselenides,[8,91] respectively. In 1996, the first catalytic approach was published by Onami and coworkers, who reported the oxidation of alcohols by using dimethyl 2,2′-diselenodibenzoate as a catalyst and *N*-chloro-4-chlorobenzenesulfonamide sodium salt as stoichiometric oxidant.[92] This approach was later extended to a more powerful catalyst, bis-[2-(2-pyridyl)phenyl] diselenide.[93] In 2009, Arends et al. reported the conversion of benzyl alcohol to benzaldehyde by using diphenyl diselenide as precatalyst and *tert*-butyl hydroperoxide as oxidant.[94] Thus, *N*-methylmorpholine *N*-oxide (NMO), tBuOOH, NaOCl, and $NaIO_4$ were employed as terminal oxidants in partnership with "piano-stool" ruthenium complexes where selenium ligands showcased a superior activity when compared to the corresponding sulfur-containing ones.[95] Recently, secondary alcohols were converted into ketones by using molecular bromine and 10 mol% of a benzisoselenazolone as catalyst in dichloromethane at room temperature.[96]

Besides, phenols were converted to quinones using organoselenium compounds as reagents[11,97] or as catalysts as demonstrated by Giurg and Młochowski when they converted *o*-aminophenols to 2-amino-3H-phenoxazin-3-ones.[98] Da Silva and coworkers published the most recent example in 2018. They reported the phenylseleninic acid–mediated conversion of variously functionalized naphthols (**41**) into the corresponding 1,4 naphthoquinones (**42**) using 40% of catalyst loading, H_2O_2 as green oxidant and O_2 as cooxidant at room temperature using organic solvents (Scheme 7.14).

Interestingly, the catalyst loading was reduced to 2% in the presence of 0.25 mol% of a rhodium-based carbon nanotube, which accelerates the reaction because the metal allows the formation of a seminaphthoquinone radical, which, in the presence of oxygen, is rapidly converted into the desired 1,4-naphthoquinones.[99]

7.2.4 Bayer–Villiger oxidations

The Bayer–Villiger (BV) oxidation was first presented in 1899 by Adolf von Baeyer and Victor Villiger. It involved the conversion of ketones into the corresponding esters in the presence of a permosulfuric acid.[100] Many years later, the BV reaction is still a highly useful and commonly employed method to perform the synthesis of esters from reduced homologs. As reported in Scheme 7.15, the acidic environment protonates the carbonyl carbon (**43**) leading to, after resonance stabilization, the carbocation (**44**) attached by the oxidant anion (**45**). The successive intramolecular rearrangement (**46**) affords the target ester (**47**). Such a mechanism was proposed by Criegee in 1948, and it is still the most accepted.[101]

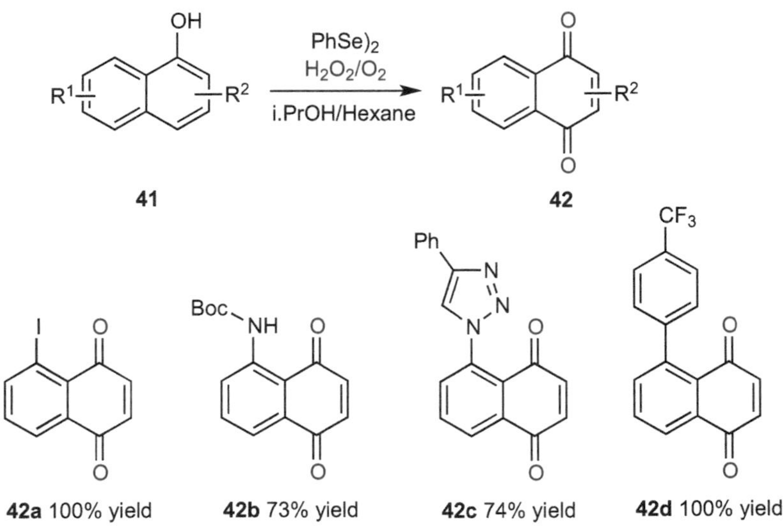

SCHEME 7.14 Synthesis of 1,4 naphthoquinones.

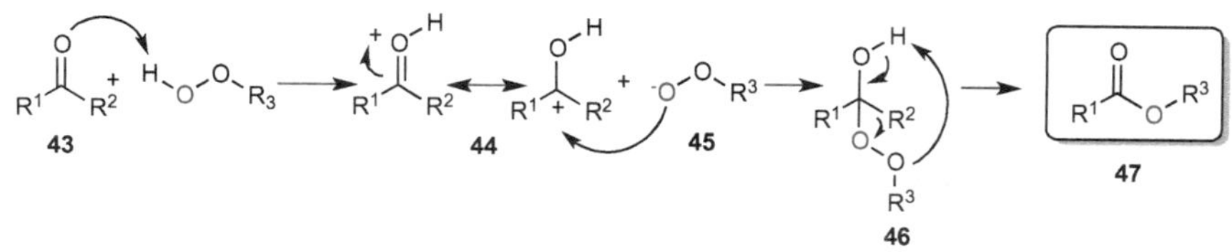

SCHEME 7.15 Mechanism of the Bayer–Villiger oxidation.

Being one of the most employed transformations, the BV reaction has been extensively studied, and several catalysts and oxidants have been developed, even recently.[102–105] In the frame of selenium chemistry, pioneering examples were reported by Syper, in 1987, who described the conversion of α,β-unsaturated aldehydes mainly into the corresponding formylesters by using several benzeneseleninic acids or their precursors.[106] More recently, Detty and Goodman reported the oxidation of trimetoxybenzaldehyde, cyclohexanone, and adamantone using catalytic amounts of compound (**2**) and hydrogen peroxide as green oxidant.[107] Other catalysts were also screened, but none reached the activity exhibited by compound (**2**). The aromatic aldehyde was converted into the formyl ester in 8 hours, while the conversion of aliphatic ketones into lactone took longer time (18 hours for adamantane, 24 hours for cyclohexanone) (Scheme 7.16). From a mechanistic point of view, the selenoxide initially reacts with H_2O_2 to give the trigonal bipyramidal dihydroxy peroxy selenane (**50**), which is the actual catalyst able to transfer oxygen to various substrates. After that, selenane (**50**) can reversibly lose water to regenerate the selenoxide and begin the catalytic cycle again. Alternatively, it can be protonated and dehydrated, leading to perhydroxyselenonium (**51**) still able to act as oxidant. Experimental evidences for the formation of the intermediate (**51**) have been reported.[108]

Diselenides, in particular dibenzyldiselenide (compound (**6**)), have also proved to be efficient catalysts in the BV oxidation of unsaturated ketones for the preparation of the corresponding vinyl esters. Compound (**6**), which was employed in 5 mol%, was recycled and reused for five consecutive cycles in the synthesis of phenyl vinyl acetate without appreciable loss of catalytic activity.[109] Analogously, cyclic ketones, such as alkylidene cyclobutanones, were converted into the corresponding lactones by using as catalyst, diphenyl diselenide.[110] As shown in Scheme 7.17, 5 mol% of diphenyl diselenide was employed as a catalyst for the conversion of isatins (**52**) into cyclic anhydrides (**53**) with hydrogen peroxide as oxidative partner. In this case also, the catalytic system was recovered and reused for seven consecutive cycles, making this waste-free method one of the most suited for large-scale preparation of isatoic anhydrides (**53**).[111] Compounds (**53**) are useful

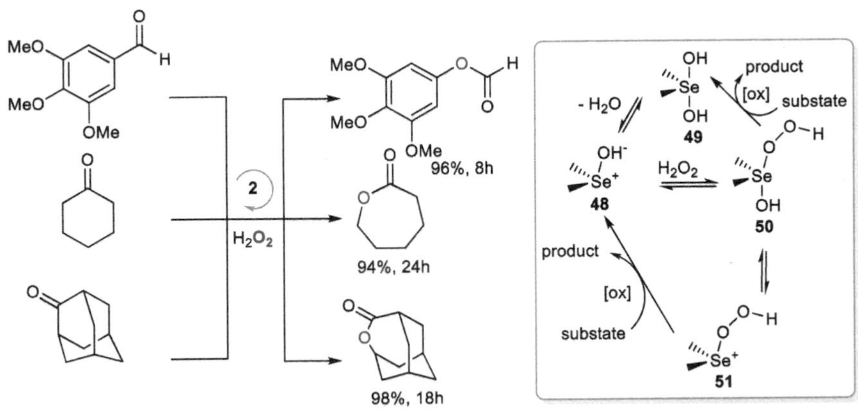

SCHEME 7.16 Bayer–Villiger oxidation catalyzed by compound (**2**).

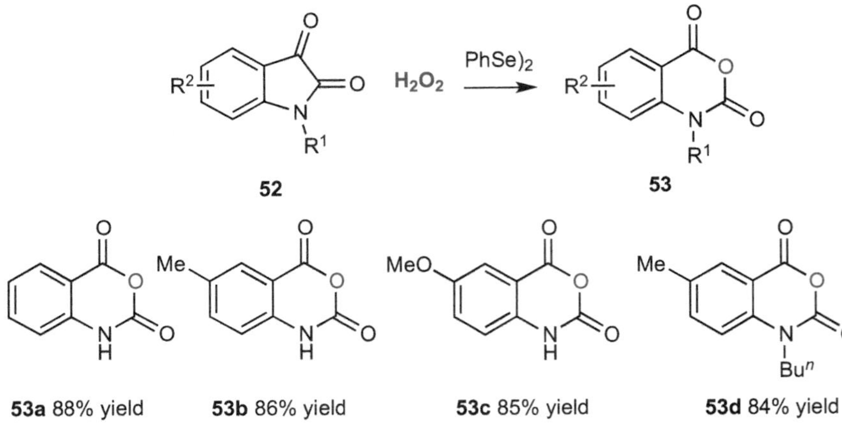

SCHEME 7.17 Oxidation of isatins (**52**) catalyzed by PhSe)$_2$.

building blocks for the synthesis of several biologically relevant compounds, such as pyrrolo-quinazolinoquinoline alkaloids,[112] antivirals[113] or anti-*Trypanosoma cruzi* agents.[114] Thus, methods devoted to their facile preparation are highly demanded.

7.2.5 Oxidation of aldehydes

The synthesis of carboxylic acids and their derivatives is of pivotal importance for the industrial preparation of a number of organic molecules, including drugs and fine chemicals. They are usually obtained by oxidation of lower oxidation-state homologs such as alcohols, aldehydes, or ketones. Aldehydes have been chosen as starting points in reactions where heavy metals were employed as catalysts and/or hazardous reagents such as potassium permanganate, chromates, and chlorite as over-stoichiometric oxidants. Examples of organoselenium-catalyzed approaches have been reported in literature, whereas the pioneering methods were reported by Choi in 1988[115] and Mlochowski in 2000,[116] when they proved that selenium-containing compounds, such as selenium dioxide, were able to convert aldehydes to carboxylic acids with hydrogen peroxide as over-stoichiometric oxidant in refluxing tetrahydrofuran. Later, Sheldon reported the development of fluorine-containing diselenides as precatalyst in aldehyde oxidation with H_2O_2 in fluorous solvents.[117] More recently, some of us, taking inspiration from Mlochowski's approach,[116] discovered that the replacement of SeO_2 with $PhSeO_2H$ allowed the oxidation to carboxylic acids (**54**) to be carried out in *on water* condition, at room temperature and with stoichiometric amounts of H_2O_2. As depicted in Scheme 7.18, by carrying out the reaction in the presence of alcohols and increasing the temperature, instead of the carboxylic acid, several esters (**55**) were prepared.[118]

Both aromatic and aliphatic aldehydes were efficiently converted into the corresponding carboxylic acids in just 6 hours with yields ranging from good to excellent. Aromatic aldehydes bearing functional groups with different electronic properties did not influence the performance of the reaction; the same was with sterically demanding substrates. The multigram synthesis of benzoic acid with pharmaceutical purity was also demonstrated by recycling and reusing the catalytic system several times. Of worth noting, the oxidation of *trans*-cinnamaldehyde (**56**) gave the target cinnamic acid (**57**) in good yield but, unexpectedly, as a *Z*/*E* isomer mixture in which the less stable *Z*-isomer was the major product.

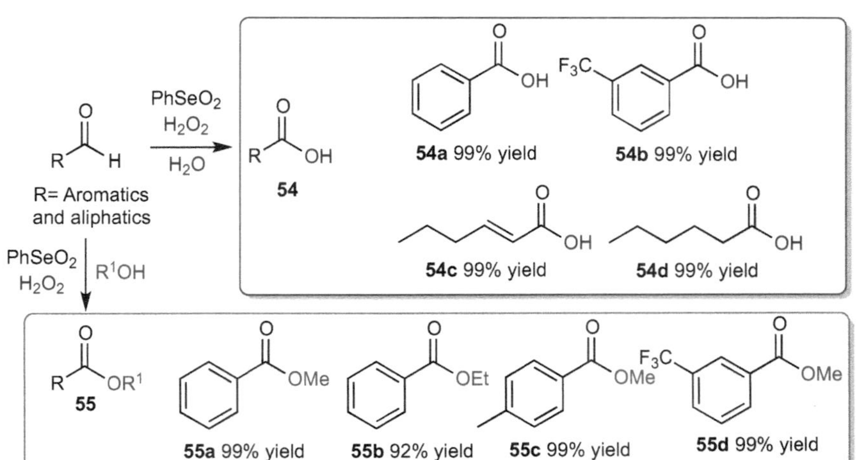

SCHEME 7.18 Oxidation of aldehydes to carboxylic acids and esters.

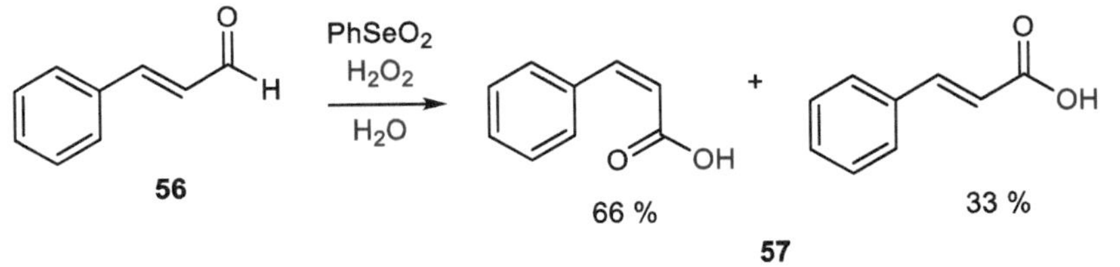

SCHEME 7.19 Oxidation of trans-cinnamaldehyde.

This is a remarkable example of divergent synthesis that still lacks a mechanistic explanation[118] (Scheme 7.19).

Examples of organoselenium-mediated carboxylic acids synthesis through heterogeneous approaches are reported in literature starting from 2002, when Huang et al. reported the development of polystyrene-bound phenylseleninic acid capable of converting several aldehydes into the corresponding oxidized products.[119] Successively, Nemati et al. built Fe_3O_4 magnetic nanoparticles covered with aminopropylsilica in turn decorated with an organodiselenide compound. Once in contact with hydrogen peroxide, the diselenide bridge is thought to be converted into a perseleninc acid moiety capable of turning aldehydes into their oxidized counterparts. The brilliance of this system is that, after the reaction, the magnetic nature of Fe_3O_4 allows the recovery of the catalyst by using an external magnet, and the reusability was tested for four consecutive cycles.[120] In 2019, Pich and colleagues developed Poly(*N*-vinylcaprolactam) microgels functionalized with seleninic acid moieties capable of converting acrolein into acrylic acid using H_2O_2 as oxidant in a water/dioxane mixture. In addition, running the reaction in presence of methanol also, the methyl ester of acrylic acid was prepared in high yield.[121]

7.2.6 Oxygen-transfer mechanism of organoselenium compounds

Given the widespread utilization of organoselenium compounds as catalysts in oxidative transformations, several mechanisms have been proposed to explain the actual oxygen-transfer reaction. A clear picture is drawn for two classes of compounds: selenides and diselenides when they activate hydrogen peroxide, with comparable activation energies. Mechanistic details inferred by studying the selenides and diselenides classes of compounds can be translated to explain the catalytic activity of other selenium-containing functional groups.

The catalytic activity of selenides is shown in Scheme 7.20. From a general point of view, selenide lacking aliphatic β-protons (**58**), is converted into the selenoxide (**59**) and then to hydroxyperoxy selenane

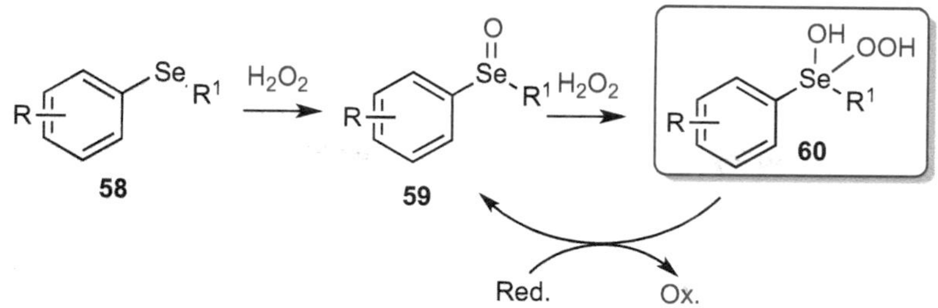

SCHEME 7.20 Conversion of selenides into the catalytically active species.

(**60**), which is the active catalytic species able to convert a reduced substrate into the oxidized counterpart. This mechanism was proven first by Braga in 2012,[122] and then, using combined mass spectrometry, nuclear magnetic resonance (NMR), and computational methods by Orian in 2017.[123] The hydroxy perhydroxy selenium derivatives were demonstrated to be worse oxidants than the sole peroxide toward selenides, still able to be catalytically active, though. It means that in the activation of selenides, an autocatalytic process should be ruled out.[123]

On the contrary, the activation of diselenides follows an autocatalytic, two-step process in which they are initially converted into seleninic acid and then, most probably, into perseleninic acid, which was believed as the actual catalytic species, before 2019, when Back et al. through high-resolution mass spectrometry, ^{77}Se NMR, and reactivity tests proved that the treatment of phenylseleninic acids with 0.5 equivalents of hydrogen peroxide and successive crystallization affords the previously unknown selenonium–selenonate salt (**61**), which upon reacting again with the oxidant gives the peroxiselinonic acid (**62**), which is hereafter thought to be the actual catalytic species when dealing with this kind of selenium-based catalysts, at least in epoxidation of C–C double bonds (Scheme 7.21).[124] This is the first example in literature where a direct engagement of a Se(VI) species in the catalysis of selenium compounds is proved.

7.2.7 Amide bond formation

Organoselenium compounds have been investigated in reactions and synthetic procedures where the oxygen transfer step is not central. An example worth mentioning is the amide synthesis catalyzed by benzisolselenazolones and diselenides developed by Liebeskind and coworkers.[125] In this procedure, an amidic bond is formed through the reaction of a carboxylic acid (**63**) and an amine in the presence of triethyl phosphite (**67**), atmospheric oxygen and 2.5 mol% of organoselenium compounds. As shown in Scheme 7.22, the *ortho*

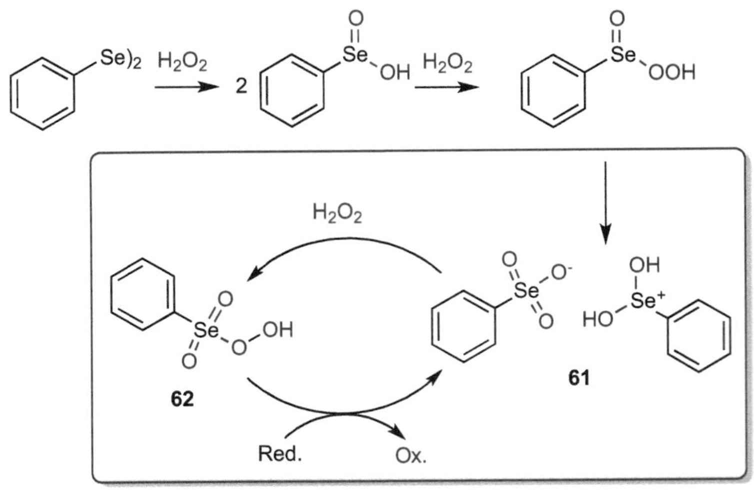

SCHEME 7.21 Activation of diphenyl diselenide.

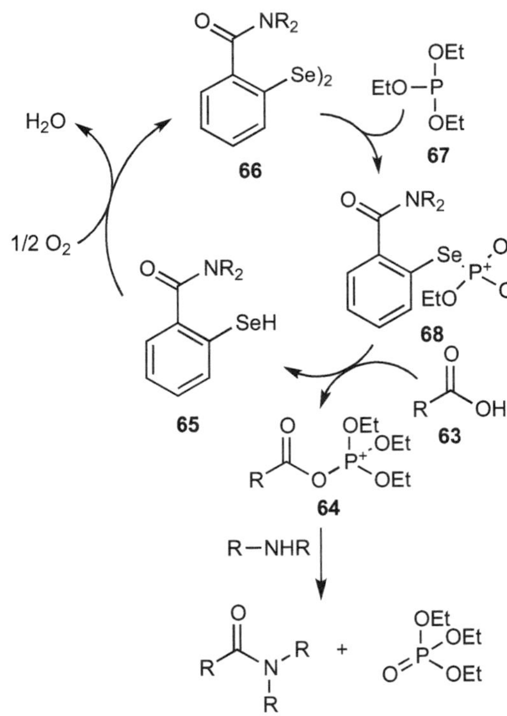

SCHEME 7.22 Selenium-catalyzed amide bond formation.

amido–substituted diselenide (**66**) reacts with triethyl phosphite (**67**) leading to the electrophilic phosphonium (**68**) rapidly attacked by the carboxylic acid (**63**) leading to the activated specie susceptible of nucleophilic attach by the amine, thus giving the target amide (**69**) and a phosphate ester (**70**) as a byproduct. The diselenide bridge is formed because of the selenol's (**65**) ability to reduce the atmospheric oxygen closing the catalytic cycle. It is important to note that the same reaction with sulfur-containing compounds requires copper salts for the catalyst to be oxidatively restored.[126]

7.3 Oxygen transfer by organoselenium glutathione peroxidase mimetics

The physiological production of reactive oxygen species (ROS) within the cellular metabolism for energy production is finely balanced in order to not overload the cytoplasm with these highly reactive molecules. Indeed, high ROS levels are responsible for DNA damage and protein denaturation. Besides, abnormal ROS concentration led to many human pathologies such as inflammation, heart, and neurodegenerative diseases.[127] In this context, glutathione peroxidases (GPx) are among the most important enzymatic systems implemented by the cell to control ROS levels and to fight oxidative damage. GPx is a family made of eight isoenzymes capable of dismantling ROS catalyzing the conversion of glutathione (GSH) to its oxidized counterpart (GSSG). Five out of eight enzymes contain in their active site a residue of Sec, which represents the hot spot, the catalytic core of the enzyme. Sec, kept in its ionized state by a hydrogen bond network with neighboring glutamine and tryptophan residues,[128] turns harmful peroxides into harmless molecules forming a selenenic acid intermediate (B), which reacts with GSH. The formed selenylsulfide (C) is then attached by another molecule of GSH restoring the selenol and giving GSSG.[51,129] In the presence of a low concentration of GSH or high concentration of oxidative species, Sec may be oxidized to the corresponding seleninic acid (D), that, however, is converted back to the selenylsulfide intermediate (C) after the reaction with further equivalents of GSH. The pool of cellular GSH is restored through the action of glutathione reductase (GR) enzyme and nicotinamide adenine dinucleotide phosphate (NADPH) (Fig. 7.1).

Considering the importance of the GPx in keeping redox homeostasis and fighting ROS damage in living systems, the scientific research addressed the need of developing small selenium-containing molecules with the ability to mimic the catalytic cycle of GPx in order to be exploited as antioxidant. However, it is important to stress that, after more than 30 years of research, the concepts of GPx mimics and antioxidant do not fully overlap and none

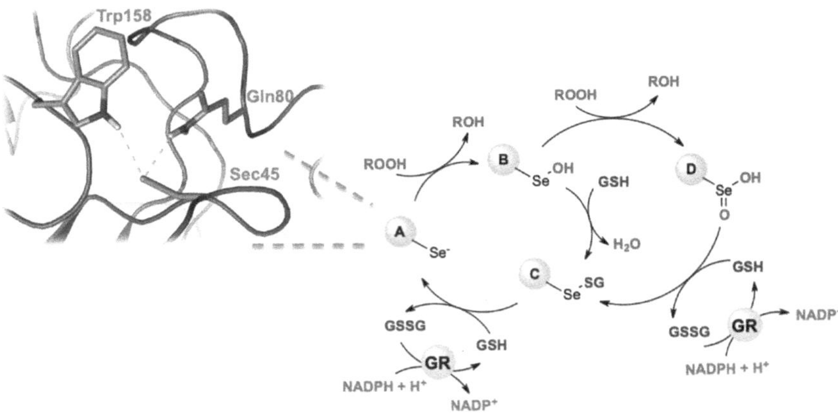

FIGURE 7.1 Active site and catalytic cycle of GPx.

of the GPx-mimic compounds have displayed the potential to be clinically utilized in oxidative stress–related diseases.[51,52]

Several methods for the evaluation of such an enzymatic-like activity have been developed over the years, and all of them have been broadly applied. The catalytic activity is evaluated according to the rate improvement the selenium-containing compound is able to exert in the reaction reported in Scheme 7.23; the faster the reaction, the better is the catalyst.

Direct and indirect methods are known and selenium compounds are employed in catalytic amounts ranging from 10 to 20 mol%. Among the indirect methods, the GR-coupled assay was among the first to be developed. It measures the consumption of NADPH by means of UV spectroscopy[130] (Fig. 7.2). This method was first adopted by Sies in the evaluation of the GPx-mimic activity of Ebselen,[131] and unfortunately, some pitfalls emerged as the selenylsulfide formed upon the reaction with the first molecule of GSH is substrate of GR, giving misleading results.[132] In the direct methods, the consumption of one of the reaction partners is monitored through different approaches. As example, 3-carboxy-4-nitrobenzenethiol was adopted as thiol probe, and the kinetic of the reaction was followed spectroscopically as a decrease of the absorbance at 410 nm.[133] The conversion of thiophenol, used as a probe, into diphenyl disulfide was followed as the increase of $(PhS)_2$ absorbance by Iwaoka and Tomoda.[134] The reduction of H_2O_2, measured through the Hildebrandt and Roots method, was taken as readout while observing the GPx-mimic activity of ebselen using GSH and lipoic acid as thiols probes.[135] NMR-based, direct methods are among the most employed as they are rapid, reliable, and easy to perform with a variety of deuterated solvents and thiol probes.[136] As a result, several papers have been published starting from the seminal work of

$$2RSH + ROOH \xrightarrow{Se} RSSR + 2\,ROH$$

SCHEME 7.23 Oxidation of thiols through a GPx-mimic approach.

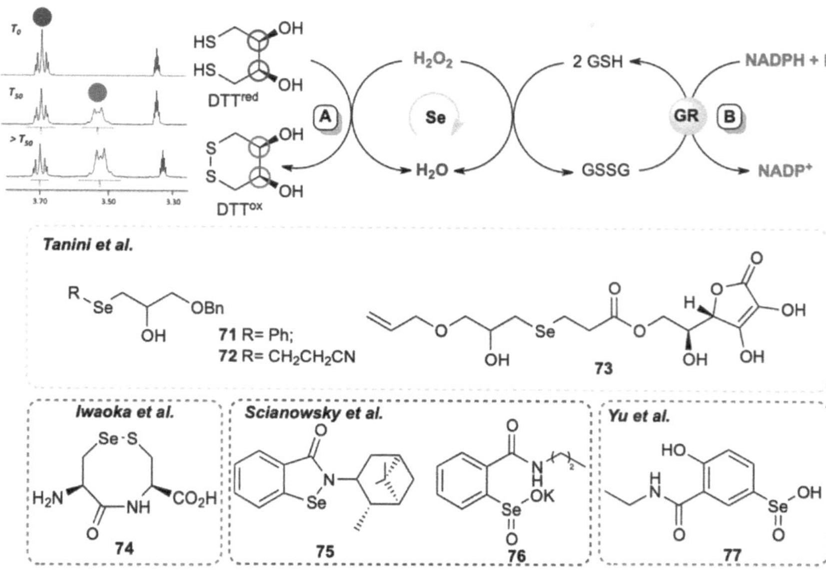

FIGURE 7.2 GPx activity evaluation methods and mimic compounds.

Engman;[132] then, Iwaoka developed a practical protocol based on the conversion of dithiothreitol (DTT^{red}) into the corresponding cyclic derivative (DTT^{ox}) in deuterated methanol (Fig. 7.2). The oxidation of the DTT^{red} can be clearly monitored by observing the disappearance of the characteristic ^1H NMR peak at 3.68 ppm; as the reaction progresses, the intensity of this peak decreases, while the signals of the oxidized thiol (DTT^{ox}) increase (exemplary at 3.52 ppm Fig. 7.2).[137] Later on, Santi and coworkers reported that even deuterium oxide is suitable, but, for certain catalysts, the kinetic is too fast to be measured correctly.[138] GSH in D_2O and DMSO was more recently adopted by Back in a series of NMR experiments meant to determine the GPx activity of water-soluble and polyethylene glycol-supported selenides.[139] High-performance liquid chromatography–based direct methods were also proposed using benzylmercaptan or thiophenol as thiol probes.[140,141]

Unfortunately, these methods have some variants and reinterpretations that created a really heterogeneous collection of data, tricky to understand and sometimes awkward to read. Here, few examples are presented, chosen among the most recent example, in consideration that review articles and book chapters precisely devoted to this topic were published quite recently.[51,52,129,142–145]

A series of really interesting compounds was developed by Tanini et al. incorporating the β-hydroxy ethylselenide scaffold into different molecules in order to obtain potent GPx-mimetic compounds. Derivative (**71**), as example, tested through the NMR method based on the conversion of DTT^{red} into DTT^{ox} (method A of Fig. 7.2) displayed a T_{50} of about 6 minutes,[146] while compound (**72**) was tested, besides the NMR-based method displaying a T_{50} of 6.5 minutes, also by the GR-based approach with a T_{50} as low as 18 seconds.[147] The same authors very recently reported the functionalization of vitamin C with selenium-containing structures obtaining amphiphilic antioxidants. Compound (**73**), synthesized through an enzymatic approach to overcome the instability of vitamin C, was tested through both methods displaying lower GPx-mimicking property but remarkable radical chain breaking activity, demonstrated by the DPPH test.[148] Cyclic selenium-sulfide peptides, such as compound (**74**),[149,150] were recently assayed by Iwaoka, while alpha-methyl selenocysteine containing peptides were developed by Hondal, who demonstrated that such peptides are more stable than the corresponding selenocysteine-containing analogues.[151] Scianowski's group contributed significantly to this research field,[152–156] and recently they developed a series of terpene-containing ebselen analogues such as compound (**75**),[157] and the seleninic acids potassium salts as compound (**76**).[158] These molecules, besides being endowed with a striking GPx-like activity measured through method A in both methanol and deuterium oxide, are noticeably unusual. Indeed, in the extant literature, there is just another example (compound (**77**)) reported by Yu et al.[159] In conclusion, worth to be mentioned are the selenium-decorated metal–organic framework UIO-66 reported by Huang, Huo and Xu[160] and a selenopeptide placed on gold nanoparticles in order to create a nanozyme,[161] both endowed with GPx-mimicking properties exploitable in heterogeneous processes.

7.4 Conclusions

Starting from the early works of Riley and the widespread utilization of selenium dioxide, through the well-known Sharpless oxidation protocols to the latest catalytic procedures, Se reagents have marked their place as highly efficient oxidants in modern organic synthesis. Catalytic systems are built up from the organoselenium compound, and the primary oxidant had been widely used in various oxygen-transfer reactions including: oxidation of alkenes to epoxides and diols; transformation of azomethine compounds to carboxylic acids, esters or nitriles; synthesis of carbonyls from alkenes, alcohols, phenols, and aldehydes; and the Bayer–Villiger oxidations. The unique reactivity of Se reagents toward oxygen species is not only utilized in organic synthesis but also in medicinal chemistry. The presence of the selenium atom, in the form of the amino acid Sec, is essential for several enzymatic systems to express their catalytic activity. Considering the possibility of application as biologically potent compounds that could efficiently mimic natural selenoenzymes, along with the synthetical potential and unresolved mechanistic problems, it needs to be highlighted that this field of knowledge creates many application opportunities with numerous issues that still require further evaluation.

References

1. Singh, F. V.; Wirth, T. Selenium Reagents as Catalysts. *Catal. Sci. Technol.* **2019,** *9,* 1073–1091.
2. Back, T. G. Oxidations Catalyzed By Seleninic Acids and Anhydrides, their Precursors and Congeners. *Curr. Green Chem.* **2016,** *3,* 76–91.
3. Młochowski, J.; Wójtowicz-Młochowska, H. Developments in Synthetic Application of Selenium(IV) Oxide and Organoselenium Compounds as Oxygen Donors and Oxygen-Transfer Agents. *Molecules* **2015,** *20,* 10205–10243.
4. Krasowska, D.; Sancineto, L.; Deska, M.; Drabowicz, J. Optically Active Selenoxides: Structural and Synthetic Aspects. *Symmetry (Basel)* **2020,** *12,* 349.
5. Riley, H. L.; Morley, J. F.; Friend, N. A. C. 255. Selenium Dioxide, a New Oxidising Agent. Part I. Its Reaction with Aldehydes and Ketones. *J. Chem. Soc.* **1932,** 1875.
6. Barton, D. H. R.; Brewster, A. G.; Hui, R. A. H. F.; Lester, D. J.; Ley, S. V.; Back, T. G. Oxidation of Alcohols Using Benzeneseleninic Anhydride. *J. Chem. Soc. Chem. Commun.* **1978,** 952.
7. Toki, N.; Satoh, T. Selective Oxidation of Alcohols at the Benzylic Position by Benzeneseleninic Anhydride. *ChemInform* **2005,** 36.
8. Kuwajima, I.; Shimizu, M.; Urabe, H. Oxidation of Alcohols with Tert-Butyl Hydroperoxide and Diaryl Diselenide. *J. Org. Chem.* **1982,** *47,* 837–842.
9. Barton, D. H. R.; Brewster, A. G.; Ley, S. V.; Read, C. M.; Rosenfeld, M. N. Oxidation of Phenols, Pyrocatechols, and Hydroquinones to Ortho-Quinones using Benzeneseleninic Anhydride. *J. Chem. Soc. Perkin Trans. 1* **1981,** 1473.
10. Barton, D. H. R.; Finet, J.-P.; Thomas, M. Comparative Oxidation of Phenols with Benzeneseleninic Anhydride and with Benzeneseleninic Acid. *Tetrahedron* **1988,** *44,* 6397–6406.
11. Henriksen, L. Competing Ene-Reactions in the P-OXIDATION and O-Phenylselenylation of Phenol with Benzeneseleninic Acid. *Tetrahedron Lett.* **1994,** *35,* 7057–7060.
12. Barton, D. H. R.; Lester, D. J.; Ley, S. V. Oxidation of Ketone and Aldehyde Hydrazones, Oximes, and Semicarbazones and of Hydroxylamines and Hydrazo-Compounds, Using Benzeneseleninic Anhydride. *J. Chem. Soc. Perkin Trans. 1* **1980,** 1212.
13. Back, T. G.; Collins, S.; Kerr, R. G. Oxidation of Hydrazines with Benzeneseleninic Acid and Anhydride. *J. Org. Chem.* **1981,** *46,* 1564–1570.
14. Back, T. G. Oxidation of 1,1-Disubstituted Hydrazines to Tetrazenes with Benzeneseleninic Acid. *J. Chem. Soc. Chem. Commun.* **1981,** 530.

15. Back, T. G.; Kerr, R. G. ChemInform Abstract: Oxidation of 1,1-Disubstituted Hydrazines with Benzeneseleninic Acid and Selenium Dioxide. Facile Preparation of Tetrazenes. *Chem. Informationsd.* **1983,** 14.
16. Barton, D. H. R.; Lusinchi, X.; Ramírez, J. S. Improved Synthesis of 1,2,4-Triazoline-3,5-Dione Derivatives of Ergosterol and a New Method for their Reconversion to Ergosterol. *Tetrahedron Lett.* **1983,** *24,* 2995–2998.
17. Back, T. G.; Collins, S. A Convenient Synthesis of Selenolsulfonates from the Oxidation of Sulfonhydrazides with Benzeneseleninic Acid. *Tetrahedron Lett.* **1980,** *21,* 2213–2214.
18. Back, T. G.; Collins, S.; Krishna, M. V. Reactions of Sulfonhydrazides with Benzeneseleninic Acid, Selenium Halides, and Sulfur Halides. A Convenient Preparation of Selenosulfonates and Thiosulfonates. *Can. J. Chem.* **1987,** *65,* 38–42.
19. Czarny, M. R. Oxidation of Amines with Diphenylseleninic Anhydride. *J. Chem. Soc. Chem. Commun.* **1976,** 81.
20. Czarny, M. R. Oxidation of Amines with Benzeneseleninyl Chloride. *Synth. Commun.* **1976,** *6,* 285–293.
21. Barton, D. H. R.; Billion, A.; Boivin, J. Oxidation of Secondary Amines to α-Cyanoamines. *Tetrahedron Lett.* **1985,** *26,* 1229–1232.
22. Barton, D. H. R.; Lusinchi, X.; Milliet, P. Studies on the Reaction of Primary and Secondary Amines with Phenylseleninic Anhydride and with Phenylseleninic Acid. *Tetrahedron* **1985,** *41,* 4727–4738.
23. Barton, D. H.; Lusinchi, X.; Milliet, P. La Transformation D'indolines en Indoles et D'autres Reactions Apparentees. *Tetrahedron Lett.* **1982,** *23,* 4949–4952.
24. Ninomiya, I.; Hashimoto, C.; Kiguchi, T.; Naito, T.; Barton, D. H. R.; Lusinchi, X.; Milliet, P. Dehydrogenation with Benzeneseleninic Anhydride in the Total Synthesis of Ergot Alkaloids. *J. Chem. Soc. Perkin Trans. 1* **1990,** 707.
25. Faehl, L. G.; Kice, J. L. Oxidation of Sulfides and Phosphines by Aromatic Selenonic and Seleninic Acids. *J. Org. Chem.* **1979,** *44,* 2357–2361.
26. Barton, D. H. R.; Cussans, N. J.; Ley, S. V. Removal of 1,3-Dithiolan Protecting Groups by Benzeneseleninic Anhydride. *J. Chem. Soc. Chem. Commun.* **1977,** 751.
27. Cussans, N. J.; Ley, S. V.; Barton, D. H. R. Removal of Thioacetal Protecting Groups by Benzeneseleninic Anhydride. *J. Chem. Soc. Perkin Trans. 1* **1980,** 1654.
28. Burton, A.; Hevesi, L.; Dumont, W.; Cravador, A.; Krief, A. Cleavage (Deprotection) of Selenoacetals to Carbonyl Compounds; A Comparative Study. *Synthesis (Stuttg.)* **1979,** *1979,* 877–880.
29. Barton, D. H. R.; Okano, T.; Parekh, S. I. On the Reactions of Benzeneseleninic Anhydride with Monosubstituted Hydrazones. *Evid. Radic. Pathways Tetrahedron* **1991,** *47,* 1823–1836.
30. Barton, D. H. R.; Choi, S.-Y.; Liu, W.; Smith, J. A. Oxidation of Phenylhydrazones with Benzeneseleninic Anhydride: A New Mechanistically Interesting Observation. *Mol. Online* **1998,** *2,* 22–28.
31. Barton, D. H. R.; Cussans, N. J.; Ley, S. V. Conversion of Thiocarbonyl Compounds into Their Corresponding Oxo Derivatives Using Benzeneseleninic Anhydride. *J. Chem. Soc. Chem. Commun.* **1978,** 393.
32. Cussans, N. J.; Ley, S. V.; Barton, D. H. R. Conversion of Thiocarbonyl Compounds into their Corresponding Oxoderivatives Using Benzeneseleninic Anhydride. *J. Chem. Soc. Perkin Trans. 1* **1980,** 1650.
33. Barton, D. H. R.; Lester, D. J.; Ley, S. V. Dehydrogenation of Steroidal and Triterpenoid Ketones Using Benzeneseleninic Anhydride. *J. Chem. Soc. Perkin Trans. 1* **1980,** 2209.
34. Back, T. G. Dehydrogenation of Hydrazines and of 4-Azacholestan-3-One with Benzeneseleninic Acid and Benzeneseleninic Anhydride. *J. Chem. Soc. Chem. Commun.* **1978,** 278.

35. Back, T. G. Oxidation of Azasteroid Lactams and Alcohols with Benzeneseleninic Anhydride. *J. Org. Chem.* **1981,** *46,* 1442–1446.
36. Barton, D. H. R.; Hui, R. A. H. F.; Ley, S. V.; Williams, D. J. Dehydrogenation of Lactones using Benzeneseleninic Anhydride. X-Ray Crystal Structure of 3β-Acetoxy-14α-Hydroxy-17a-oxa-D-Homo-5α-Androst-15-en-17-One. *J. Chem. Soc. Perkin Trans. 1* **1982,** 1919–1922.
37. Yamakawa, K.; Satoh, T.; Takita, S.; Iida, T.; Iwasaki, M. Studies on the Terpenoids and Related Alicyclic Compounds. XXX. An Application of the Angular Hydroxylation Using Benzeneseleninic Anhydride to the Syntheses of 10.BETA.-Hydroxyfuranoeremophilane Derivatives. *Chem. Pharm. Bull. (Tokyo).* **1983,** *31,* 3544–3552.
38. Barton, D. H. R.; Hui, R. A. H. F.; Lester, D. J.; Ley, S. V. Preparation of Aldehydes and Ketones by Oxidation of Benzylic Hydrocarbons with Benzeneseleninic Anhydride. *Tetrahedron Lett.* **1979,** *20,* 3331–3334.
39. Barton, D. H. R.; Hui, R. A. H. F.; Ley, S. V. Oxidation of Benzylic Hydrocarbons with Benzeneseleninic Anhydride and Related Reactions. *J. Chem. Soc. Perkin Trans. 1* **1982,** 2179.
40. Clayton, M. D.; Marcinow, Z.; Rabideau, P. W. Benzeneseleninic Anhydride Oxidation of 1,2-Diarylethanes and 1,2-Diarylethylenes to 1,2-Diaryldiketones. *Tetrahedron Lett.* **1998,** *39,* 9127–9130.
41. Grieco, P. A.; Yokoyama, Y.; Gilman, S.; Nishizawa, M. Organoselenium Chemistry. Epoxidation of Olefins with Benzeneseleninic acid and Hydrogen Peroxide ("Benzeneperoxyseleninic Acid"). *J. Org. Chem.* **1977,** *42,* 2034–2036.
42. Hori, T.; Sharpless, K. B. Synthetic Applications of Arylselenenic and Arylseleninic Acids. Conversion of Olefins to Allylic Alcohols and Epoxides. *J. Org. Chem.* **1978,** *43,* 1689–1697.
43. Kametani, T.; Nemoto, H.; Fukumoto, K. A New Method for an Epoxidation of Olefins and Its Application to a Biomimetic Type Synthesis of Monoterpenes, Linalyloxides. *Heterocycles* **1977,** *6,* 1365.
44. Kametani, T.; Nemoto, H.; Fukumoto, K. A New Method for Selective Epoxidation and a Biogenetic-Type Synthesis of Linalyloxides. *Bioorg. Chem.* **1978,** *7,* 215–220.
45. Grieco, P. A.; Yokoyama, Y.; Gilman, S.; Ohfune, Y. Conversion of Ketones into Lactones with Benzeneseleninic acid and Hydrogen Peroxide (Benzeneperoxyseleninic acid): A New Reagent for the Baeyer-Villiger Reaction. *J. Chem. Soc. Chem. Commun.* **1977,** 870.
46. Syper, L. The Baeyer-Villiger Oxidation of Aromatic Aldehydes and Ketones with Hydrogen Peroxide Catalyzed by Selenium Compounds. A Convenient Method for the Preparation of Phenols. *Synthesis (Stuttg.)* **1989,** *1989,* 167–172.
47. Syper, L.; Młochowski, J. Benzeneperoxyseleninic Acids-Synthesis and Properties. *Tetrahedron* **1987,** *43,* 207–213.
48. Sancineto, L.; Palomba, M.; Bagnoli, L.; Marini, F.; Santi, C. Advances in Electrophilic Organochalcogen Reagents. *Curr. Org. Chem.* **2015,** *20,* 122–135.
49. Perin, G.; Nobre, P.; Mailahn, D.; Silva, M.; Barcellos, T.; Jacob, R.; Lenardão, E.; Santi, C.; Roehrs, J. Synthesis of 4-Organoselanyl-1H-pyrazoles: Oxone®-Mediated Electrophilic Cyclization of α,β-Alkynyl Hydrazones by Using Diorganyl Diselenides. *Synthesis (Stuttg.)* **2019,** *51,* 2293–2304.
50. Flohe, L.; Günzler, W. A.; Schock, H. H. Glutathione Peroxidase: A Selenoenzyme. *FEBS Lett.* **1973,** *32,* 132–134.
51. Santi, C.; Tidei, C.; Scalera, C.; Piroddi, M.; Galli, F. Selenium Containing Compounds from Poison to Drug Candidates: A Review on the GPx-like Activity. *Curr. Chem. Biol.* **2013,** *7,* 25–36.
52. Pacuła, A. J.; Mangiavacchi, F.; Sancineto, L.; J. Lenardão, E.; Ścianowski, J.; Santi, C. An Update on "Selenium Containing Compounds from Poison to Drug Candidates: A Review on the GPx-like Activity. *Curr. Chem. Biol.* **2016,** *9,* 97–112.

53. Parnham, M. J.; Sies, H. The Early Research and Development of Ebselen. *Biochem. Pharmacol.* **2013,** *86*, 1248–1253.
54. Nogueira, C. W.; Rocha, J. B. T. Toxicology and Pharmacology of Selenium: Emphasis on Synthetic Organoselenium Compounds. *Arch. Toxicol.* **2011,** *85*, 1313–1359.
55. Krasowska, D.; Iraci, N.; Santi, C.; Drabowicz, J.; Cieslak, M.; Kaźmierczak-Barańska, J.; Palomba, M.; Królewska-Golińska, K.; Magiera, J.; Sancineto, L. Diselenides and Benzisoselenazolones as Antiproliferative Agents and Glutathione-S-Transferase Inhibitors. *Molecules* **2019,** *24*, 2914.
56. Nascimento, V.; Ferreira, N. L.; Canto, R. F. S.; Schott, K. L.; Waczuk, E. P.; Sancineto, L.; Santi, C.; Rocha, J. B. T.; Braga, A. L. Synthesis and Biological Evaluation of New Nitrogen-Containing Diselenides. *Eur. J. Med. Chem.* **2014,** *87*, 131–139.
57. Sancineto, L.; Mariotti, A.; Bagnoli, L.; Marini, F.; Desantis, J.; Iraci, N.; Santi, C.; Pannecouque, C.; Tabarrini, O. Design and Synthesis of DiselenoBisBenzamides (DISeBAs) as Nucleocapsid Protein 7 (NCp7) Inhibitors with Anti-HIV Activity. *J. Med. Chem.* **2015,** *58*, 9601–9614.
58. Iraci, N.; Tabarrini, O.; Santi, C.; Sancineto, L. NCp7: Targeting a Multitask Protein for Next-Generation Anti-HIV Drug Development Part 2. Noncovalent Inhibitors and Nucleic Acid Binders. *Drug Discov. Today* **2018,** *23*, 687–695.
59. Sancineto, L.; Iraci, N.; Tabarrini, O.; Santi, C. NCp7: Targeting a Multitasking Protein for Next-Generation Anti-HIV Drug Development Part 1: Covalent Inhibitors. *Drug Discov. Today* **2018,** *23*, 260–271.
60. Reich, H. J.; Chow, F.; Peake, S. L. Seleninic Acids as Catalysts for Oxidations of Olefins and Sulfides Using Hydrogen Peroxide. *Synthesis (Stuttg.)* **1978,** *1978*, 299–301.
61. Betzemeier, B.; Lhermitte, F.; Knochel, P. A Selenium Catalyzed Epoxidation in Perfluorinated Solvents with Hydrogen Peroxide. *Synlett* **1999,** *1999*, 489–491.
62. Goodman, M. A.; Detty, M. R. Selenoxides as Catalysts for Epoxidation and Baeyer—Villiger Oxidation with Hydrogen Peroxide. *ChemInform* **2006,** 37.
63. ten Brink, G.-J.; Fernandes, B. C. M.; van Vliet, M. C. A.; Arends, I. W. C. E.; Sheldon, R. A. Selenium Catalysed Oxidations with Aqueous Hydrogen Peroxide. Part I: Epoxidation Reactions in Homogeneous Solution. *J. Chem. Soc. Perkin Trans. 1* **2001,** 224–228.
64. García-Marín, H.; van der Toorn, J. C.; Mayoral, J. A.; García, J. I.; Arends, I. W. C. E. Glycerol-Based Solvents as Green Reaction Media in Epoxidations with Hydrogen Peroxide Catalysed by Bis[3,5-bis(trifluoromethyl)-diphenyl] Diselenide. *Green Chem.* **2009,** *11*, 1605.
65. De Torres, M.; Arends, I. W. C. E.; Mayoral, J. A.; Pires, E.; Jiménez-Osés, G. A Highly Efficient, Green and Recoverable Catalytic System for the Epoxidation of Fatty Esters and Biodiesel with H_2O_2. *Appl. Catal. A Gen.* **2012,** *425–426*, 91–96.
66. Yu, L.; Bai, Z.; Zhang, X.; Zhang, X.; Ding, Y.; Xu, Q. Organoselenium-Catalyzed Selectivity-Switchable Oxidation of β-Ionone. *Catal. Sci. Technol.* **2016,** *6*, 1804–1809.
67. Yang, Y.; Fan, X.; Cao, H.; Chu, S.; Zhang, X.; Xu, Q.; Yu, L. Fabrication of Se/C Using Carbohydrates as Biomass Starting Materials: An Efficient Catalyst for Regiospecific Epoxidation of β-Ionone with Ultrahigh Turnover Numbers. *Catal. Sci. Technol.* **2018,** *8*, 5017–5023.
68. Santoro, S.; Santi, C.; Sabatini, M.; Testaferri, L.; Tiecco, M. Eco-Friendly Olefin Dihydroxylation Catalyzed by Diphenyl Diselenide. *Adv. Synth. Catal.* **2008,** *350*, 2881–2884.
69. Santi, C.; Di Lorenzo, R.; Tidei, C.; Bagnoli, L.; Wirth, T. Stereoselective Selenium Catalyzed Dihydroxylation and Hydroxymethoxylation of Alkenes. *Tetrahedron* **2012,** *68*, 10530–10535.
70. Sancineto, L.; Mangiavacchi, F.; Tidei, C.; Bagnoli, L.; Marini, F.; Gioiello, A.; Scianowski, J.; Santi, C. Selenium-Catalyzed Oxacyclization of Alkenoic Acids and Alkenols. *Asian J. Org. Chem.* **2017,** *6*, 988–992.

71. Cerra, B.; Mangiavacchi, F.; Santi, C.; Lozza, A. M.; Gioiello, A. Selective Continuous Flow Synthesis of Hydroxy Lactones from Alkenoic Acids. *React. Chem. Eng.* **2017,** *2,* 467–471.
72. Yu, L.; Wang, J.; Chen, T.; Ding, K.; Pan, Y. Access to Cyclohexane-1,2-diol through the Diphenyldiselenide Catalyzed Oxidation of Cyclohexene by Hydrogen Peroxide. *Chin. J. Org. Chem.* **2013,** *33,* 1096.
73. Yu, L.; Wang, J.; Chen, T.; Wang, Y.; Xu, Q. Recyclable 1,2-Bis[3,5-Bis(trifluoromethyl) phenyl]Diselane-Catalyzed Oxidation of Cyclohexene with H 2 O 2: A Practical Access to Trans -1,2-Cyclohexanediol. *Appl. Organomet. Chem.* **2014,** *28,* 652–656.
74. Wang, Y.; Yu, L.; Zhu, B.; Yu, L. Design and Preparation of a Polymer Resin-Supported Organoselenium Catalyst with Industrial Potential. *J. Mater. Chem. A* **2016,** *4,* 10828–10833.
75. Santi, C.; Jacob, R.; Monti, B.; Bagnoli, L.; Sancineto, L.; Lenardão, E. Water and Aqueous Mixtures as Convenient Alternative Media for Organoselenium Chemistry. *Molecules* **2016,** *21,* 1482.
76. Penteado, F.; Monti, B.; Sancineto, L.; Perin, G.; Jacob, R. G.; Santi, C.; Lenardão, E. J. Ultrasound-Assisted Multicomponent Reactions, Organometallic and Organochalcogen Chemistry. *Asian J. Org. Chem.* **2018,** *7,* 2368–2385.
77. Tidei, C.; Sancineto, L.; Bagnoli, L.; Battistelli, B.; Marini, F.; Santi, C. A Recyclable Biphasic System for Stereoselective and Easily Handled Hydrochalcogenations. *Eur. J. Org. Chem.* **2014,** *2014,* 5968–5975.
78. Begini, F.; Krasowska, D.; Jasiak, A.; Drabowicz, J.; Santi, C.; Sancineto, L. Continuous Flow Synthesis of 2,2′-Diselenobis(benzoic acid) and Derivatives. *React. Chem. Eng.* **2020.**
79. Gogoi, P.; Sharma, S. D.; Konwar, D. SeO2/H2O2/H2O-Dioxane: A New Catalytic System for Trans Dihydroxylation of Olefins. *Lett. Org. Chem.* **2007,** *4,* 249–252.
80. Wang, T.; Jing, X.; Chen, C.; Yu, L. Organoselenium-Catalyzed Oxidative C=C Bond Cleavage: A Relatively Green Oxidation of Alkenes into Carbonyl Compounds with Hydrogen Peroxide. *J. Org. Chem.* **2017,** *82,* 9342–9349.
81. Jing, X.; Chen, C.; Deng, X.; Zhang, X.; Wei, D.; Yu, L. Design and Preparation of Poly-Selenides: Easily Fabricated and Efficient Organoselenium Materials for Heavy Metal Removing and Recycling. *Appl. Organomet. Chem.* **2018,** *32,* e4332.
82. Yu, L.; Chen, F.; Ding, Y. Organoselenium-Catalyzed Oxidative Ring Expansion of Methylenecyclopropanes with Hydrogen Peroxide. *ChemCatChem* **2016,** *8,* 1033–1037.
83. Giurg, M.; Said, S. B.; Syper, L.; Młochowski, J. One-Pot Oxidation of Azomethine Compounds into Arenecarboxylic Acids. *Synth. Commun.* **2001,** *31,* 3151–3159.
84. Giurg, M.; Mlochowski, J.; Ambrozak, A. Hydrogen Peroxide Oxidation of N,N-Dimethylhydrazones Promoted by Selenium Compounds, Titanosilicalites or Acetonitrile. *Pol. J. Chem.* **2002,** *76,* 1713–1720.
85. Sheng, S.-R.; Huang, X. Oxidative Conversion of Aldoximes into Carboxylic Acid Esters Catalyzed by Polystyrene-Bound Phenylseleninic Acid. *ChemInform* **2004,** 35.
86. Yu, L.; Li, H.; Zhang, X.; Ye, J.; Liu, J.; Xu, Q.; Lautens, M. Organoselenium-Catalyzed Mild Dehydration of Aldoximes: An Unexpected Practical Method for Organonitrile Synthesis. *Org. Lett.* **2014,** *16,* 1346–1349.
87. Zhang, X.; Sun, J.; Ding, Y.; Yu, L. Dehydration of Aldoximes Using PhSe(O)OH as the Pre -Catalyst in Air. *Org. Lett.* **2015,** *17,* 5840–5842.
88. Jing, X.; Wang, T.; Ding, Y.; Yu, L. A Scalable Production of Anisonitrile Through Organoselenium-Catalyzed Dehydration of Anisaldoxime Under Solventless Conditions. *Appl. Catal. A Gen.* **2017,** *541,* 107–111.
89. Jing, X.; Yuan, D.; Yu, L. Green and Practical Oxidative Deoximation of Oximes to Ketones or Aldehydes with Hydrogen Peroxide/Air by Organoselenium Catalysis. *Adv. Synth. Catal.* **2017,** *359,* 1194–1201.

90. Chen, C.; Zhang, X.; Cao, H.; Wang, F.; Yu, L.; Xu, Q. Iron-Enabled Utilization of Air as the Terminal Oxidant Leading to Aerobic Oxidative Deoximation by Organoselenium Catalysis. *Adv. Synth. Catal.* **2019,** *361*, 603–610.
91. Shimizu, M.; Kuwajima, I. Highly Efficient Method for Oxidation of Alcohol with -Butyl Hydroperoxide Catalyzed by Diaryl Diselenide. *Tetrahedron Lett.* **1979,** *20*, 2801–2804.
92. Onami, T.; Ikeda, M.; Woodard, S. S. Catalytic Oxidation of Alcohols to Carbonyl Compounds Mediated by N -(Arylseleno)-4-chlorobenzenesulfonamide. *Bull. Chem. Soc. Jpn.* **1996,** *69*, 3601–3605.
93. Ehara, H.; Noguchi, M.; Sayama, S.; Onami, T. Bis[2-(2-Pyridyl)phenyl] Diselenide, a More Effective Catalyst for Oxidation of Alcohols to Carbonyl Compounds. *J. Chem. Soc. Perkin Trans. 1* **2000,** 1429–1431.
94. van der Toorn, J. C.; Kemperman, G.; Sheldon, R. A.; Arends, I. W. C. E. Diphenyldiselenide-Catalyzed Selective Oxidation of Activated Alcohols with tert -Butyl Hydroperoxide: New Mechanistic Insights. *J. Org. Chem.* **2009,** *74*, 3085–3089.
95. Singh, P.; Singh, A. K. "Piano-Stool" Complexes of Ruthenium(II) Designed with Arenes and N-[2-(Arylchalcogeno)ethyl]morpholines: Highly Active Catalysts for the Oxidation of Alcohols with N-Methylmorpholine N-Oxide, tert-Butyl Hydroperoxide and Sodium Periodate and Oxychloride. *Eur. J. Inorg. Chem.* **2010,** *2010*, 4187–4195.
96. Balkrishna, S. J.; Prasad, C. D.; Panini, P.; Detty, M. R.; Chopra, D.; Kumar, S. Isoselenazolones as Catalysts for the Activation of Bromine: Bromolactonization of Alkenoic Acids and Oxidation of Alcohols. *J. Org. Chem.* **2012,** *77*, 9541–9552.
97. Sukumaran, K. B.; Harvey, R. G. Synthesis of the Non-K-region o-Quinones and Dihydrodiols of Polycyclic Aromatic Hydrocarbons from the Corresponding Phenols. *J. Org. Chem.* **1980,** *45*, 4407–4413.
98. Giurg, M.; Piekielska, K.; Gębala, M.; Ditkowski, B.; Wolański, M.; Peczyńska-Czoch, W.; Młochowski, J. Catalytic Oxidative Cyclocondensation of o-Aminophenols to 2-Amino-3 H -Phenoxazin-3-Ones. *Synth. Commun.* **2007,** *37*, 1779–1789.
99. de Carvalho, R. L.; Jardim, G. A. M.; Santos, A. C. C.; Araujo, M. H.; Oliveira, W. X. C.; Bombaça, A. C. S.; Menna-Barreto, R. F. S.; Gopi, E.; Gravel, E.; Doris, E., et al. Combination of Aryl Diselenides/Hydrogen Peroxide and Carbon-Nanotube/Rhodium Nanohybrids for Naphthol Oxidation: An Efficient Route towards Trypanocidal Quinones. *Chem.—A Eur. J.* **2018,** *24*, 15227–15235.
100. Krow, G. R. *The Baeyer-VilligerOxidation of Ketones and Aldehydes. Organic Reactions;* John Wiley & Sons, Inc: Hoboken, NJ, 1993251–798.
101. Criegee, R. Die Umlagerung der Dekalin-peroxydester als Folge von kationischem Sauerstoff. *Justus Liebigs Ann. Chem.* **1948,** *560*, 127–135.
102. Olah, G. A.; Wang, Q.; Trivedi, N. J.; Surya Prakash, G. K. Baeyer-Villiger Oxidation of Ketones to Esters with Sodium Percarbonate/Trifluoroacetic Acid. *Synthesis (Stuttg.)* **1991,** *1991*, 739–740.
103. Ochiai, M.; Yoshimura, A.; Miyamoto, K.; Hayashi, S.; Nakanishi, W. Hypervalent λ 3-Bromane Strategy for Baeyer – Villiger Oxidation: Selective Transformation of Primary Aliphatic and Aromatic Aldehydes to Formates, Which is Missing in the Classical Baeyer – Villiger Oxidation. *J. Am. Chem. Soc.* **2010,** *132*, 9236–9239.
104. Poladura, B.; Martínez-Castañeda, Á.; Rodríguez-Solla, H.; Llavona, R.; Concellón, C.; del Amo, V. General Metal-Free Baeyer–Villiger-Type Synthesis of Vinyl Acetates. *Org. Lett.* **2013,** *15*, 2810–2813.
105. Zhou, L.; Liu, X.; Ji, J.; Zhang, Y.; Hu, X.; Lin, L.; Feng, X. Enantioselective Baeyer–Villiger Oxidation: Desymmetrization of Meso Cyclic Ketones and Kinetic Resolution of Racemic 2-Arylcyclohexanones. *J. Am. Chem. Soc.* **2012,** *134*, 17023–17026.

106. Syper, L. Reaction of α,β-Unsaturated Aldehydes with Hydrogen Peroxide Catalysed by Benzeneseleninic Acids and Their Precursors. *Tetrahedron* **1987,** *43,* 2853–2871.
107. Goodman, M.; Detty, M. Selenoxides as Catalysts for Epoxidation and Baeyer-Villiger Oxidation with Hydrogen Peroxide. *Synlett* **2006,** *2006,* 1100–1104.
108. Detty, M. R.; Friedman, A. E.; Oseroff, A. R. A Mechanism for the Oxidation of Glutathione to Glutathione Disulfide with Organotellurium(IV) and Organoselenium(IV) Compounds. A Stepwise Process with Implications for Photodynamic Therapy and Other Oxidative Chemotherapy. *J. Org. Chem.* **1994,** *59,* 8245–8250.
109. Zhang, X.; Ye, J.; Yu, L.; Shi, X.; Zhang, M.; Xu, Q.; Lautens, M. Organoselenium-Catalyzed Baeyer-Villiger Oxidation of α,β-Unsaturated Ketones by Hydrogen Peroxide to Access Vinyl Esters. *Adv. Synth. Catal.* **2015,** *357,* 955–960.
110. Yu, L.; Wu, Y.; Cao, H.; Zhang, X.; Shi, X.; Luan, J.; Chen, T.; Pan, Y.; Xu, Q. Facile Synthesis of 2-Methylenecyclobutanones via Ca(OH) 2-Catalyzed Direct Condensation of Cyclobutanone with Aldehydes and (PhSe) 2-Catalyzed Baeyer–Villiger Oxidation to 4-Methylenebutanolides. *Green Chem.* **2014,** *16,* 287–293.
111. Yu, L.; Ye, J.; Zhang, X.; Ding, Y.; Xu, Q. Recyclable (PhSe) 2-Catalyzed Selective Oxidation of Isatin by H_2O_2: A Practical and Waste-Free Access to Isatoic Anhydride Under Mild and Neutral Conditions. *Catal. Sci. Technol.* **2015,** *5,* 4830–4838.
112. Rao, K. R.; Mekala, R.; Raghunadh, A.; Meruva, S. B.; Kumar, S. P.; Kalita, D.; Laxminarayana, E.; Prasad, B.; Pal, M. A Catalyst-Free Rapid, Practical and General Synthesis of 2-Substituted Quinazolin-4(3H)-Ones Leading to Luotonin B and E, Bouchardatine and 8-Norrutaecarpine. *RSC Adv.* **2015,** *5,* 61575–61579.
113. Sancineto, L.; Iraci, N.; Massari, S.; Attanasio, V.; Corazza, G.; Barreca, M. L.; Sabatini, S.; Manfroni, G.; Avanzi, N. R.; Cecchetti, V., et al. Computer-Aided Design, Synthesis and Validation of 2-Phenylquinazolinone Fragments as CDK9 Inhibitors with Anti-HIV-1 Tat-Mediated Transcription Activity. *ChemMedChem* **2013,** *8,* 1941–1953.
114. Bollini, M.; Bruno, A. M.; Niño, M. E.; Casal, J. J.; Sasiambarrena, L. D.; Valdez, D. A. G.; Battini, L.; Puente, V. R.; Lombardo, M. E. Synthesis, 2D-QSAR Studies and Biological Evaluation of Quinazoline Derivatives as Potent Anti-Trypanosoma Cruzi Agents. *Med. Chem. (Los. Angeles)* **2019,** *15,* 265–276.
115. Choi, J.-K.; Chang, Y.-K.; Yeap Hong, S. Catalytic Oxidation of Aldehydes to Carboxylic Acids with Hydrogen Peroxide as Oxidant. *Tetrahedron Lett.* **1988,** *29,* 1967–1970.
116. Brzaszcz, M.; Kloc, K.; Maposah, M.; Mlochowski, J. Selenium(IV) Oxide Catalyzed Oxidation of Aldehydes to Carboxylic Acids with Hydrogen Peroxide. *Synth. Commun.* **2000,** *30,* 4425–4434.
117. ten Brink, G.-J.; Vis, J. M.; Arends, I. W. C.; Sheldon, R. A. Selenium Catalysed Oxidations with Aqueous Hydrogen Peroxide. Part 3: Oxidation of Carbonyl Compounds Under Mono/Bi/Triphasic Conditions. *Tetrahedron* **2002,** *58,* 3977–3983.
118. Sancineto, L.; Tidei, C.; Bagnoli, L.; Marini, F.; Lenardão, E.; Santi, C. Selenium Catalyzed Oxidation of Aldehydes: Green Synthesis of Carboxylic Acids and Esters. *Molecules* **2015,** *20,* 10496–10510.
119. Qian, H.; Shao, L.-X.; Huang, X. Polystyrene-Bound Phenylseleninic Acid: Catalytic Oxidation of Aldehydes to Carboxylic Acids with Hydrogen Peroxide. *J. Chem. Res.* **2002,** *2002,* 514–515.
120. Rangraz, Y.; Nemati, F.; Elhampour, A. Diphenyl Diselenide Immobilized on Magnetic Nanoparticles: A Novel and Retrievable Heterogeneous Catalyst in the Oxidation of Aldehydes Under Mild and Green Conditions. *J. Colloid Interface Sci.* **2018,** *509,* 485–494.
121. Tan, K. H.; Xu, W.; Stefka, S.; Demco, D. E.; Kharandiuk, T.; Ivasiv, V.; Nebesnyi, R.; Petrovskii, V. S.; Potemkin, I. I.; Pich, A. Selenium-Modified Microgels as Bio-Inspired Oxidation Catalysts. *Angew. Chem. Int. (Ed.)* **2019,** *58,* 9791–9796.

122. Nascimento, V.; Alberto, E. E.; Tondo, D. W.; Dambrowski, D.; Detty, M. R.; Nome, F.; Braga, A. L. GPx-Like Activity of Selenides and Selenoxides: Experimental Evidence for the Involvement of Hydroxy Perhydroxy Selenane as the Active Species. *J. Am. Chem. Soc.* **2012,** *134,* 138–141.
123. Ribaudo, G.; Bellanda, M.; Menegazzo, I.; Wolters, L. P.; Bortoli, M.; Ferrer-Sueta, G.; Zagotto, G.; Orian, L. Mechanistic Insight into the Oxidation of Organic Phenylselenides by H_2O_2. *Chem. - A Eur. J.* **2017,** *23,* 2405–2422.
124. Sands, K. N.; Mendoza Rengifo, E.; George, G. N.; Pickering, I. J.; Gelfand, B. S.; Back, T. G. The Unexpected Role of Se VI Species in Epoxidations with Benzeneseleninic Acid and Hydrogen Peroxide. *Angew. Chem. Int. (Ed.)* **2020,** *59,* 4283–4287.
125. Akondi, S. M.; Gangireddy, P.; Pickel, T. C.; Liebeskind, L. S. Aerobic, Diselenide-Catalyzed Redox Dehydration: Amides and Peptides. *Org. Lett.* **2018,** *20,* 538–541.
126. Liebeskind, L. S.; Gangireddy, P.; Lindale, M. G. Benzoisothiazolone Organo/Copper-Cocatalyzed Redox Dehydrative Construction of Amides and Peptides from Carboxylic Acids using (EtO) 3 P as the Reductant and O_2 in Air as the Terminal Oxidant. *J. Am. Chem. Soc.* **2016,** *138,* 6715–6718.
127. Manoharan, S.; Guillemin, G. J.; Abiramasundari, R. S.; Essa, M. M.; Akbar, M.; Akbar, M. D. The Role of Reactive Oxygen Species in the Pathogenesis of Alzheimer's Disease, Parkinson's Disease, and Huntington's Disease: A Mini Review. *Oxid. Med. Cell. Longev.* **2016,** *2016,* 1–15.
128. Epp, O.; Ladenstein, R.; Wendel, A. The Refined Structure of the Selenoenzyme Glutathione Peroxidase at 0.2-nm Resolution. *Eur. J. Biochem.* **1983,** *133,* 51–69.
129. Bortoli, M.; Torsello, M.; Bickelhaupt, F. M.; Orian, L. Role of the Chalcogen (S, Se, Te) in the Oxidation Mechanism of the Glutathione Peroxidase Active Site. *ChemPhysChem* **2017,** *18,* 2990–2998.
130. Yasuda, K.; Watanabe, H.; Yamazaki, S.; Toda, S. Glutathione Peroxidase Activity of D,L-Selenocystine and Selenocystamine. *Biochem. Biophys. Res. Commun.* **1980,** *96,* 243–249.
131. Müller, A.; Cadenas, E.; Graf, P.; Sies, H. A Novel Biologically Active Seleno-Organic Compound-1. Glutathione Peroxidase-Like Activity In Vitro and Antioxidant Capacity of PZ 51 (Ebselen). *Biochem. Pharmacol.* **1984,** *33,* 3235–3239.
132. Engman, L.; Stern, D.; Cotgreave, I. A.; Andersson, C. M. Thiol Peroxidase Activity of Diaryl Ditellurides as Determined by a Proton NMR Method. *J. Am. Chem. Soc.* **1992,** *114,* 9737–9743.
133. Bell, I. M.; Hilvert, D. Peroxide Dependence of the Semisynthetic Enzyme Selenosubtilisin. *Biochemistry* **1993,** *32,* 13969–13973.
134. Iwaoka, M.; Tomoda, S. A Model Study on the Effect of an Amino Group on the Antioxidant Activity of Glutathione Peroxidase. *J. Am. Chem. Soc.* **1994,** *116,* 2557–2561.
135. Haenen, G. R.; De Rooij, B. M.; Vermeulen, N. P.; Bast, A. Mechanism of the Reaction of Ebselen with Endogenous Thiols: Dihydrolipoate Is a Better Cofactor than Glutathione in the Peroxidase Activity of Ebselen. *Mol. Pharmacol.* **1990,** *37,* 412–422.
136. Lenardão, E. J.; Santi, C.; Sancineto, L. *Bioactive Organoselenium Compounds and Therapeutic Perspectives. New Frontiers in Organoselenium Compounds;* Springer International Publishing: Cham, 2018, 99–143.
137. Iwaoka, M.; Kumakura, F. Applications of Water-Soluble Selenides and Selenoxides to Protein Chemistry. *Phosphorus, Sulfur. Silicon Related Elements* **2008,** *183,* 1009–1017.
138. Tidei, C.; Piroddi, M.; Galli, F.; Santi, C. Oxidation of Thiols Promoted by PhSeZnCl. *Tetrahedron Lett.* **2012,** *53,* 232–234.
139. McNeil, N. M. R.; Press, D. J.; Mayder, D. M.; Garnica, P.; Doyle, L. M.; Back, T. G. Enhanced Glutathione Peroxidase Activity of Water-Soluble and Polyethylene Glycol-Supported Selenides, Related Spirodioxyselenuranes, and Pincer Selenuranes. *J. Org. Chem.* **2016,** *81,* 7884–7897.

140. Back, T. G.; Dyck, B. P. A Novel Camphor-Derived Selenenamide that Acts as a Glutathione Peroxidase Mimetic. *J. Am. Chem. Soc.* **1997,** *119,* 2079–2083.
141. Elsherbini, M.; Hamama, W. S.; Zoorob, H. H.; Bhowmick, D.; Mugesh, G.; Wirth, T. Synthesis and Antioxidant Activities of Novel Chiral Ebselen Analogues. *Heteroat. Chem.* **2014,** *25,* 320–325.
142. Tiezza, M. D.; Ribaudo, G.; Orian, L. Organodiselenides: Organic Catalysis and Drug Design Learning from Glutathione Peroxidase. *Curr. Org. Chem.* **2019,** *23,* 1381–1402.
143. Ruberte, A. C.; Sanmartin, C.; Aydillo, C.; Sharma, A. K.; Plano, D. Development and Therapeutic Potential of Selenazo Compounds. *J. Med. Chem.* **2020,** *63,* 1473–1489.
144. Barbosa, N. V.; Nogueira, C. W.; Nogara, P. A.; de Bem, A. F.; Aschner, M.; Rocha, J. B. T. Organoselenium Compounds as Mimics of Selenoproteins and Thiol Modifier Agents. *Metallomics* **2017,** *9,* 1703–1734.
145. Santi, C.; Marini, F.; Lenardão, E. J. *Synthetic Advances on Bioactive Selenium CompoundsChapter 2. Looking Beyond the Traditional Idea of Glutathione Peroxidase Mimics as Antioxidants. Organoselenium Compounds in Biology and Medicine;* Royal Society of Chemistry: Cambridge, 2017, 35–76.
146. Tanini, D.; Lupori, B.; Lo Nostro, P.; Capperucci, A. Synthesis and Catalytic Antioxidant Activity of Functionalized Chalcogen-Containing GPx Mimics. *Phosphorus. Sulfur. Silicon Relat. Elem.* **2019,** *194,* 746–749.
147. Tanini, D.; Scarpelli, S.; Ermini, E.; Capperucci, A. Seleno-Michael Reaction of Stable Functionalised Alkyl Selenols: A Versatile Tool for the Synthesis of Acyclic and Cyclic Unsymmetrical Alkyl and Vinyl Selenides. *Adv. Synth. Catal.* **2019,** *361,* 2337–2346.
148. Tanini, D.; Lupori, B.; Malevolti, G.; Ambrosi, M.; Nostro, P. L.; Capperucci, A. Direct Biocatalysed Synthesis of First Sulfur-, Selenium- and Tellurium-Containing L-Ascorbyl Hybrid Derivatives with Radical Trapping and GPx-like Properties. *Chem. Commun.* **2019,** *55,* 5705–5708.
149. Shimodaira, S.; Takei, T.; Hojo, H.; Iwaoka, M. Synthesis of Selenocysteine-Containing Cyclic Peptides via Tandem N -to- S Acyl Migration and Intramolecular Selenocysteine-Mediated Native Chemical Ligation. *Chem. Commun.* **2018,** *54,* 11737–11740.
150. Shimodaira, S.; Iwaoka, M. Synthesis of Selenocysteine-Containing Dipeptides Modeling the Active Site of Thioredoxin Reductase. *Phosphorus. Sulfur. Silicon Relat. Elem.* **2019,** *194,* 750–752.
151. Wehrle, R. J.; Ste Marie, E. J.; Hondal, R. J.; Masterson, D. S. Synthesis of Alpha-Methyl Selenocysteine and Its Utilization as a Glutathione Peroxidase Mimic. *J. Pept. Sci.* **2019,** *25,* e3173.
152. Pacuła, A. J.; Ścianowski, J.; Aleksandrzak, K. B. Highly Efficient Synthesis and Antioxidant Capacity of N-Substituted Benzisoselenazol-3(2H)-Ones. *RSC Adv.* **2014,** *4,* 48959–48962.
153. Pacuła, A. J.; Kaczor, K. B.; Wojtowicz, A.; Antosiewicz, J.; Janecka, A.; Długosz, A.; Janecki, T.; Ścianowski, J. New Glutathione Peroxidase Mimetics—Insights into Antioxidant and Cytotoxic Activity. *Bioorg. Med. Chem.* **2017,** *25,* 126–131.
154. Pacuła, A. J.; Kaczor, K. B.; Antosiewicz, J.; Janecka, A.; Długosz, A.; Janecki, T.; Wojtczak, A.; Ścianowski, J. New Chiral Ebselen Analogues with Antioxidant and Cytotoxic Potential. *Molecules* **2017,** *22,* 492.
155. Pacuła, A. J.; Obieziurska, M.; Ścianowski, J.; Kaczor, K. B.; Antosiewicz, J. Water-Dependent Synthesis of Biologically Active Diaryl Diselenides. *Arkivoc* **2018,** *2018,* 153–164.
156. Obieziurska, M.; Pacuła, A.; Juhas, U.; Antosiewicz, J.; Ścianowski, J. The Influence of O/S Exchange on the Biocatalytical Activity of Benzisoselenazol-3(2H)-Ones. *Catalysts* **2018,** *8,* 493.

157. Obieziurska, M.; Pacuła, A. J.; Długosz-Pokorska, A.; Krzemiński, M.; Janecka, A.; Ścianowski, J. Bioselectivity Induced by Chirality of New Terpenyl Organoselenium Compounds. *Mater. (Basel)* **2019,** *12*, 3579.
158. Obieziurska, M.; Pacuła, A. J.; Laskowska, A.; Długosz-Pokorska, A.; Janecka, A.; Ścianowski, J. Seleninic Acid Potassium Salts as Water-Soluble Biocatalysts with Enhanced Bioavailability. *Mater. (Basel)* **2020,** *13*, 661.
159. Yu, S.-C.; Ri, D.-M.; Kühn, H. Hydrophobicity and Glutathione Peroxidase-Like Activity of Substituted Salicyloyl-5-Seleninic Acids: Re-Investigations on Aromatic Selenium Compounds based on their Hydrophobicity. *J. Organomet. Chem.* **2018,** *862*, 86–94.
160. Zhou, W.; Li, H.; Xia, B.; Ji, W.; Ji, S.; Zhang, W.; Huang, W.; Huo, F.; Xu, H. Selenium-Functionalized Metal-Organic Frameworks as Enzyme Mimics. *Nano Res.* **2018,** *11*, 5761–5768.
161. Zhang, D.; Shen, N.; Zhang, J.; Zhu, J.; Guo, Y.; Xu, L. A Novel Nanozyme Based on Selenopeptide-Modified Gold Nanoparticles with a Tunable Glutathione Peroxidase Activity. *RSC Adv.* **2020,** *10*, 8685–8691.

PART III

Biomass/Natural Products as a Precursor of Organochalcogen Compounds

CHAPTER

8

Semisynthetic bioactive organoselenium and organotellurium compounds

Daniela Hartwig[1], *Raquel G. Jacob*[1], *Eder J. Lenardão*[1], *José E.R. Nascimento*[1], *Laura Abenante*[1], *Liane K. Soares*[1] *and Carl H. Schiesser*[2]

[1]**Laboratório de Síntese Orgânica Limpa—LASOL—CCQFA—Universidade Federal de Pelotas—UFPel, Pelotas, Brazil,** [2]**Seleno Therapeutics Pty. Ltd., Brighton East, VIC, Australia**

8.1 Introduction

It is well-known that compounds present in nature exhibit a multitude of properties, with some natural compounds having been used in herbal remedies since ancient times.[1,2] Modifying a natural compound is a strategy for producing a more active molecule. In some cases, starting from a compound that already exists in nature avoids the tedious and multistep syntheses of complex molecular scaffolds.[3,4]

The biological activities of organochalcogen compounds have been well documented in several review articles and books,[5–10] which reveal that the insertion of a chalcogen atom into a natural bioactive compound often greatly increases its biological activity.

Considering the number of reports collated, for ease of discussion, this chapter is divided into five sections according to the starting functionalized material, namely: (1) terpenes and essential oils, (2) phenolic compounds, (3) carbohydrates, (4) nitrogen-containing heterocycles, and (5) miscellaneous.

In essence, this chapter is not intended to provide a comprehensive account of semisynthetic bioactive organoselenium and organotellurium

Organochalcogen Compounds
DOI: https://doi.org/10.1016/B978-0-12-819449-2.00003-3

compounds. It is rather intended to furnish an overview that highlights some of the key elements of their chemistries and potential biological and medical applications.

8.2 Terpenes and essential oils

Terpenes or terpenoids are the largest and most important class of natural products. They are characterized as being constructed from five-carbon isoprene units and are classified on the basis of the number of isoprene (C_5H_8) units present in their structure, as hemiterpenes (C_5), monoterpenes (C_{10}), sesquiterpenes (C_{15}), diterpenes (C_{20}), sesterterpenes (C_{25}), triterpenes (C_{30}), and polyterpenes ($C_n \geq 40$) (Fig. 8.1).[11–13] When they contain oxygenated functionalities, they are referred to as "terpenoids."

Essential oils, which are composed of terpenes, are natural liquids extracted from plants, flowers, fruits, herbs, or leaves, and are obtained mainly through steam distillation.[14] They are volatile or ethereal compounds used in the flavor and fragrance industries and are also known for their antimicrobial,[15] antiviral,[16] and insecticidal[17] properties. Essential oils have been extensively used in the food,[18] perfume,[19] and pharmaceutical industries.[20]

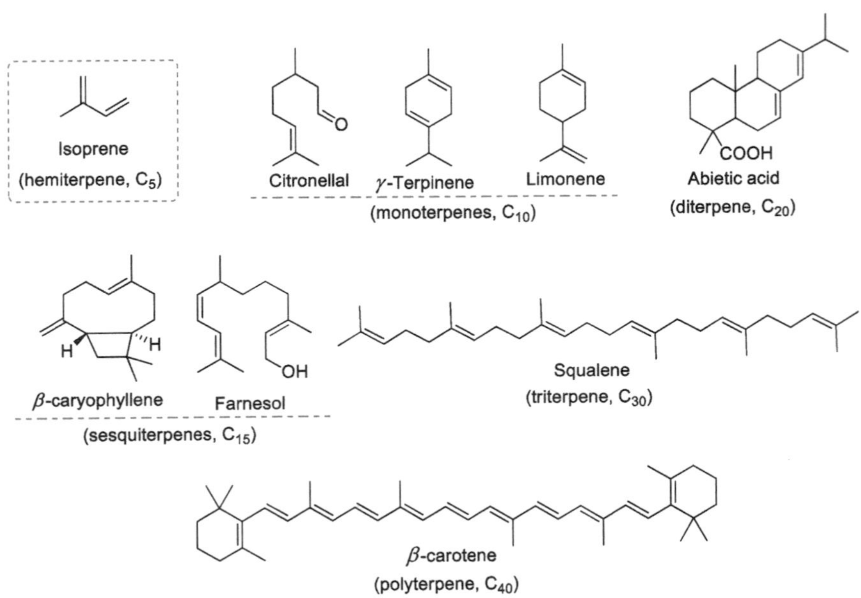

FIGURE 8.1 Chemical structures of some natural terpenes.[10–12]

Terpenes have been the targets of several studies aimed at incorporating chalcogen atoms into their structures. The addition of an organochalcogen group was observed to increase biological activity compared to that of the natural terpene.[21] Based on this, Back and coworkers[22] reported the synthesis of *endo,endo*-dicamphoryl diselenide (**1**) and selenide (**2**) through the reaction of camphor enolate with elemental selenium, followed by aerobic oxidation; that is, (1 *R*)-(+)-camphor (**3**) was treated with lithium diisopropylamide (LDA) in tetrahydrofuran (THF) as the solvent, followed by the addition of elemental selenium at −40°C, to afford selenolate (**4**) in situ. Aerobic oxidation of the reaction mixture subsequently afforded diselenide (**1**) and selenide (**2**) in yields of 76% and 14%, respectively (Scheme 8.1).

When the authors attempted to alkylate in situ−generated selenolate (**4**) with methyl iodide, the expected methyl selenide product (**5**) was formed in only 12% yield, together with the selenoacetal (**6**), which was obtained in 15% yield. However, the main product of this reaction was a mixture of the corresponding *syn*- and *anti*-1,3-diselenetanes (**7**) and (**8**), which were produced in yields of 32% and 10%, respectively. Furthermore, traces of diselenide (**3**) and selenide (**4**) were obtained under these reaction conditions (Scheme 8.1).

Subsequently, Pietrella and coworkers analyzed the in vitro ability of camphor diselenide (**1**) to inhibit biofilm formation and to disperse the preformed microbial biofilms of some microorganisms involved in infected wounds.[23] Camphor diselenide (**1**) demonstrated good antibiofilm activities against *Staphylococcus aureus, Staphylococcus epidermidis* and was active against *Streptococcus pyogenes* and yeast *Candida albicans* biofilms.

Another convenient and interesting strategy for the synthesis of camphor-based selenides was developed by Martínez and coworkers.[24]

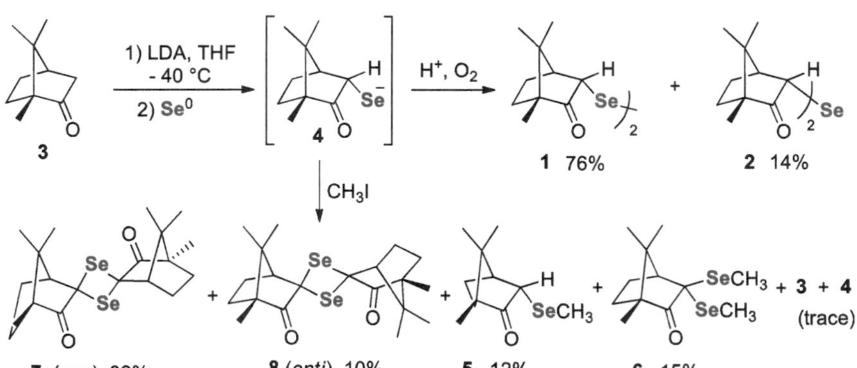

SCHEME 8.1 Synthesis of *endo,endo*-dicamphoryl diselenide (1), selenide (2), *syn*- and *anti*-1,3-diselenetanes (7) and (8) from (1 *R*)-(+)-camphor (3).[22]

These authors prepared (1 *S*)-10-(phenylselanyl)camphor (**9**) from 3,3-dimethyl-2-methylenenorbornan-1-ol (**10**), which was obtained from (1 *R*)-camphor (**3**) in two easy steps (Scheme 8.2). The key step of this protocol was the enantiospecific Wagner–Meerwein rearrangement of **10**, promoted by electrophilic treatment with phenylselanyl chloride (**11**) at room temperature to give the corresponding product **9** in 82% yield (Scheme 8.2).

In 2016, Rodríguez and coworkers[25] reported the synthesis of isoselenocyanate derivatives from (-)-8,15-diisocyano-11(20)-amphilectene (**12**), a marine natural product isolated from the Caribbean sponge *Svenzea flava*, and evaluated their antimalarial and antituberculosis activities. The insertion of two selenium atoms into the molecule was achieved by the isoselenocyanation of diisocyanide (**12**) with elemental selenium in the presence of triethanolamine in THF as the solvent at 25°C for 24 hours. The respective product, 15-diisoselenocyano-11(20)-amphilectene (**13**), was obtained in 78% yield (Scheme 8.3, Procedure A).

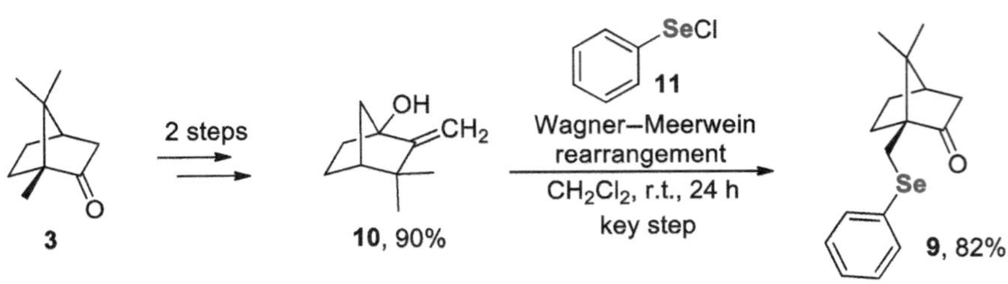

SCHEME 8.2 Synthesis of (1 *S*)-10-(phenylselanyl)camphor (**9**).[24]

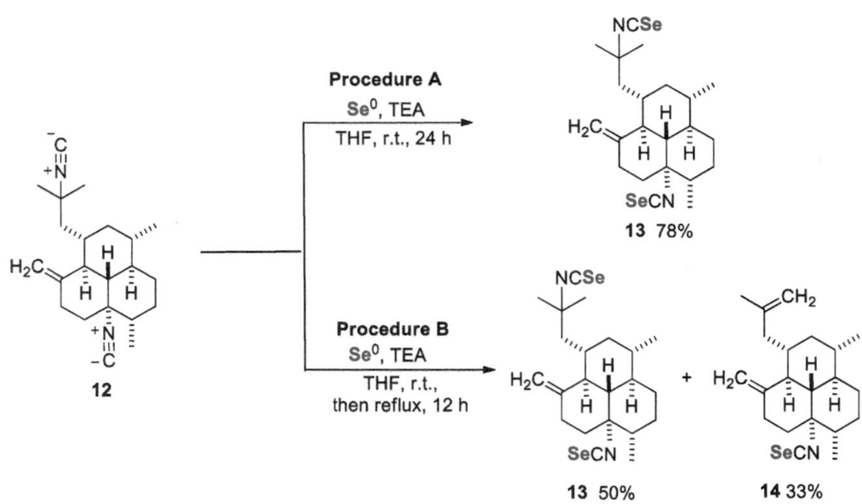

SCHEME 8.3 Synthesis of isoselenocyanate analogs **13** and **14**.[25]

However, when the same reaction was conducted under reflux for 12 hours, a mixture of two products, namely, diisoselenocyanate (**13**) and isoselenocyanate (**14**), was obtained in yields of 50% and 33%, respectively (Scheme 8.3, Procedure B).

The isoselenocyanate derivatives **13** and **14** were examined for their in vitro antimalarial and antituberculosis activities by evaluating their abilities to inhibit the growth of two malaria parasites, namely *Plasmodium falciparum* Dd2 (drug resistant) and 3D7 (chloroquine-sensitive), and *Mycobacterium tuberculosis* $H_{37}Rv$. Se compound **13** presented similar IC_{50} values (0.0066 μM and 0.0025 μM) as the parent natural product **12** against malaria parasites Dd2 and 3D7, respectively. These tuberculosis bacterium (*Mycobacterium tuberculosis* $H_{37}Rv$) studies revealed that the monoselenylated compound **14** is more potent (MIC of 2.1 μM) and less toxic (selectivity index of 45.3) than the natural terpenoid **12**.

El Houssame and coworkers[26] reported an efficient method for the preparation of new Se(β-pinene)$_2$ (**15**) starting from β-pinene (**16**). The selenide (**15**) was easily prepared in good yield (71%) at room temperature by reacting β-pinene with an equivalent amount of selenium oxide in CH_2Cl_2 in the presence of triethylamine at room temperature for 72 hours (Scheme 8.4).

In 2007, Tiecco and coworkers[27] described the easy conversions of several unsaturated natural terpenes (**16**) into the corresponding enantiomerically pure β-azidoselenides (**17a–f**) in 60%–90% yield (Scheme 8.5). These azidoselenenylation reactions are stereospecific anti-additions, with each involving the Markovnikov addition of electrophilic PhSeOTf (**18**) in the presence of sodium azide (**19**) in dry MeCN at 0°C for 24–30 hours.

Possible synthetic uses of the β-azidoselenides (**17**) were explored in classical reactions. For instance, (**17a**) was transformed into the enantiomerically pure β-aminoselenide (**20**) (60% yield) by reduction of the azido group in the presence of Ph_3P and H_2O at 50°C for 5 hours. A similar procedure using acetyl chloride or benzoyl chloride (**21**) and Et_3N provided the respective β-amidoselenides (**22a**) and (**22b**) in 64% yield (Scheme 8.6).

Stereoselective selenenylation reactions are other reactions that are important and of significant interest in organic synthesis. In this sense, Ścianowski and coworkers[28] described a convenient method for the

SCHEME 8.4 Synthesis of Se(β-pinene)$_2$ (**15**).[26]

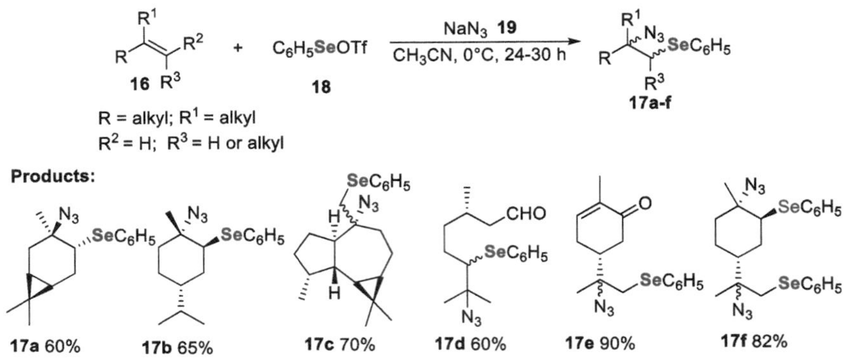

SCHEME 8.5 Synthesis of β-azidoselenides (**17a–f**) from unsaturated natural terpenes.[27]

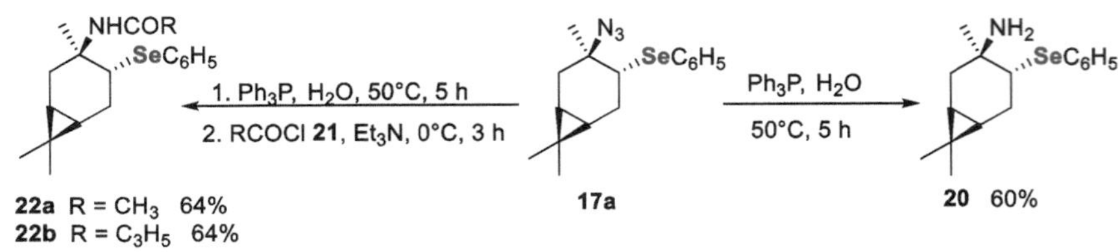

SCHEME 8.6 Synthetic applications of β-azidoselenides (**17**).[27]

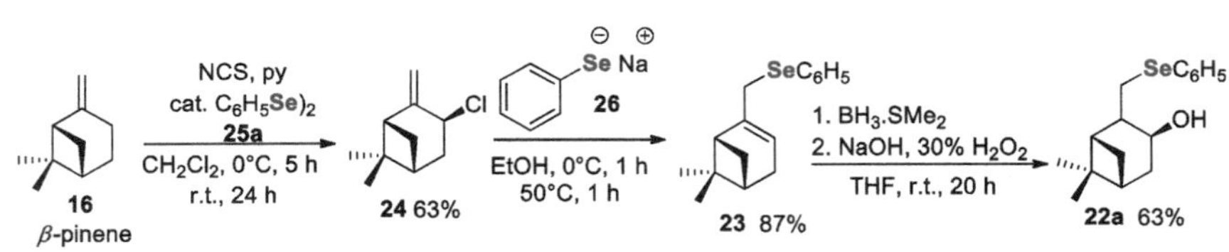

SCHEME 8.7 Synthesis of 10-phenylselanylisopinocampheol (**22a**).[28]

synthesis of optically active dialkyl diselenides (**21**) derived from terpenes and functionalized with hydroxy, alkoxy, sulfanyl, and selanyl groups. First, the authors prepared 10-phenylselanylisopinocampheol (**22a**) in 63% yield by the hydroboration-oxidation of 10-phenylselanyl-2-pinene (**23**) (87%) prepared from chlorinated β-pinene (**24**), itself prepared in 63% yield from **16** by the action of *N*-chlorosuccinimide in the presence of PhSeSePh (**25a**) as the catalyst, by substitution with sodium benzeneselenolate (**26**) (Scheme 8.7).

In the same paper,[28] the authors report the use of the Se-functionalized terpenic alcohols (**22**) obtained in Scheme 8.7 to prepare diselenides (**21a–f**) by the reaction of the respective alkyl tosylates and halides with disodium diselenide generated in situ (Scheme 8.8).

The reaction proceeded in two different ways (routes **A** and **B**) to produce the desired dialkyl diselenides (**21**). In route **A**, terpenic alcohols (**22**) are first reacted with TsCl to give several chiral tosylates (**23a–f**) in 67%–91% yield. The tosylates (**23a–f**) were then reacted with disodium diselenide, leading to the desired diselenides (**21a–f**) with single inversions in 8%–81% yield. In route **B**, double inversion was observed during the transformation of alcohols (**22**) in the presence of Ph_3P and CCl_4. The corresponding inverted chlorides (**24**) were formed in 41%–73% yield. Another inversion occurred when chloride (**24**) was reacted with Na_2Se_2 in DMF, and the respective diaryl diselenides (**21a–h**) were obtained in 19%–55% yield (Scheme 8.8).

In 2009, Lenardão and coworkers[29] α-selenated (*R*)-citronellal (**27**) using KF/Al_2O_3 (40%) and PEG-400 as a recyclable basic system, with α-phenylselenocitronellal (**28**) obtained in 80% yield after 3 hours (Scheme 8.9). 2-Phenylselanyl citronellol (**29**) was prepared in 73% yield

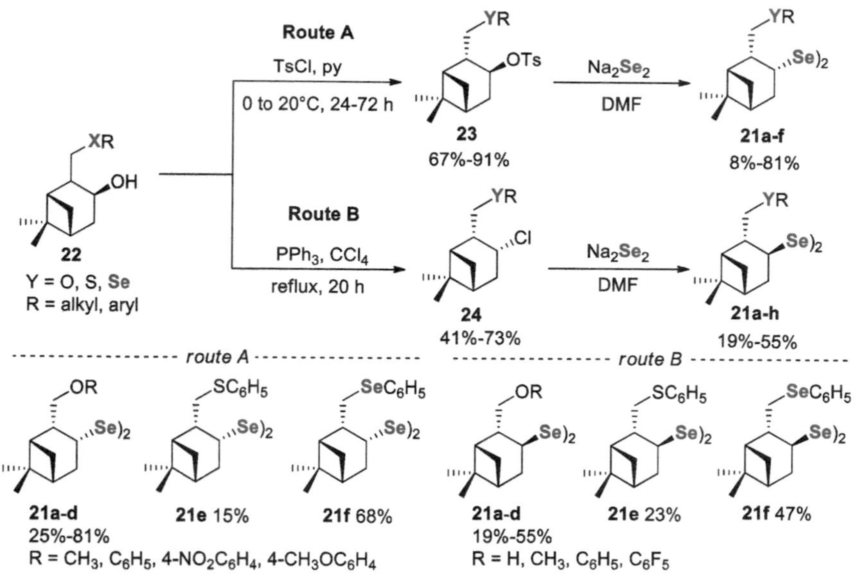

SCHEME 8.8 Syntheses of functionalized diselenides.[28]

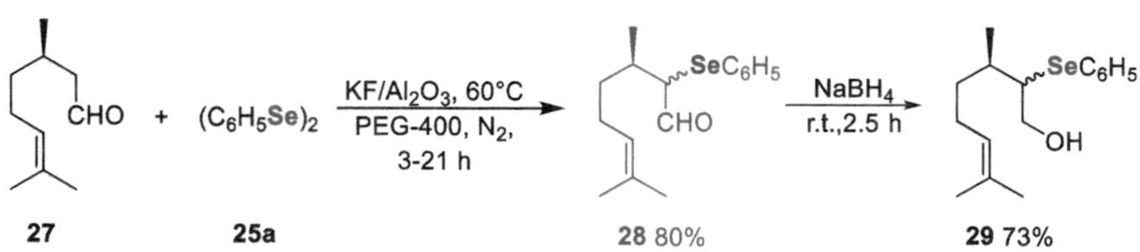

SCHEME 8.9 Synthesis of α-phenylselenocitronellal (**28**) and 2-phenylselenocitronellol (**29**).[29]

from **28** by reduction with $NaBH_4$. The antibacterial activities of both products **28** and **29** were examined; these compounds were found to be active against *Listeria monocytogenes*, *S. aureus*, and *Salmonella enteridis*.

Starting with α-phenylselenocitronellal (**28**), Jacob and coworkers[30] synthesized seleno-containing nitrone (**30**), which was formed in 65% yield by reaction with *N*-methylhydroxylamine hydrochloride in the presence of aqueous sodium carbonate. When hydroxylamine hydrochloride was used instead, oxime (**31**) was obtained in 90% yield. Compounds **30** and **31** were evaluated for their antimicrobial and antioxidant activities *in vitro* against foodborne pathogenic bacteria, such as Gram-positive *Listeria monocytogenes*, *S. aureus*, *Bacillus cereus*, and Gram-negative *Salmonella typhimurium* and *Escherichia coli*, and the clinically important bacterium *Pseudomonas aeruginosa*. Both derivatives **30** and **31** were found to be more active than the (*R*)-citronellal precursor (**27**) (Scheme 8.10).

8.3 Phenolic compounds

Phenolic and polyphenolic compounds are found in many fruits and vegetables and are commonly present in the average human diet.[31] Recently, several research groups have modified the structures of naturally occurring phenolic compounds and have found that some modifications boost biological activity. With this in mind, the insertion of organochalcogen groups into phenolic structures, with the aim of improving their pharmacological properties, is described in this chapter.

For example, chrysin (**33**), which is commonly found in a variety of plant extracts, honey, fruits, and vegetables, such as passion fruit (*Passiflora edulis*) and propolis, among others,[32,33] has been reported to exhibit multiple pharmacological properties, including antiviral,[34]

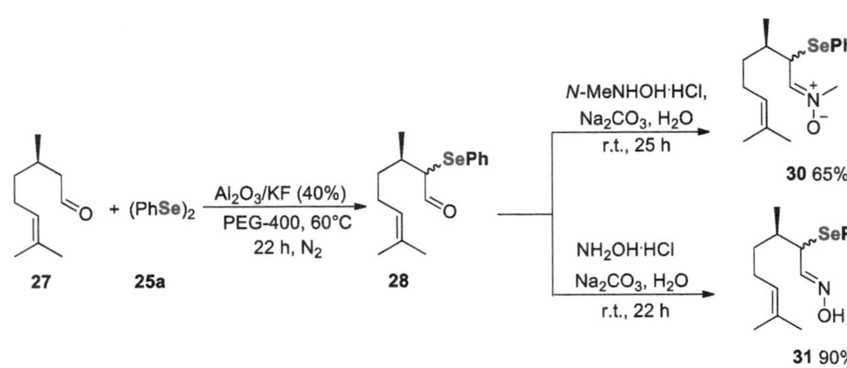

SCHEME 8.10 Synthesis of nitrone (**30**) and oxime (**31**) from α-phenylselenocitronellal **28**.[30]

anticancer,[35] and antioxidant[36] activities. In 2015, the organoselenium and organotellurium derivatives (**36**) and (**37**) of chrysin were synthesized by Lenardão and coworkers.[37] The first step in the synthesis involved the reaction of chrysin with 1,2-dibromoethane (**34**) in the presence of potassium carbonate as the base and acetone as the solvent in a N_2 atmosphere and under reflux for 21 hours. The brominated chrysin (**32**) was reacted with diorganyl dichalcogenides (**25**) and (**35**) in the presence of $NaBH_4$ and ethanol as the solvent. After stirring under reflux (N_2 atmosphere) for 1.5–6.0 hours, eight selenium- and three tellurium-containing chrysin derivatives (**36**) and (**37**) were obtained in 57%–96% yield (Scheme 8.11). The semisynthetic chrysin derivatives were evaluated for their antioxidant activities in vitro. The tellurium-containing derivatives (**37**) were found to be generally more active than the selenium analogs, with the butyltelluro-derivative (**37**) ($R = C_4H_9$) being the most active.

The same research group used ultrasonic irradiation (US) to synthesize semisynthetic selenium-containing chrysin derivatives by directly functionalizing chrysin (**33**) with diaryl diselenides (**25**) and CuI (5 mol%) in dimethyl sulfoxide for 3–5 hours. The authors prepared six new selenochrysins (**38**) in 60%–89% yield (Scheme 8.12).[38]

The selenocrysins (**38**) were subjected to antioxidant- and anticancer-activity testing in vitro, with the organoselanyl group in the A-ring of chrysin observed to enhance both antioxidant and anticancer properties, compared to the parent compound. Semisynthetic Se-chrysins **38a** and **38b** exhibited the best antioxidant results. In an antiproliferative assay against lung adenocarcinoma A-549 cells, derivative **38c** was the more effective, with an IC_{50} of 19.9 μM after 72 hours of treatment, while compound **38b** presented a higher antiproliferative activity after 48 hours of treatment ($IC_{50} = 41.4$ μM).

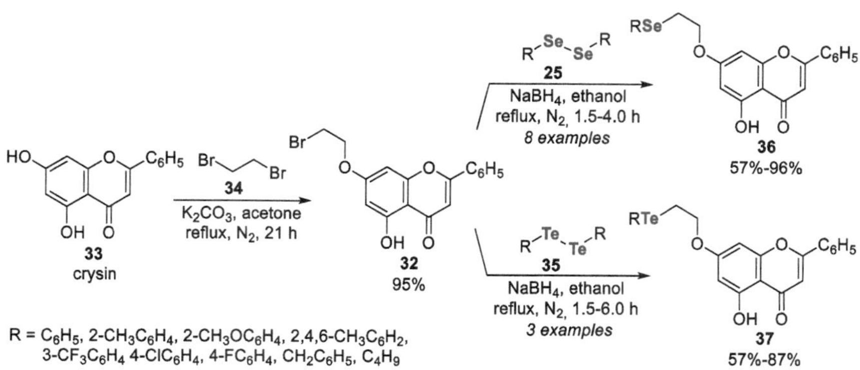

SCHEME 8.11 Preparation of selenium- and tellurium-containing chrysin derivatives (**36**) and (**37**).[37]

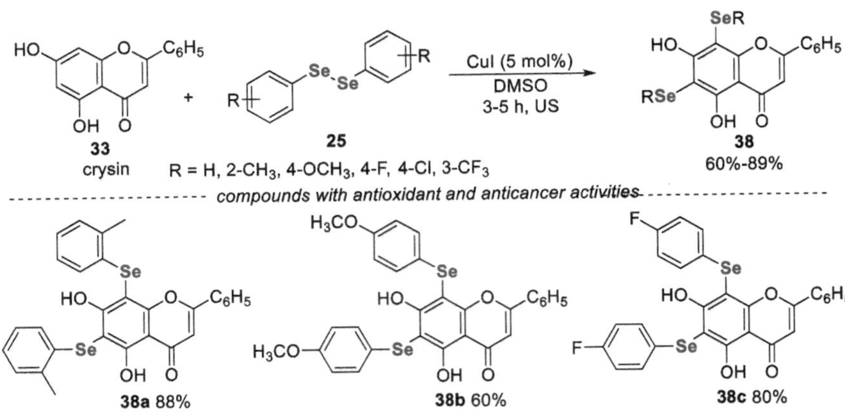

SCHEME 8.12 Ultrasound-promoted copper-catalyzed synthesis of bis-arylselanyl chrysin derivatives with boosted antioxidant and anticancer activities.[38]

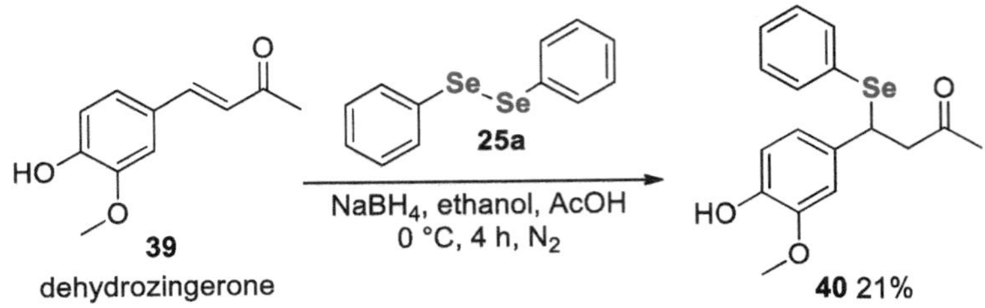

SCHEME 8.13 Synthesis of selenium-containing zingerone derivative (40).[43]

Dehydrozingerone (DHZ) and zingerone are phenolic compounds structurally related to curcumin, and are present as minor components in ginger, the rhizomes of *Zingiber officinale* Rosco.[39] DHZ has been reported to have radioprotective,[40] antimicrobial,[41] and antioxidant[42] properties. In 2015, Lenardão and coworkers[43] synthetized a selenium-containing zingerone derivative with antioxidant properties. The strategy involved generating the nucleophilic selenium species in situ (PhSeSePh **25a**/$NaBH_4$/ethanol) in the presence of acetic acid at room temperature, to avoid the reduction of the carbonyl group. Subsequent Michael-type addition of the selenol to dehydrozingerone (**39**) gave derivative **40** in 21% yield after 4 hours at 0°C (Scheme 8.13). 4-(4-Hydroxy-3-methoxyphenyl)-4-(phenylseleno)-2-butanone (**40**) presented significantly improved antioxidant activity compared to that of dehydrozingerone (**39**), exhibiting twice the antioxidant activity in the ABTS assay. The improved activity of **40** was attributed to the ability of the phenylselanyl group to eliminate phenylselanyl radicals.

The tocopherols are important phenolic compounds. As a mixture of their α- to δ- forms, they constitute vitamin E, a well-known lipid-soluble antioxidant, with α-tocopherol (**41**) playing the most important chain-breaking antioxidant role.[44] With the aim of enhancing the antioxidant capacity of α-tocopherol, and recognizing that organoselenium compounds react rapidly with free radicals,[45] Schiesser and Engman synthesized tocopherol analog (**42**) using intramolecular homolytic substitution chemistry (Scheme 8.14).[46] Farnesol (**43**), a natural sesquiterpenoid,

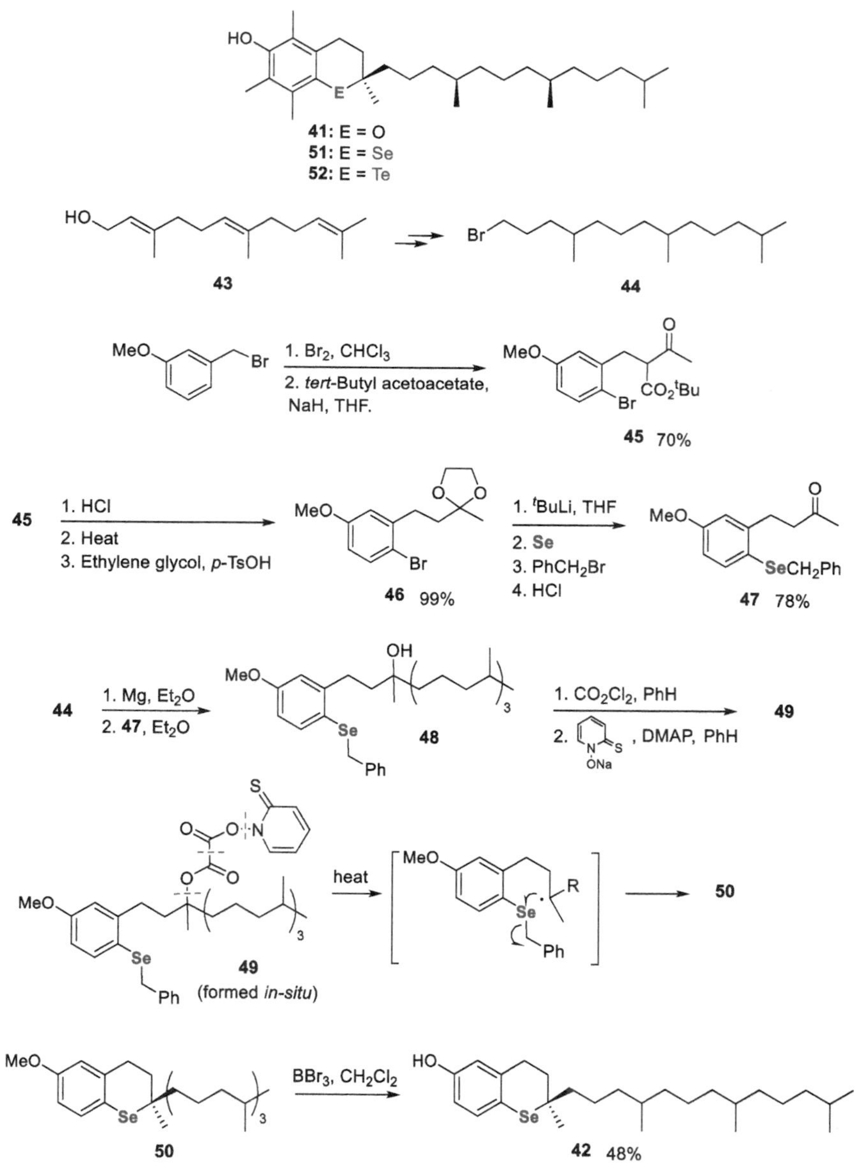

SCHEME 8.14 Synthesis of a selenium-containing tocopherol analog.[46]

was first reduced and then converted into 1-bromo-4,8,12-trimethyltridecane (**44**) as reported by Cohen and Schaer.[47] Concurrently, 3-methoxybenzyl bromide was electrophilically reacted with Br_2, and then with the enolate of *tert*-butyl acetoacetate to afford ketoester (**45**) in 70% yield. Acid hydrolysis, thermal decarboxylation, and deprotection gave 1,3-dioxolane (**46**) in almost quantitative yield, which was then reacted with *tert*-butyllithium, elemental selenium, and benzylbromide, to afford benzylselenide (**47**) in 78% from **46**. Further reaction with 4,8,12-trimethyltridecylmagnesium bromide (from **44**) afforded the tertiary alcohol **48** in 74% yield.

In a one-pot process, **48** was reacted with oxalyl chloride in benzene, followed by the sodium salt of *N*-hydroxypyridine-2-thione to generate the active radical precursor **49** in situ. Subsequent heating afforded the methyl-protected tocopherol analog (**50**) through cleavage of the weak N–O bond, double-decarboxylation, and intramolecular homolytic substitution of the tertiary carbon radical at selenium, with expulsion of the benzyl radical as the leaving group; **50** was converted into **42** by the action of BBr_3.

A similar strategy was later used by Engman and coworkers to prepare α-selenotocopherol (**51**), albeit in a low overall yield (6.6%).[48] Interestingly, **51** proved to be slightly less effective as an antioxidant than α-tocopherol itself.[48] Shanks also reported tentative steps toward the preparation of α-tellurotocopherol **52**[49]; however, to the best of our knowledge, **52** was never prepared.

8.4 Carbohydrates

Carbohydrates play a vital role in many biological processes and are important energy sources and crucial for the transport of energy. Moreover, they have essential structural functions and are important for the immune system, fertility, and biological development.[50]

The pharmacological activities of selenium-containing carbohydrates have been recently described. For instance, the octylselenoxylofuranoside (OSX) exhibits free-radical-scavenging activity in mice,[51] preventing damage to proteins and lipids, and is a promising antidepressant (Fig. 8.2).[52]

In 2000, Schiesser and coworkers[53] reported the thermolytic synthesis of 2,3,4-tri-*O*-benzyl-1,5-dideoxy-5-seleno-D-pentopyranose (**53**). The first step in the reaction sequence involved the conversion of D-arabinose into 2,3,4-tri-*O*-benzyl-5-*O*-methanesulfonylarabinose (**54**) in 59% yield, which was then converted into the benzylselenoarabitol (**55**) in 61% yield (Scheme 8.15). The alcohol (**55**) was then transformed to the selenocarbonate (**56**) (71% yield, three steps), and finally, when

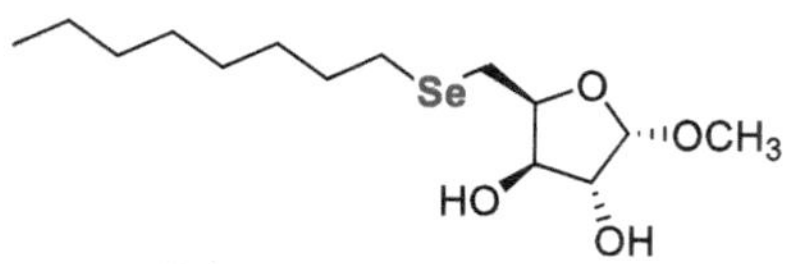

FIGURE 8.2 Structure of methyl octylselenoxylofuranoside (OSX).[51, 52]

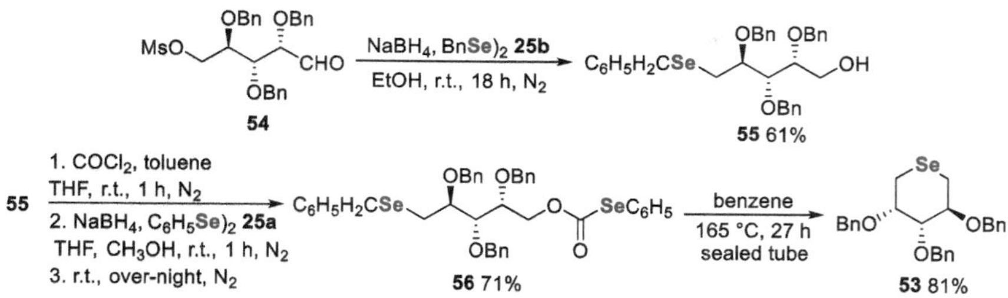

SCHEME 8.15 Synthesis of 2,3,4-tri-*O*-benzyl-1,5-dideoxy-5-seleno-D-arabinopyranose (53).[53]

heated at 165°C in benzene, **56** was converted into 2,3,4-tri-*O*-benzyl-1,5-dideoxy-5-seleno-D-arabinopyranose (**53**) in 81% yield.

In the same way, L-arabinose was converted into 2,3,4-tri-*O*-benzyl-1,5-dideoxy-5-seleno-L-arabinopyranose in 74% yield, while D-ribose and D-xylose afforded compounds **57** and **58**, respectively. These substrates were converted into aldehydes **59** and **60** (3 steps, 99% and 72% yield), which were the intermediates for the synthesis of selenocarbonates **61** and **62**. The protected cyclic selenosugars, namely, 2,3,4-tri-*O*-benzyl-1,5-dideoxy-5-seleno-D-ribopyranose (**63**) and 2,3,4-tri-*O*-benzyl-1,5,-dideoxy-5-seleno-D-xylopyranose (**64**) were obtained in yields of 56% and 96% from the respective selenocarbonates **61** and **62** (Scheme 8.16).[54]

Nguyen and Schiesser also described the synthesis of tellurosugar (**67**), starting from D-arabinose.[55] The sequence first involved the synthesis of 2,3,4-tri-*O*-benzyl-5-*O*-methanesulfonyl-D-arabitol (**65**), which was dimesylated to **66** in 89% yield. After treatment with Na_2Te, 2,3,4-tri-*O*-benzyl-1,5-dideoxy-5-telluro-D-arabinose (**67**) was obtained in 82% yield as a single enantiomer (Scheme 8.17). In a similar way, L-arabinose effectively afforded 2,3,4-tri-*O*-benzyl-1,5-dideoxy-5-telluro-L-arabinose, the enantiomer of **67**.

In 2010, Lüdtke and coworkers[56] described the synthesis of 5′-selenoxylofuranosides (**68a–h**), which were prepared from D-xylose. First, the starting material was converted into the tosylate (**69**) that, upon substitution with a nucleophilic selenium species, afforded seleno-substituted products (**68**) after 24 hours of reaction in yields of between 72% and 88%. The reaction was observed to be tolerant to a variety of

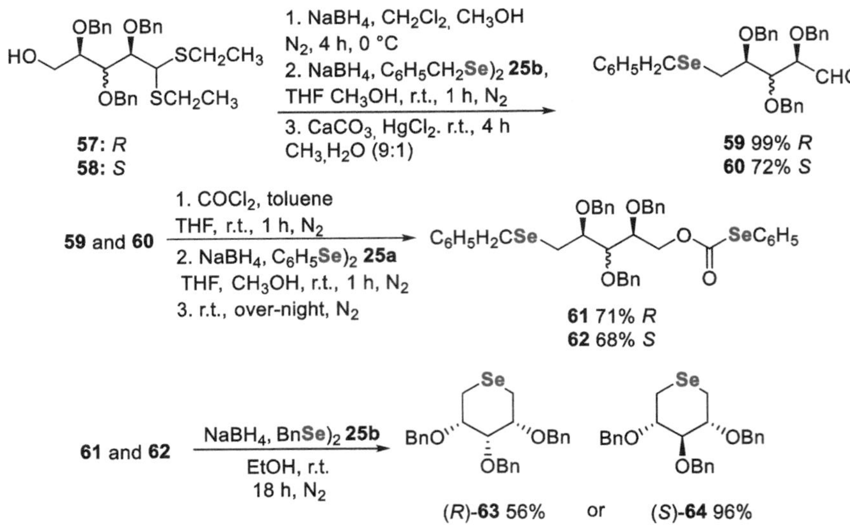

SCHEME 8.16 Synthesis of cyclic selenosugars **63** and **64**.[54]

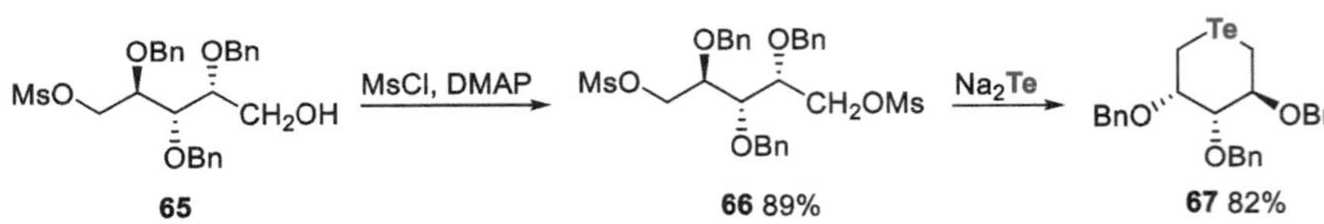

SCHEME 8.17 Synthesis of 2,3,4-tri-*O*-benzyl-1,5-dideoxy-5-telluro-D-arabinose (**67**).[55]

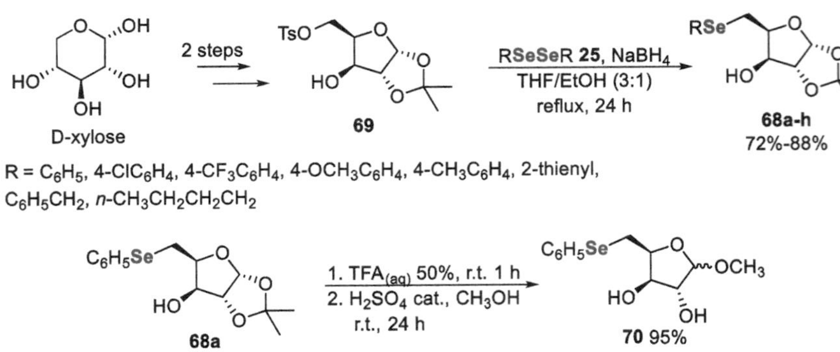

SCHEME 8.18 Synthesis of selenoxylofuranoses (**68a–h**).[56]

substituents on the organoselenium moiety. Additionally, product **68a** (R = C_6H_5) was transformed into the methyl seleno-xylofuranoside (**70**) in 95% yield after deprotection, followed by treatment with methanol in an acidic medium (Scheme 8.18). In vivo studies revealed that **68a** (R = C_6H_5) exhibited protective effects on long-term memory and on

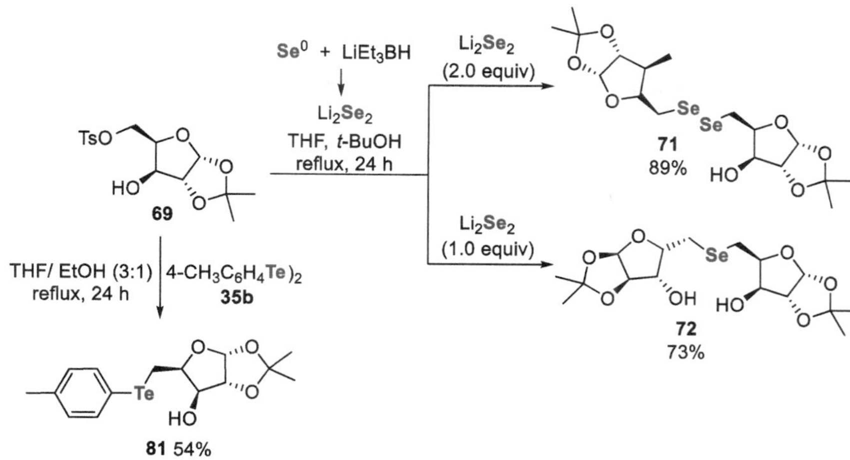

SCHEME 8.19 Synthesis of sugar-derived diselenide (**71**), selenide (**72**), and telluride (**81**).[56]

oxidative stress in mice by acting on the sodium–potassium pump and on acetylcholinesterase.[57]

Moreover, by treating tosylate (**69**) with lithium diselenide, the sugar-derived diselenide (**71**) and selenide (**72**) were obtained in yields of 89% and 73%, respectively (Scheme 8.19). The tellurocarbohydrate (**81**) was prepared in 54% yield (Scheme 8.19) by the reaction of (**69**) with *p*-tolylditelluride under reflux for 24 hours. This telluride showed the ability to reduce Mn-induced oxidative stress in worms,[58] while compounds **60a** and **71** demonstrated antioxidative activities in vitro, inhibiting the enzyme δ-aminolevulinate dehydratase (δ-ALA-D).

Diselenide (**71**) was subjected to reductive cleavage with $NaBH_4$, followed by alkylation with propargyl bromide (**75**) and allyl bromide (**76**) to afford the respective nonsymmetrical selenides (**73**) and (**74**) in yields of 92% and 94%, respectively. When benzoyl chloride (**77**) was used as the electrophile, the selenoester (**78**) was obtained in 95% yield, while the reaction with aziridine (**79a**), derived from the amino acid L-valine, afforded the seleno-glycoconjugate (**80**) in 83% yield (Scheme 8.20).

This strategy was used by the same group in 2010[59] to prepare seleno-carbohydrates **82a–d** in excellent yields (91%–98%) from D-galactose, which was first converted into the key tosylate and mesylate intermediates **83** and **84** (Scheme 8.21).

When tosylate (**83**) was reacted with lithium diselenide, the diseleno galactopyranoside (**85**) was obtained in 87% yield, which was then reductively cleaved with $NaBH_4$ and subsequently treated with several electrophiles, to afford nonsymmetrical selenides (**86a–d**) in yields of 75%, 76%, 83%, and 83%, respectively (Scheme 8.22).

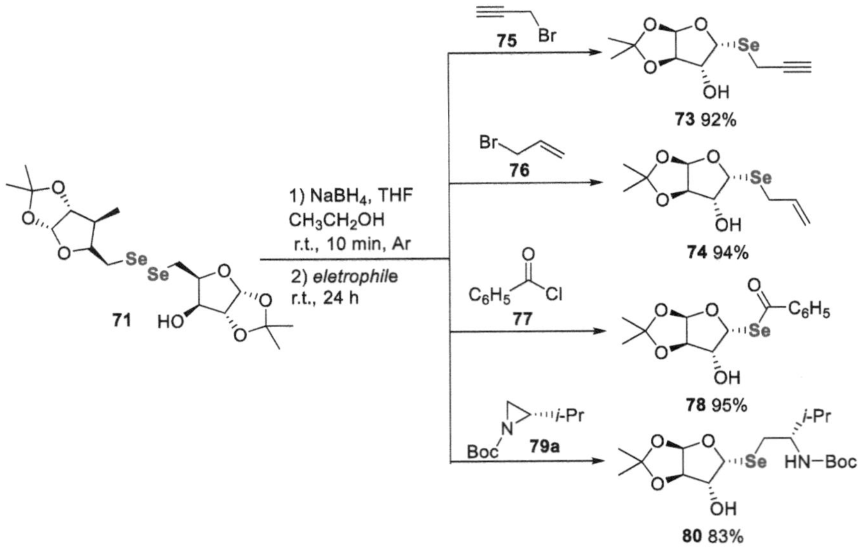

SCHEME 8.20 Synthesis of functionalized seleno-furanosides **73, 74, 78,** and **80.**[56]

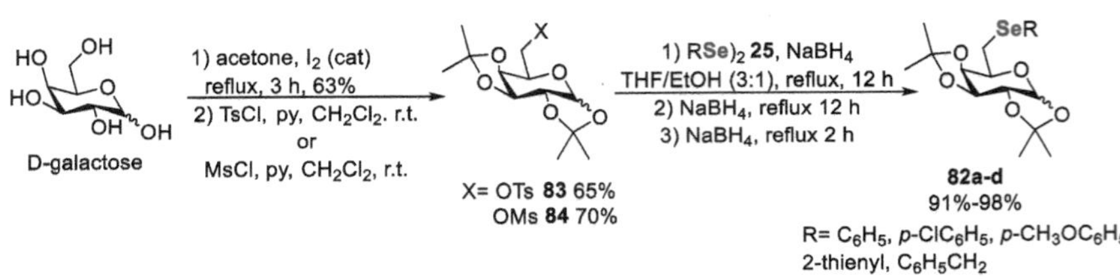

SCHEME 8.21 Synthesis of seleno-carbohydrates **82a−d.**[59]

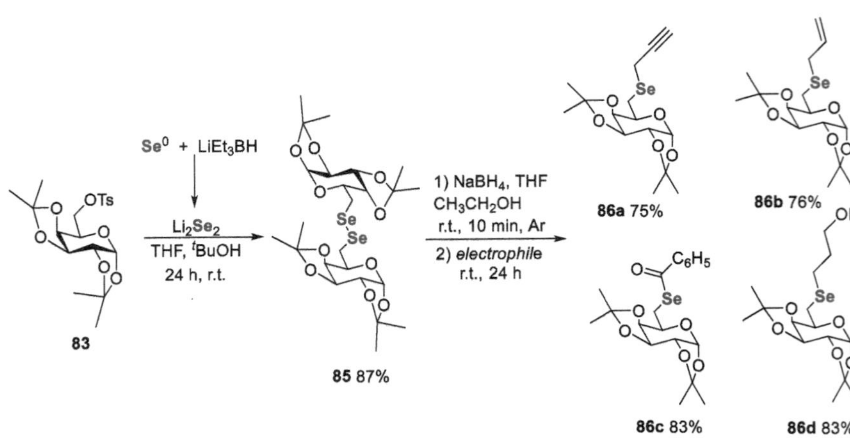

SCHEME 8.22 Synthesis of galactopyranoside (**85**) and functionalized seleno-galactopyranosides (**86a−d**).[59]

In 2012, Lüdtke and coworkers[60] reported the preparation of selenium-linked neoglycoconjugates **87a–h** in yields that ranged from 30% to 85% by reacting the previously prepared selenium-containing D-xylose, D-galactose, and D-ribose carbohydrates (**88a–c**) and various *N*-Boc aziridines (**89a–e**) (Scheme 8.23).

Moreover, the authors reported a method for preparing pseudodisaccharides (**91a–c**) in 50%–95% yield by the reactions of diselenides (**88a–c**) with tosylates (**90a**) and (**90c**). Interestingly, the reaction of diselenide **89a** with tosylate **90c** was more effective than the reaction of **89c** with **90a** as a strategy for preparing selenide **91a** (Scheme 8.24).

While the above-mentioned examples reveal that numerous selenium-containing carbohydrate derivatives can be prepared, including endocyclic ones, unprotected endocyclic selenosugars have remained largely unexplored since some early reports by van Es and Whistler.[61,62] Recognizing that such selenosugars would be potent water-soluble antioxidants useful for the treatment of cardiovascular disease,[54,63] Davies and Schiesser prepared a number of 1,5-anhydro-5-selenohexitols. Inspired by the work of Pinto and coworkers, who prepared selenonium sulfates (e.g., **92**) as potential glycosidase inhibitors[64] for the treatment of diabetes, 1,5-anhydro-5-seleno-L-gulitol (**93**), 1,5-anhydro-5-seleno-D-mannitol (**94**), and 1,5-anhydro-5-seleno-L-iditol (**95**) were straightforwardly prepared (Scheme 8.25).[54,65]

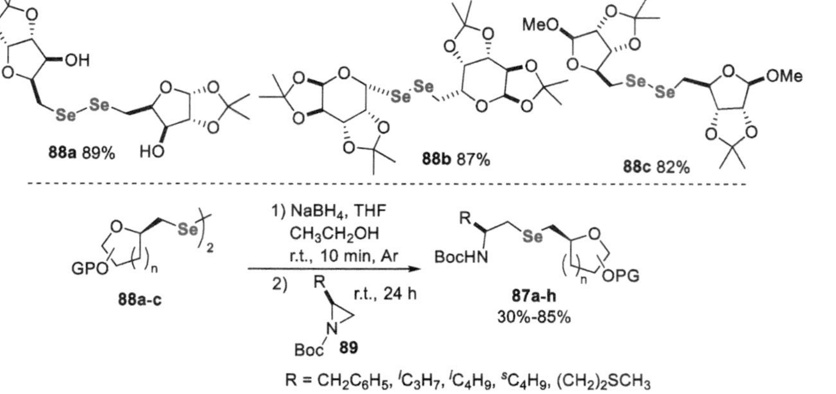

SCHEME 8.23 Synthesis of selenium-linked neoglycoconjugates (**87a–h**).[60]

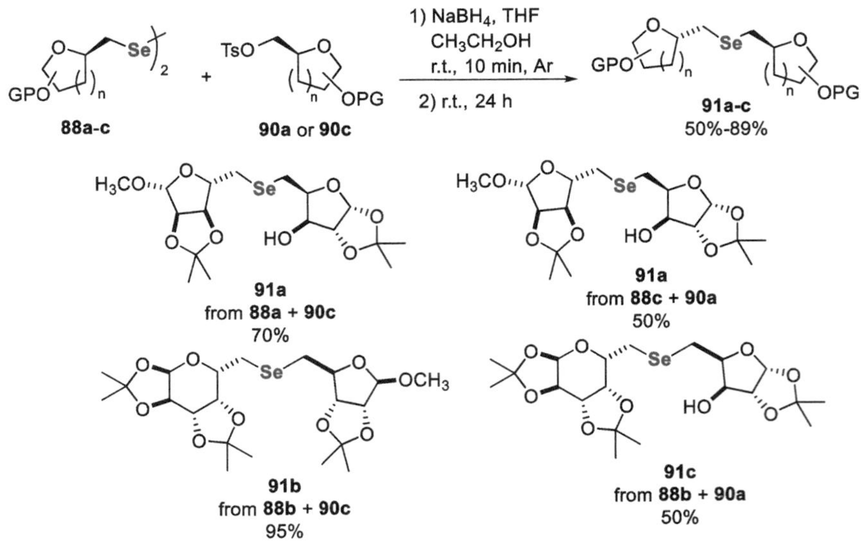

SCHEME 8.24 Synthesis of pseudodisaccharides (**91a–c**).[60]

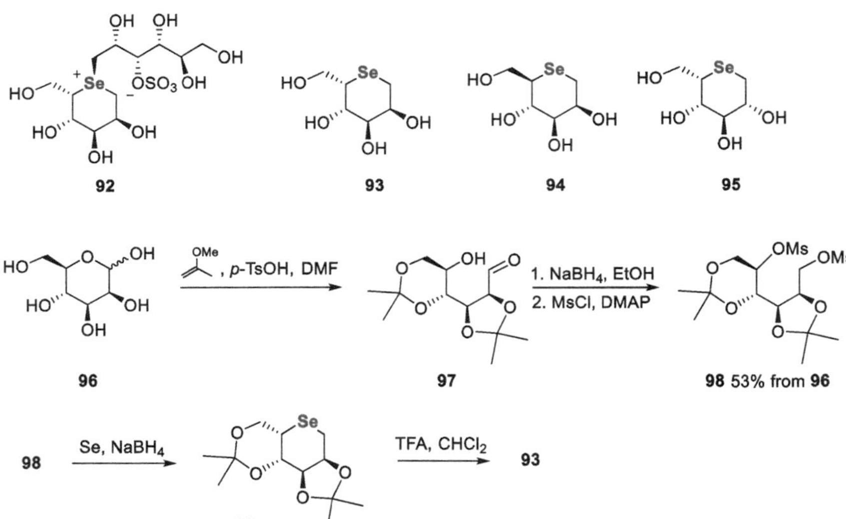

SCHEME 8.25 Selenonium sulfate (**92**) and the preparation of 1,5-anhydro-5-selenohexitols (**93–95**).[54,65]

The synthesis of **93** began with D-mannose (**92**), which was protected with 2-methoxypropene to afford aldehyde **97**; this aldehyde was immediately reduced with sodium borohydride and then dimesylated with methanesulfonyl chloride under standard conditions to afford **98** in 53% yield over three steps. Subsequent reaction with disodium selenide, prepared by the reduction of elemental selenium with $NaBH_4$, afforded the protected selenosugar **99** in 61% yield; careful deprotection with TFA provided 1,5-anhydro-5-seleno-L-gulitol (**93**) in moderate yield. The other selenohexitols (**94**) and (**95**) were prepared in a similar manner.[54,65]

Because selenosugars **93**–**95** were slightly cumbersome to prepare, the same authors prepared the corresponding 1,4-anhydro-4-selenohexitols and pentitols, namely, 1,4-anhydroseleno-D-tallitol (**100**, SeTal), 1,4-anhydroseleno-L-tallitol (**101**, *ent*-SeTal), 1,4-anhydroseleno-DL-glucitol (**102**), and 1,4-anyhdro-4-seleno-L-lyxitol (**103**), which were more straightforwardly prepared and purified by adapting Pinto's protocol,[54,63,65,66] along with 1,6-anhydro-6-selenogalacticol (**104**), a seven-membered-ring analog, as shown in Scheme 8.26.[54,63]

SeTal (**100**) was prepared by first reacting D-mannose (**96**) with 2,2-dimethoxypropane followed by reduction with $NaBH_4$, which afforded acetonide **105** in good yield. Dimesylation, reaction with disodium

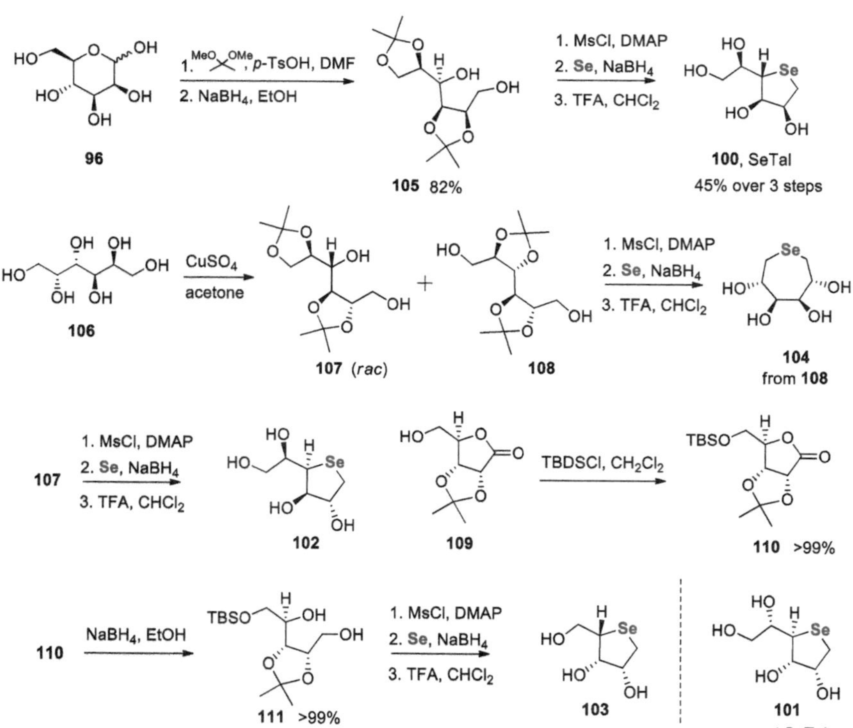

SCHEME 8.26 Preparation of 1,4-anhydro-4-selenohexitols and pentitols **100**–**103** and 1,6-anhydro-6-selenogalacticol (**104**).[54,63,65,66]

selenide, and careful deprotection with TFA, as described above, yielded **100** in 45% yield over three steps as a white crystalline solid. The same procedure applied to L-mannose afforded the enantiomeric SeTal (**101**). Diastereomeric 1,4-anhydroseleno-DL-glucitol (**102**) was prepared from achiral galactitol (**106**), which was reacted with acetone and anhydrous $CuSO_4$ (as the dehydrating agent) to afford a mixture of acetonides **107** and **108**.[67] Interestingly, **107** and **108** were efficiently separated by "old-school" fractional crystallization, which was found not to be possible when other dehydrating agents were used.[68]

Selenopentitol **103** was prepared from the commercially available acetonide-protected lactone **109**, which was further TBS-protected under standard conditions to afford **110** in quantitative yield. Reduction of **110** with $NaBH_4$ gave diol **111**, which was then transformed into 1,4-anyhdro-4-seleno-L-lyxitol (**103**) under standard conditions.

Selenosugars **93–95** and **100–104** were patented as novel water-soluble cardioprotective agents,[54] while **100–104** were further patented as novel skin-repair molecules.[69] Extensive biological testing revealed that the above-mentioned selenosugars are potent water-soluble antioxidants and are about two orders of magnitude more reactive than their sulfur counterparts toward HOCl.[66] In addition, SeTal (**100**) has been shown to react rapidly with both amino-acid- and protein-derived chloramines and detoxify both HOCl, the primary oxidant produced by myeloperoxidase, a mammalian heme enzyme whose function is to kill invading pathogens, and secondary chloramines produced by the reactions of HOCl with amino residues, such as lysine in proteins.[70]

Quite fortuitously, 1,4-anhydroseleno-D-tallitol (**100**, SeTal) was found to significantly increase the rate of wound-healing in both wildtype (C57BL/6 J) and diabetic (db/db) mice, with the rate of healing of the db/db mice increased to a level similar to that of the wildtype mice, but from a significantly lower level.[69,71] Apparently, **100** can overcome the wound-healing disruptions often associated with diabetes. Unexpectedly, SeTal would appear to be a privileged structure; in "in vitro" studies, it prevented pyrogallol and high-glucose-induced endothelial dysfunction in mouse aorta in a manner that its isomers and other related structures **101–104** are unable to, suggesting that this selenosugar is uniquely capable of repairing endothelial dysfunction brought on by the diabetic condition.[63,72] In addition, it is important to note that **100** is a stable molecule, with no discernible toxicity in a number of cell studies. SeTal does not appear to affect cell viability or metabolic function in mouse studies, nor does it appear to upregulate the protein levels of (Se-dependent) thioredoxin reductase or glutathione peroxidase-1,[73] nor does it elicit an inflammatory response when applied to mouse wounds.[69,71] Indeed, SeTal appears to operate in a catalytic cycle involving its (stable) selenoxide.[74]

In order to complete the vertical series, Borges, Schiesser, and Davies recently prepared the tellurium analogs **112–116** of selenosugars **100–104**, with the synthesis of TeTal **(112)**, the tellurium analog of SeTal, presented as an example in Scheme 8.27.[75] The use of NaHTe in PEG-400 as the solvent proved to offer an operational advantage over the use of ethanol, presumably because oxygen, which rapidly oxidizes NaHTe, is less diffusible in the PEG solvent. Not unexpectedly, these tellurosugars proved to be excellent water-soluble antioxidants.[75]

8.5 Nitrogen-containing heterocycles

The five nitrogen-containing bases used to construct nucleosides are adenine, guanine, uracil, cytosine, and thymine. They are classified as purine and pyrimidine derivatives and are the bases present in the structures of DNA and RNA.[76] In addition, a variety of nucleoside analogs derived from purine and pyridine systems have been approved for cancer and viral-disease therapies.[77,78] In this sense, nitrogenous bases functionalized with organoselenium groups constitute a class of molecule with potential biological applications, some of which are reported below.

In 2011, Minakawa and coworkers[79] described the synthesis of 4′-selenouridine **117** and 4′-selenocytidine **118**, substrates for the synthesis of 4′-selanyl-RNA. Initially, D-ribose was converted into the protected 4-selenosugar **119** in several steps, after which it was subjected to a Pummerer-like reaction with iodosylbenzene, trifluoromethane sulfonate (TMSOTf), 2,6-lutidine as a base, and silylated uracil in dichloroethane at 50°C, to give intermediate **120** in 64% yield. In this sequence, **120** was converted into the 4′-selenopyrimidine nucleosides **117** and **118**. The 4′-selenouridine **121**, was prepared in 91% yield by the

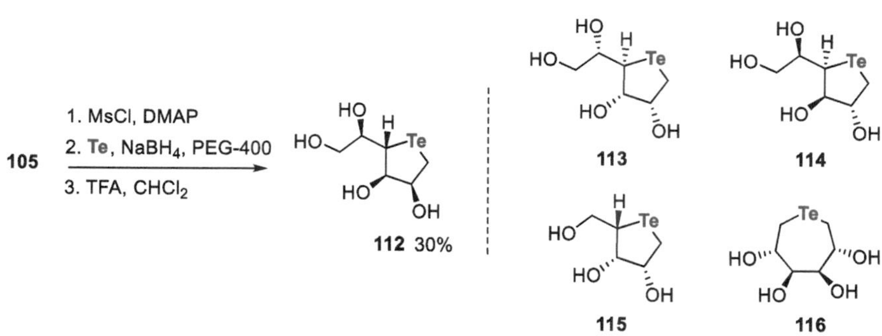

SCHEME 8.27 1,4-Anhydro-4-tellurohexitols and pentitols, and 1,6-anhydro-6-telluro-galacticol (112–116).[75]

treatment of **120** with tetrabutylammonium fluoride (TBAF), followed by the reaction with methylamine. To synthesize 4′-selenocytidine **(118)**, intermediate **120** was first treated with triisopropylbenzenesulfonyl chloride (TPSCl), followed by ammonium hydroxide, to give the protected cytidine intermediate **121** in 91% yield. Deprotection of **121** as described above for **119**, afforded 4′-selenocytidine **118** in 64% yield (Scheme 8.28).

Alternatively, 4′-selenopurine nucleosides can be prepared from purine derivatives, such as 6-chloropurine, which facilitated the introduction of the purine unit into the final product. In 2013, Jeong and coworkers[80] reported the synthesis of new RNA purine building blocks, namely 4′-selenoadenosine (**122a**) and 4′-selenoguanosine (**122b**), starting with D-ribose and 6-chloropurine. The key step of the synthesis was the reaction between the protected selenosugar acetate **123** and 6-chloropurine in the presence of TMSOTf, which afforded a mixture of the N7 (**124**, 15% yield) and N9 (**125**, 49% yield) regioisomers, through the N7 to N9 rearrangement (Scheme 8.29).

Using a similar strategy, the 2-amino-6-chloropurine derivatives **126** and **127** were also prepared from selenosugar **123**, as shown in Scheme 8.29. These protected purine derivatives were converted in a few steps into 4′-selenoadenosine **122a** and 4′-selenoguanosine **122b**. These new

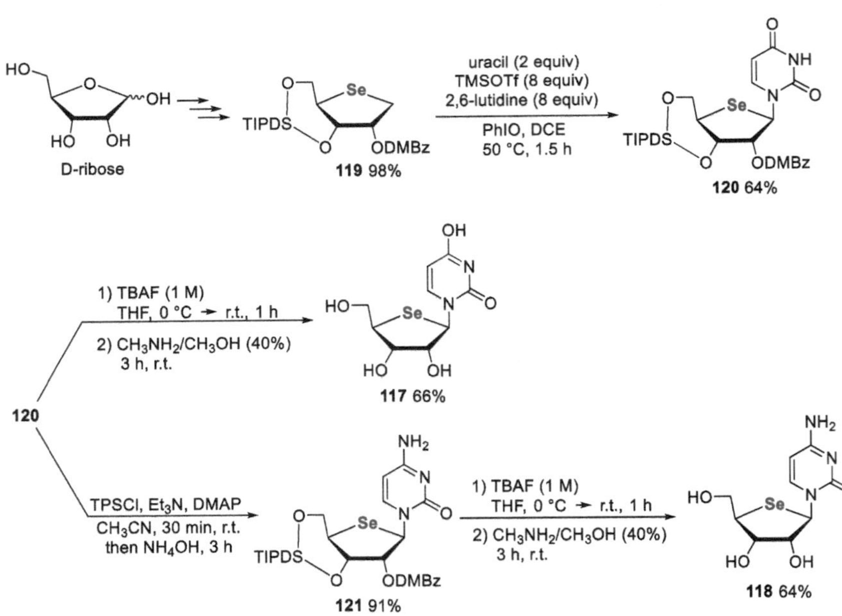

SCHEME 8.28 Preparation of 4′-selenouridine **117** and 4′-selenocytidine **118**.[58]

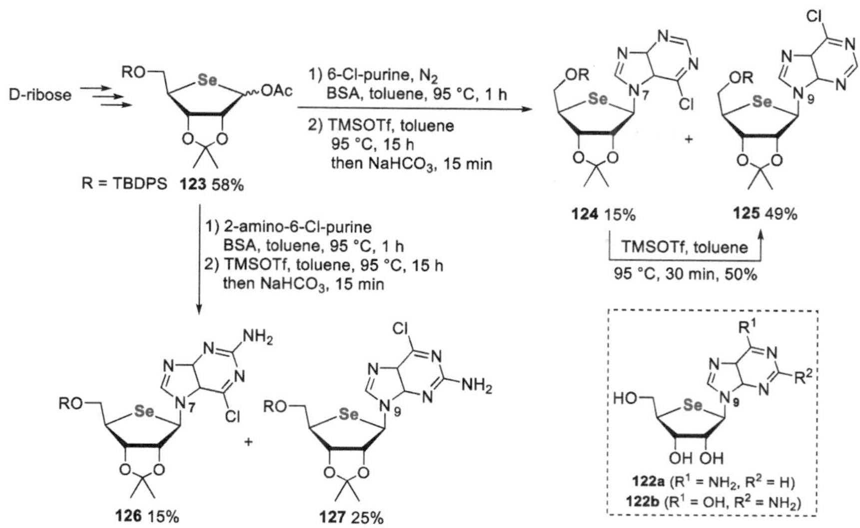

SCHEME 8.29 Rearrangement of the N7 isomers **124** and **129** to the N9 ones **125** and **127**.[80]

RNA purine building blocks may be valuable for the design of therapeutically as well as biochemically useful antimetabolites and oligonucleotides.

In 2016, Minakawa and coworkers[81] developed a similar protocol that involved combining purine bases (6-chloropurine and 2,6-dichloropurine) and 4-selenosugars bearing electron-donating groups at their 2-positions. Through a one-pot Pummerer-like reaction, several 4′-selenopurine nucleosides were obtained in moderate to good yields starting from selenosugar **128**. The reaction of the hypervalent iodine intermediate derived from **128** with 6-chloropurine provided the N7–6-chloropurine isomer **129** (31% yield), which isomerized to the desired N9 isomer **130** (53% yield) after treatment with trimethylsilyl TMSOTf in toluene at 90°C.

In a similar manner, 2-amino-6-chloropurine formed a mixture of the desired N9 isomer **131** (16% yield) and the N7 isomer **132** (31% yield) after heating for 8.5 hours. In the 2,6-dichloropurine base, the coupling reaction proceeded smoothly to give a mixture of the N7 isomer **133** and the N9 isomer **134** in 64% yield as a 1:1 inseparable mixture. This mixture was subsequently treated with TMSOTf in toluene at 90°C to exclusively afford the desired product **134** in 62% yield. Two more steps afforded guanosine derivative **135** in 77% isolated yield from **134** (Scheme 8.30).

A different strategy was adopted by Alves and coworkers[82] for the preparation of 6-arylselanylpurines diaryl diselenides **25** and

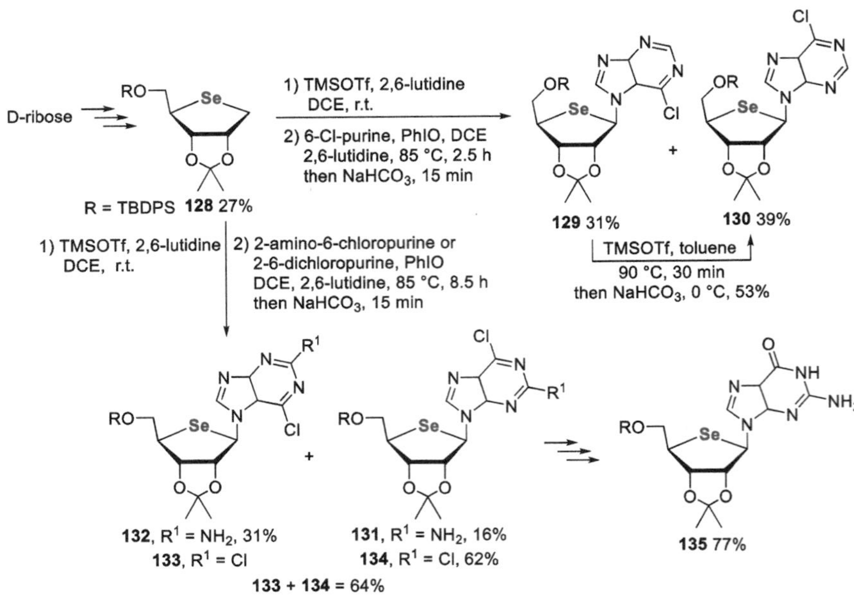

SCHEME 8.30 Synthesis of 4′-selenoguanosine derivatives.[81]

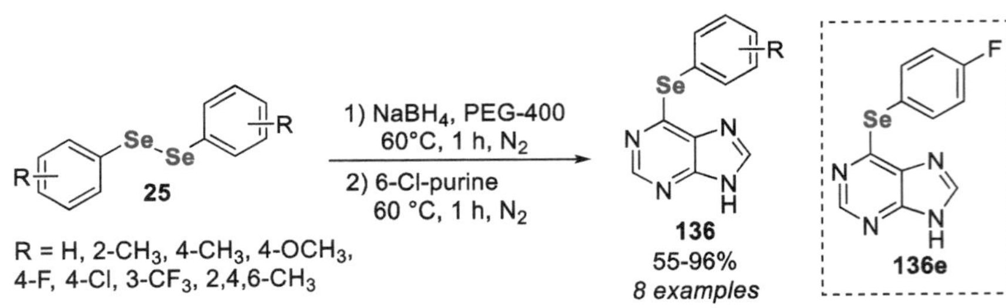

SCHEME 8.31 Synthesis of 6-arylselanylpurines **136**.[82]

6-chloropurine. In a typical procedure, diaryl diselenides **25** were reduced with $NaBH_4$ in PEG-400 as the solvent (1 hour, 60°C under N_2) after which 6-chloropurine was added to the solution of the arylselenolate anion generated in situ, and the mixture was stirred at 60°C for an additional 1 hour. Using this protocol, the authors were able to prepare eight samples of 6-arylselanylpurines in yields of between 55% to 96% (Scheme 8.31).

It should be noted that under the same reaction conditions, other *N*-heterocycles (chloropyrimidine and chloroquinoxaline) also performed well to provide the desired selenylated products in excellent yields.[82] Moreover, fluorine-substituted **136e** behaved as a cerebral acetylcholinesterase (AChE) inhibitor. When administered to mice, this 6-arylselanylpurine caused cognitive enhancement during the memory phase and is considered to be a promising therapeutic agent for the treatment of Alzheimer's disease.

Tecadenoson (**137a**) was developed as a novel selective adenosine receptor type 1 (A_1AR) agonist for the conversion of tachycardia (paroxysmal supraventricular tachycardia) to sinus rhythm, and along with ethylene n-butyl acrylate (ENBA) (**138a**) is among the most-effective A_1R agonists reported.[83,84] In 2007, Schiesser and Scammells reported the preparation of the selenium analog **137b** of tecadenoson (Scheme 8.32).[85] Starting with boc-protected D-asparaginol **139**, a (standard) protocol involving dimesylation and selenation afforded boc-protected (*R*)-3-aminotetrahydroselenophene **140**; subsequent careful treatment with hydrogen chloride in ether afforded (*R*)-3-aminotetrahydroselenophene which was isolated as its hydrochloride salt **141** for stability reasons (87% from **139**). The final step in the synthesis of **137b** involved its reaction with 6-chloropurine riboside (**142**) in the presence of base, which afforded **137b** in 96% yield. Interestingly, selenium analog **137b** proved to be a more potent A_1AR agonist than tecadenoson itself, with a K_i value of 8 ± 2 nM and an IC_{50} of 1.9 ± 0.1 nM (c.f., **137a**: $K_i = 65 \pm 26$ nm; $IC_{50} = 8.2 \pm 1.2$ nm).

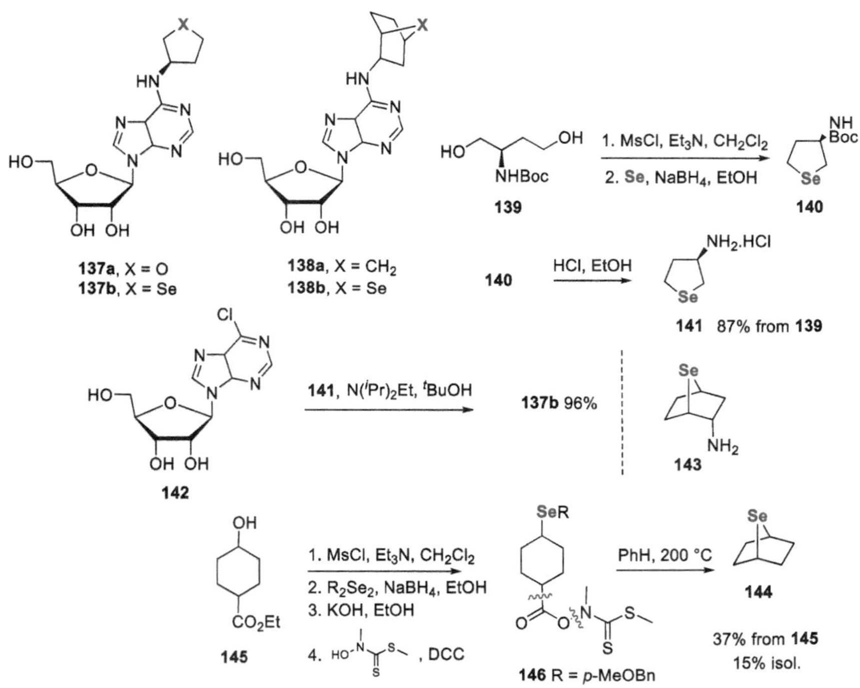

SCHEME 8.32 Synthesis of selenium-containing A_1AR agonists.[85,86]

The preparation of the selenium-containing analog of ENBA, that is, **138b**, proved to be quite challenging, since it required the synthesis of 2-amino-7-selenabicyclo[2.2.1]heptane (**143**), the parent skeleton of which had not been prepared prior to 2012. In order to explore the effectiveness of radical chemistry for the preparation of the bicyclic framework, the same authors demonstrated that 7-selenabicyclo[2.2.1] heptane (**144**) could be prepared straightforwardly from ethyl 4-hydroxycyclohexanecarboxylate (**145**).[86] Mesylation of **145** under standard conditions followed by reaction with sodium *p*-methoxybenzylselenoate (prepared from (p-MeOBn)$_2$Se$_2$ and NaBH$_4$ in EtOH), hydrolysis of the product ester, and coupling with *N,S*-dimethyl-*N*-hydroxydithiocarbamate, afforded the radical precursor **146**, which was not purified, but heated in benzene at 200°C (sealed vessel) to afford **144** in 37% conversion (Scheme 8.32). Owing to its volatility, **144** was difficult to isolate. It should be noted that (p-MeOBn)$_2$Se$_2$ was used as the selenating agent rather than the more-traditional Bn$_2$Se$_2$ (e.g., see Scheme 8.14) because benzylated byproducts from the radical reaction could not be separated from the desired product. It is interesting to note that this hitherto unknown parent heterocycle could be so easily prepared, albeit in low yield.

8.6 Miscellaneous

L-Ascorbic acid (vitamin C) is a powerful water-soluble antioxidant, essential for the correct functioning of the body. It is involved in a number of biological processes, ranging from the biosynthesis of collagen[87] and catecholamine to the modulation of neurotransmission.[88]

With this in mind, seleno- and telluro-containing 6-*O*-L-ascorbyl esters **147** were directly prepared by reacting vitamin C with suitable functionalized chalcogenoesters through a lipase-catalyzed transesterification reaction.[89] The novel L-ascorbyl derivatives **147** were obtained in yields of 53%–77% (Y = Se), and 67%–74% (Y = Te) by reacting three equivalents of β-seleno- or telluroester **148** and **149** with L-ascorbic acid in the presence of Lipase B, molecular sieves, and acetone at 45°C (Scheme 8.33). Eight of the synthesized selenium- and tellurium-containing vitamin C derivatives behave as catalysts that promote the reduction of hydrogen peroxide in the presence of a thiol cofactor (*in vitro* GPx-like activity). Moreover, these compounds exhibited catalytic thiol-peroxidase like properties, promoting the oxidation of both DTT and GSH.

The use of polysaccharides, which are claimed to be potential substitutes for oil-derived polymers, has been explored in order to elaborate several kinds of material that are used in a variety of areas. In

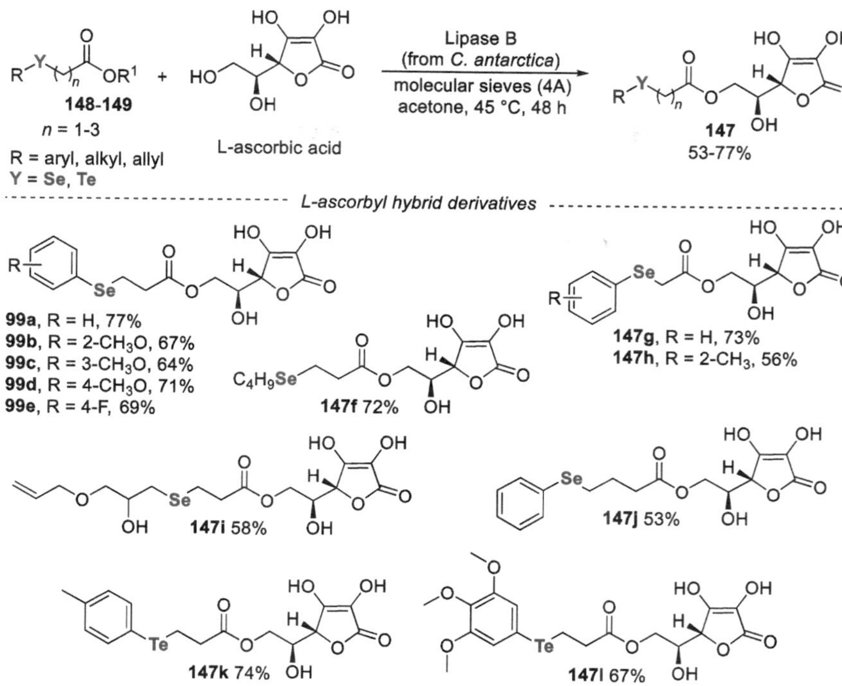

SCHEME 8.33 Synthesis of selenium- and tellurium-containing 6-*O*-ascorbyl esters **147**.[89]

particular, chitosan (CS), a chitin derivative, exhibits interesting properties, such as biodegradability, biocompatibility, nontoxicity, and antimicrobial activity, among others.[90] Aiming to improve some of these intrinsic properties and widen its applicability, Fajardo and coworkers[90] chemically modified chitosan. Selenium-containing CS derivative **150** was synthesized through the reaction of raw CS with α-phenylselanyl citronellal (**28**) in aqueous AcOH at 55°C for 3 hours. Using this strategy, CS derivative **150** was obtained in 23% conversion (calculated from ^{1}H NMR data) (Scheme 8.34).

The applicability of the CS derivative was explored through the preparation of films based on the modified CS blended with poly(vinyl alcohol) (PVA), using a solvent casting method. The authors performed in vitro and in vivo experiments in order to investigate the antimicrobial activities and potential use of these CS derivative/PVA films for the treatment of atopic dermatitis-like skin lesions. The results showed that the selenium-containing CS derivative **103** was effective in controlling the tested bacteria and fungi. The matrix formed by CS-derived organochalcogen compounds are less susceptible to hydrolysis than the parent unmodified CS. Additionally, this new film also reduced the severity of skin lesions induced by DNCB (2,4-dinitrochlorobenzene) in mice.[91]

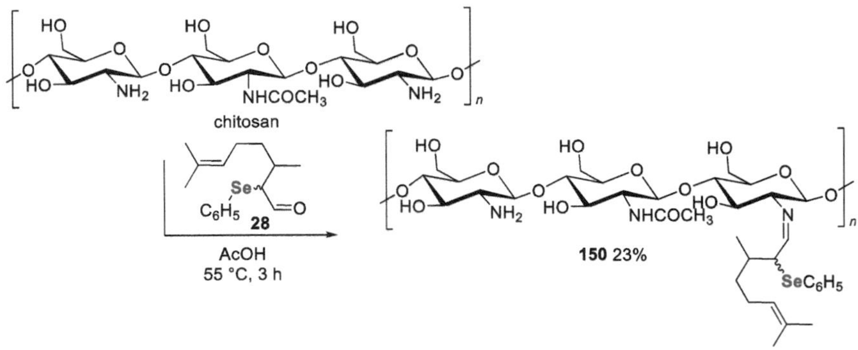

SCHEME 8.34 Synthesis of selenium-containing chitosan derivative **150**.[91]

Nucleic acids are macromolecules composed of nucleotides and are important for the storage of genetic information[92] during expression and translation.[93,94] Moreover, nucleoside analogs are recognized to have antiviral,[95] antitumor,[96,97] and antifungal properties.[98,99] Chalcogen-modified nucleotides have been developed for therapeutic gene-silencing purposes or to prepare anti-HIV drugs.[98,99] Some of this work is concisely summarized below.

In 2006, Serganov, Micura, and coworkers[100] reported the synthesis of site-specifically modified 2′-methylseleno RNAs. The key intermediate, 2′-methylselenoguanosine **151**, was prepared in several steps from 9-[β-D-arabinofuranosyl]guanine **152**, *via* the triflate intermediate **153**. The insertion of the methylselenium group was achieved by nucleophilic substitution with the nucleophilic selenium being generated *in situ* from MeSeSeMe **(25c)**, to afford intermediate **154** in 87% yield. The desired 2′-methylselenoguanosine **151** was obtained in 6% overall yield (nine steps) and was incorporated into RNA by solid-phase synthesis. The Se-modified RNA is a powerful tool in RNA X-ray structure analysis (Scheme 8.35).

The same selenoguanosine **151** (without the NPE-protecting group) was prepared few years later by Huang and coworkers using a similar strategy.[101] In this case, however, the seleno compound was used to prepare a 2′-Se-modified DNA octamer for X-ray crystallography studies.

A close-related strategy was used by Huang and coworkers in 2007 to prepare 2′-methylselenothymidine phosphoramidite **155**, itself used to prepare Se-derivative oligonucleotides.[102] In this case, the D-ribose epimer and thymidine were used instead of D-arabinose and guanine, respectively, as starting materials. The key selenium-containing intermediate **156** was prepared in 90% yield by the reaction of **157** with methylselenolate, generated in situ from **25c** and $NaBH_4$/EtOH (Scheme 8.36).[102]

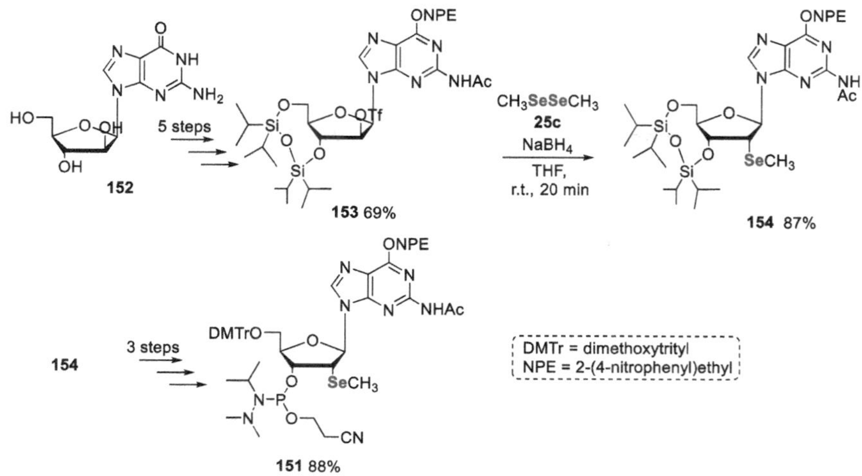

SCHEME 8.35 Preparation of 2′-methylselenoguanosine phosphoramidite **151**.[100]

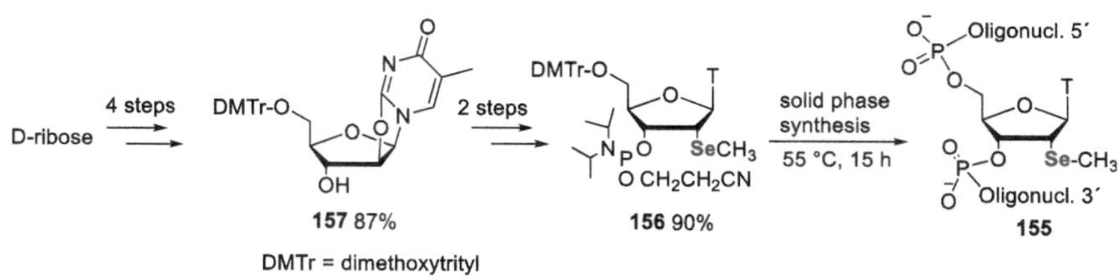

SCHEME 8.36 Synthesis of 2′-methylselenothymidine phosphoramidite **155**.[102]

In 2008, Pinto and coworkers[103] described the use of the Pummerer glycosylation reaction to prepare 4′-β-selenonucleosides derived from D-ribose and four different bases: adenosine, cytidine, thymidine, and uridine. The reaction involved the preparation of protected 1,4-anhydro-4-seleno-D-ribitol **158**, as a key intermediate, which was prepared by the reaction of **159** with Na_2Se, generated in situ from elemental selenium and $NaBH_4$. The oxidation of the seleno-D-ribitol derivative **158** with *m*CPBA gave selenoxide **160** in 95% yield, which after the Pummerer reaction using the silylated purine or pyrimidine base, stereoselectively the afforded the corresponding 4′-β-selenonucleosides **161**–**164** in yields of 59%–80% (Scheme 8.37).

In 2009, Braga, Lüdkte, and coworkers[104] described the synthesis of selenium- and tellurium-functionalized nucleosides **165a**–**i** derived from uridine. After protection with $HC(OMe)_3$ and tosylation of the 5′-hydroxyl group, intermediate **166** was obtained in 95% yield (two steps). The tosyl group then underwent nucleophilic substitution with selenolate or tellurolate, generated in situ from the respective diorganyl

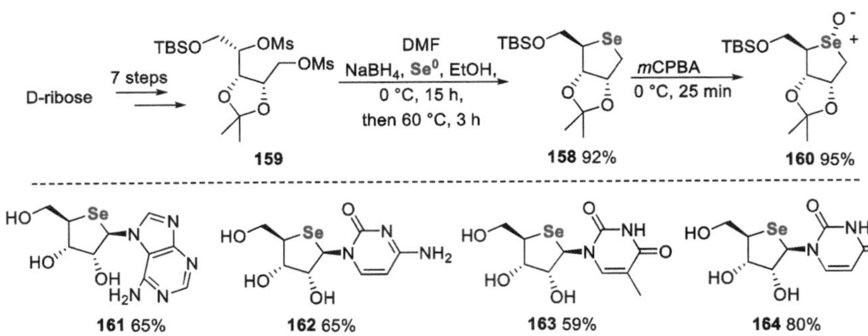

SCHEME 8.37 Synthesis of 4′-β-selenonucleosides **161–164**.[103]

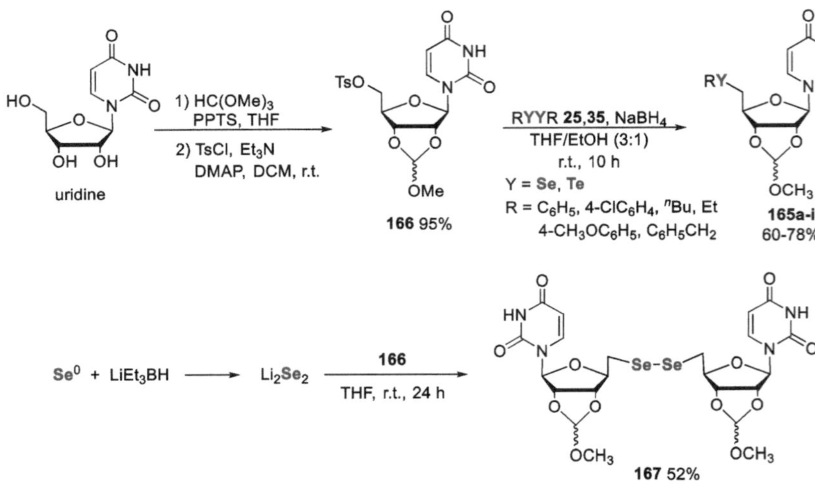

SCHEME 8.38 Synthesis of selenium- and tellurium-containing nucleosides.[104]

dichalcogenide ($NaBH_4$/THF/EtOH), to afford the respective 5′-chalcogenyl substituted nucleosides in yields of 60%–78%. When Li_2Se_2 was used instead, compound **166** reacted to form diselenide **167** in 52% yield after 24 hours (Scheme 8.38).

In 2009, Fernández-Bolaños and coworkers[105] described the synthesis of selenium-containing acyclic nucleosides and anomeric spironucleosides starting from D-mannosamine hydrochloride **168** and aryl isoselenocyanate **169**. In this sequence, D-glucosamine hydrochloride was first converted into 5-hydroxy-4-polyhydroxyalkylimidazolidine-2-selone **170** in 99% yield after 30 minutes under reflux in a mixture EtOH/H_2O. Subsequently, **170** underwent acid-catalyzed dehydration to give the bicyclic glucofuranosoimidazolidine-2-selone **171** in 89% yield. The same article reported the synthesis of the

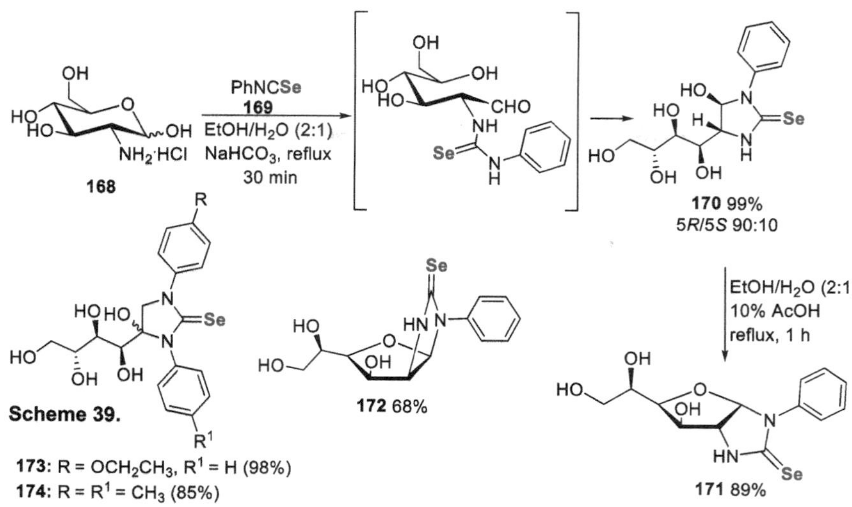

SCHEME 8.39 Synthesis of acyclic nucleosides and anomeric spironucleosides.[105]

bicyclic mannofuranosoimidazoline-2-selone **172**, in 68% yield, and 4-hydroxy-4-polyhydroxyalkylimidazolidine-2-selones **173** and **174** in yields of 98% and 85%, respectively (Scheme 8.39).

8.7 Summary and outlook

In this chapter, we described various organoselenium and organotellurium compounds synthesized from natural-product starting materials. Several of the new semisynthetic molecules exhibit superior biological activities and are anticipated to find applications in areas such as materials science as well as medicinal and food chemistry, while others are already along the road to commercialization for important health conditions.

Nature provides a significant number of unexplored possibilities as well as opportunities to create a broad library of chalcogen-containing semisynthetic compounds. New hybrid compounds that are environment-friendly offer alternatives to oil-based synthetic molecules.

Acknowledgements

This study was financed in part by the Coordenação de Aperfeiçoamento de Pessoal de Nível Superior - Brasil (CAPES) - Finance Code 001. FAPERGS CNPq and FINEP are acknowledged for financial support. Support from the Australian Research Council and Seleno Therapeutics Pty. Ltd. is also gratefully acknowledged.

References

1. Ghosh, S.; Playford, R. J. Bioactive Natural Compounds for the Treatment of Gastrointestinal Disorders. *Clin. Sci.* **2003,** *104*, 547–556.
2. Paterson, I.; Anderson, E. A. Chemistry. The Renaissance of Natural Products as Drug Candidates. *Science* **2005,** *310*, 451–453.
3. Guo, Z. The Modification of Natural Products for Medical Use. *Acta Pharm. Sin. B* **2017,** *7*, 119–137.
4. Lawal, B.; Shittu, O. K.; Kabiru, A. A.; Umar, M. B.; Berinyuy, E. B.; Alozieuwa, B. U. Potential antimalarials from African natural products: A review. *J. Intercult. Ethnopharmacol.* **2015,** *4*, 318–343.
5. Nogueira, C. W.; Rocha, J. B. T. Toxicology and Pharmacology of Selenium: Emphasis on Synthetic Organoselenium Compounds. *Arch. Toxicol.* **2011,** *85*, 1313–1359.
6. Ninomiya, M.; Garud, D. R.; Koketsu, M. Biologically Significant selenium-Containing Heterocycles. *Coord. Chem. Rev.* **2011,** *255*, 2968–2990.
7. Nogueira, C. W.; Zeni, G.; Rocha, J. B. T. Organoselenium and Organotellurium Compounds: Toxicology and Pharmacology. *Chem. Rev.* **2004,** *104*, 6255–6285.
8. Lenardão, E. J.; Santi, C.; Sancineto, L. *New Frontiers in Organoselenium Compounds;* Springer: Cham, Switzerland, 2018.
9. Jain, V. K.; Priyadarsini, K. I. *Organoselenium Compounds in Biology and Medicine: Synthesis, Biological and Therapeutic Treatments;* RSC Publishing: Cambridge, UK, 2018.
10. May, S. W. Selenium-Based Drug Designed: Rationale and Therapeutic Potential. *Expert. Opin. Invest. Drugs* **1998,** *8*, 1017–1028.
11. Rubulotta, G.; Quadrelli, E. A. Terpenes: A Valuable Family of Compounds for the Production of Fine Chemicals. *Horiz. Sustain. Ind. Chem. Catal.* **2019,** *178*, 215–229.
12. Ruzicka, L. The Isoprene Rule and the Biogenesis of Terpenic Compounds. *Experientia* **1953,** *9*, 357–367.
13. Pacuła, A. J.; Ścianowski, J. Terpenes as Green Starting Materials for New Organoselenium and Organotellurium Compounds. *Curr. Green. Chem.* **2016,** *3*, 36–50.
14. Guenther, E. *The Essential Oils.* Van Nostrand, D. vol. I, New York, 1948.
15. Masia, M. D.; Deidda, S.; Deriu, G. M.; Are, B. M.; Chessa, M.; Petretto, G.; Foddai, M.; Maida, G.; Pintore, G.; Piana, A. Antimicrobial Activities of Essential Oils Against Common Hospital Fungi Species. *Open. J. Prev. Med.* **2014,** *4*, 801–807.
16. Bishop, C. D. Antiviral Activity of the Essential Oil of *Melaleuca alternifolia* (Maiden amp; Betche) Cheel (Tea Tree) Against Tobacco Mosaic Virus. *J. Essent. Oil. Res.* **1995,** *7*, 641–644.
17. Konstantopoulou, I.; Vassilopoulou, L.; Mavragani-Tsipidou, P.; Scouras, Z. G. Insecticidal Effects of Essential Oils. A Study of the Effects of Essential Oils Extracted From Eleven Greek Aromatic Plants on *Drosophila auraria. Experientiam* **1992,** *48*, 616–619.
18. Hyldgaard, M.; Mygind, T.; Meyer, R. L. Essential Oils in Food Preservation: Mode of Action, Synergies, and Interactions with Food Matrix Components. *Front. Microbiol.* **2012,** *3*, 1–24.
19. De Godoy, L. A. F.; Hantao, L. W.; Pedroso, P. M.; Poppi, R. J.; Augusto, F. Quantitative Analysis of Essential Oils in Perfume Using Multivariate Curve Resolution Combined with Comprehensive Two-Dimensional Gas Chromatography. *Anal. Chim. Acta* **2011,** *699*, 120–125.
20. Edris, A. E. Pharmaceutical and Therapeutic Potentials of Essential Oils and Their Individual Volatile Constituents: A Review. *Phytother. Res.* **2007,** *21*, 308–323.
21. Victoria, F. N.; Radatz, C. S.; Sachini, M.; Jacob, R. G.; Alves, D.; Savegnago, L.; Perin, G.; Motta, A. S.; Silva, W. P.; Lenardão, E. J. Further Analysis of the Antimicrobial Activity of α-Phenylseleno Citronellal and α-Phenylseleno Citronellol. *Food Control.* **2012,** *23*, 95–99.

22. Back, T. G.; Dyck, B. P.; Parvez, M. 1,3-Diselenetanes and 1,3-Dithietanes Derived from Camphor. Formation, Structure, Stereochemistry, and Oxidation to Selenoxide and Sulfoxide Products. *J. Org. Chem.* **1995,** *60,* 703–710.
23. Sancineto, L.; Piccioni, M.; De Marco, S.; Pagiotti, R.; Nascimento, V.; Braga, A. L.; Santi, C.; Pietrella, D. Diphenyl Diselenide Derivatives Inhibit Microbial Biofilm Formation Involved in Wound Infection. *BMC Microbiol.* **2016,** *16,* 220–229.
24. Martínez, A. G.; Vilar, E. T.; Fraile, A. G.; Cerero, S. M.; Maroto, B. L. New Convenient Procedure for the Preparation of Enantiopure C10-S- and C10-Se-Substituted Camphor-Derived Sulfides and Selenides. *Tetrahedron Lett.* **2001,** *42,* 5017–5019.
25. Nieves, K.; Prudhomme, J.; Le Roch, K. G.; Franzblau, S. G.; Rodríguez, A. D. Natural Product-Based Synthesis of Novel Anti-Infective Isothiocyanate- and Isoselenocyanate-Functionalized Amphilectane Diterpenes. *Bioorg. Med. Chem. Lett.* **2016,** *26,* 854–857.
26. Boualy, B.; El Houssame, S.; Sancineto, L.; Santi, C.; Ali, M. A.; Stoeckli-Evans, H.; El Firdoussi, L. A. Mild and Efficient Method for the Synthesis of a New Optically Active Diallyl Selenide and Its Catalytic Activity in the Allylic Chlorination of Natural Terpenes. *N. J. Chem.* **2016,** *40,* 3395–3399.
27. Tiecco, M.; Testaferri, L.; Santi, C.; Tomassini, C.; Santoro, S.; Marini, F.; Bagnoli, L.; Temperini, A. Synthesis of Enantiomerically Pure β-azidoselenides Starting from Natural Terpenes. *Tetrahedron* **2007,** *63,* 12373–12378.
28. Ścianowski, J.; Rafiński, Z.; Szuniewicz, A.; Wojtczak, A. New Chiral Selenium Electrophiles Derived from Functionalized Terpenes. *Tetrahedron* **2009,** *65,* 10162–10174.
29. Victoria, F. N.; Radatz, C. S.; Sachini, M.; Jacob, R. J.; Perin, G.; da Silva, W. P.; Lenardão, E. J. KF/Al_2O_3 and PEG-400 as a Recyclable Medium for the Selective α-Selenation of Aldehydes and Ketones, Preparation of Potential Antimicrobial Agents. *Tetrahedron Lett.* **2009,** *50,* 6761–6763.
30. Ferraz, M. C.; Mano, R. A.; Oliveira, H. C.; Maia, D. S. V.; Silva, W. P.; Savegnago, L.; Lenardão, E. J.; Jacob, R. G. Synthesis, Antimicrobial, and Antioxidant Activities of Chalcogen-Containing Nitrone Derivatives from (*R*)-Citronellal. *Medicines* **2017,** *4,* 39–49.
31. Pandey, K. B.; Rizvi, S. I. Plant Polyphenols as Dietary Antioxidants in Human Health and Disease. *Oxid. Med. Cell. Longev.* **2009,** *2,* 270–278.
32. Dhawan, K.; Dhawan, S.; Sharma, A. Passiflora: A Review Update. *J. Ethnopharmacol.* **2004,** *94,* 1–23.
33. Bankova, V.; Popova, M.; Bogdanov, S.; Sabatini, A.-G. Chemical Composition of European Propolis: Expected and Unexpected Results. *Z. Naturforsch* **2002,** *57c,* 530–533.
34. Bae, E. A.; Han, M. J.; Lee, M.; Kim, D. H. In vitro inhibitory effect of some flavonoids on rotavirus infectivity. *Biol. Pharm. Bull.* **2000,** *23,* 1122–1124.
35. Han, C. Screening of Anticarcinogenic Ingredients in Tea Polyphenols. *Cancer Lett.* **1997,** *114,* 153–158.
36. Miura, S.; Watanabe, J.; Sano, M.; Tomita, T.; Osawa, T.; Hara, Y.; Tomita, I. Effects of Various Natural Antioxidants on the Cu^{2+}-Mediated Oxidative Modification of Low Density Lipoprotein. *Biol. Pharm. Bull.* **1995,** *18,* 1–4.
37. Fonseca, S. F.; Lima, D. B.; Alves, D.; Jacob, R. G.; Perin, G.; Lenardão, E. J.; Savegnago, L. Synthesis, Characterization and Antioxidant Activity of Organoselenium and Organotellurium Compounds Derivatives of Chrysin. *N. J. Chem.* **2015,** *39,* 3043–3050.
38. Fonseca, S. F.; Padilha, N. B.; Thurow, S.; Roehrs, J. A.; Savegnago, L.; de Souza, M. N.; Fronza, M. G.; Collares, T.; Buss, J.; Seixas, F. K., et al. Ultrasound-Promoted

Copper-Catalyzed Synthesis of Bis-Arylselanyl Chrysin Derivatives with Boosted Antioxidant and Anticancer Activities. *Ultrason. Sonochem.* **2017,** *39,* 827–836.

39. de Bernardi, M.; Vidari, G.; Vita-Finzi, P. Dehydrozingerone from Aframomum Giganteum. *Phytochemistry.* **1976,** *15,* 1785–1786.
40. Baliga, M. S.; Haniadka, R.; Pereira, M. M.; Thilakchand, K. R.; Rao, S.; Arora, R. Radioprotective Effects of Zingiber Officinale Roscoe (Ginger): Past, Present and Future. *Food Funct.* **2012,** *3,* 714–723.
41. Kubra, I. R.; Bettadaiah, N. K.; Murthy, P. S.; Rao, L. J. M. J. Structure-Function Activity of Dehydrozingerone and its Derivatives as Antioxidant and Antimicrobial Compounds. *J. Food Sci. Technol.* **2014,** *51,* 245–255.
42. Marchianni, A.; Mammi, S.; Siligardi, G.; Hussain, R.; Tessari, L.; Bubacco, L.; Delogu, G.; Fabbri, D.; Dettori, M. A.; Sanna, D., et al. Small Molecules Interacting with a-Synuclein: Antiaggregating and Cytoprotective Properties. *Amino Acids* **2013,** *45,* 327–338.
43. Martinez, D. M.; Barcellos, A. M.; Casaril, A. M.; Savegnago, L.; Perin, G.; Schiesser, C. H.; Callaghan, K. L.; Lenardão, E. J. Twice Acting Antioxidants: Synthesis and Antioxidant Properties of Selenium and Sulfur-Containing Zingerone Derivatives. *Tetrahedron Lett.* **2015,** *16,* 2243–2246.
44. Burton, G. W.; Ingold, K. U. Vitamin E: Application of the Principles of Physical Organic Chemistry to the Exploration of its Structure And Function. *Acc. Chem. Res.* **1986,** *19,* 194–201.
45. Kyne, S. H.; Schiesser, C. H. In *Encyclopedia of Radicals in Chemistry, Biology and Materials;* Chatgilialoglu, C.; Studer, A., Eds.; John Wiley and Sons: Chichester, UK, 2012.
46. Al-Maharik, N.; Engman, L.; Malmström, J.; Schiesser, C. H. Intramolecular Homolytic Substitution at Selenium: Synthesis of Novel Selenium-Containing Vitamin E Analogues. *J. Org. Chem.* **2001,** *66,* 6286–6290.
47. Cohen, N.; Schaer, B.; Scalone, M. Synthesis of (2RS, 4′R, 8′R)-α-Tocopherol and Related Compounds via a 2-Chlorochroman. *J. Org. Chem.* **1992,** *57* 5783-4785.
48. Shanks, D.; Amorati, R.; Fumo, M. G.; Pedulli, G. F.; Valgimigli, L.; Engman, L. Synthesis and Antioxidant Profile of All-Rac-A-Selenotocopherol. *J. Org. Chem.* **2006,** *71,* 1033–1038.
49. Shanks, D. Design, Synthesis and Evaluation of Catalytic Chalcogenide Antioxidants. Doctoral Thesis, Upsalla University, Sweden, 2005.
50. Hargreaves, M. Carbohydrates and Exercise. *J. Sports Sci.* **1991,** *9,* 17–28.
51. Lüdtke, D. S.; Savegnago, L.; Pinto, L. M.; Vargas, J. P. Synthesis of Alkylseleno-Carbohydrates and Evaluation of their Antioxidant Properties. *J. Braz. Chem. Soc.* **2015,** *26,* 810–815.
52. Brod, L. M. P.; Fronza, M. G.; Vargas, J. P.; Lüdtke, D. S.; Luchese, C.; Wilhelm, E. A.; Savegnago, L. Involvement of Monoaminergic System in the Antidepressant-Like Effect of (Octylseleno)-Xylofuranoside in the Mouse Tail Suspension Test. *Prog. Neuro-Psychoph* **2016,** *65,* 201–207.
53. Lucas, M. A.; Nguyen, O. T. K.; Schiesser, C. H.; Zheng, S. L. Preparation of 5-Selenopentopyranose Sugars from Pentose Starting Materials by Samarium (II) Iodide or (Phenylseleno)Formate Mediated Ring Closures. *Tetrahedron.* **2000,** *56,* 3995–4000.
54. Schiesser, C.H.; Davies, M.J.; Storkey, C. Seleno-Compounds and Therapeutic Uses Thereof. Int. Patent Appl. WO2012054988, 2012.
55. Nguyen, O. T. K.; Schiesser, C. H. Preparation of 5-Telluropentopyranose Sugars From Common Pentose Starting Materials. *Tetrahedron Lett.* **2002,** *43,* 3799–3800.
56. Braga, H. C.; Stefani, H. A.; Paixa, M.; Santos, F. W.; Lüdtke, D. S. Synthesis of 5′-Seleno-Xylofuranosides. *Tetrahedron.* **2010,** *66,* 3441–3446.

57. Prigol, M.; Santos, F. W.; Pinton, S.; Lüdtke, D. S.; Affeldt, R. F.; Pavin, N. F.; Spiazzi, C. C.; Soares, M. B.; Izaguirry, A. P.; Ramalho, J. B. Selenofuranoside improves long-term memory deficits in rats after exposure to monosodium glutamate: Involvement of Na^+, K^+-ATPase activity. *Physiol. Behav.* **2018,** *184,* 27–33.
58. Avila, D. S.; Aschner, M.; Denardin, C. C.; Santos, F. W.; Lüdtke, D. S.; Affelt, R. F.; Soares, F. A.; Gubert, P.; Viana, C.; Reis, G., et al. Seleno- and Telluro-Xylofuranoside Attenuate Mn-Induced Toxicity in *C. elegans* via the DAF-16/FOXO Pathway. *Food Chem. Toxicol.* **2014,** *64,* 192–199.
59. Braga, H. C.; Wouters, A. D.; Zerillo, F. B.; Lüdtke, D. S. Synthesis of Seleno-Carbohydrates Derived from D-Galactose. *Carbohydr. Res.* **2010,** *345,* 2328–2333.
60. Affeldt, R. F.; Braga, H. C.; Baldassari, L. L.; Lüdtke, D. S. Synthesis of Selenium-Linked Neoglycoconjugates and Pseudodisaccharides. *Tetrahedron* **2012,** *68,* 10470–10475.
61. van Es, T.; Whistler, R. L. Derivatives of 5-Deoxy-5-Seleno-D-Xylose. *Tetrahedron* **1967,** *23,* 2849–2853.
62. Blumberg, K.; Fuccello, A.; van Es, T. Selenium Derivatives of L-Arabinose, D-Ribose, and D-Xylose. *Carbohydr. Res.* **1977,** *59,* 351–362.
63. Davies, M. J.; Schiesser, C. H. 1,4-Anhydro-4-Seleno-D-Talitol (SeTal): A Remarkable Selenium-Containing Therapeutic Molecule. *N. J. Chem.* **2019,** *43,* 9759–9765.
64. Lui, H.; Pinto, B. M. Synthesis of Zwitterionic Selenonium and Sulfonium Sulfates from D-Mannose as Potential Glycosidase Inhibitors. *Can. J. Chem.* **2006,** *84,* 497–505.
65. Storkey, C.; Davies, M. J.; White, C. H.; Schiesser, C. H. Synthesis and Antioxidant Capacity of 5-Selenopyranose Derivatives. *Chem. Commun.* **2011,** *47,* 9693–9695.
66. Storkey, C.; Pattison, D. I.; White, J. M.; Schiesser, C. H.; Davies, M. J. Preventing Protein Oxidation with Sugars: Scavenging of Hypohalous Acids by 5-Selenopyranose and 4-Selenofuranose Derivatives. *Chem. Res. Toxicol.* **2012,** *25,* 2589.
67. Ohle, H.; Kohler, I. Über die Aceton-Verbindungen der Zucker und ihre Derivate, II: Die Aceton-Verbindungen der Fructose. *Ber. Deut. Chem. Ges.* **1924,** *57,* 1566.
68. Velichenko, Y.; Callaghan, K.L.; Schiesser, C.H. Unpublished Observations.
69. Schiesser, C.H.; Storkey, C.; Davies, M.J. Selenosugars and Therapeutic Uses Thereof. Int. Patent Appl. WO2016054671, 2016.
70. Carroll, L.; Pattison, D. I.; Fu, S.; Schiesser, C. H.; Davies, M. J.; Hawkins, C. L. Reactivity of Selenium-Containing Compounds with Myeloperoxidase-Derived Chlorinating Oxidants: Second-Order Rate Constants and Implications for Biological Damage. *Free. Radic. Biol. Med.* **2015,** *84,* 279–288.
71. Schiesser, C.H.; Storkey, C.; Davies, M.J. Selenosugars for skin tissue repair. US Patent 9895309, 2018.
72. Ng, H. G.; Leo, C. H.; O'Sullivan, K.; Alexander, S.-A.; Davies, M. J.; Schiesser, C. H.; Parry, L. J. 1,4-Anhydro-4-seleno-D-talitol (SeTal) Protects Endothelial Function in the Mouse Aorta by Scavenging Superoxide Radicals Under Conditions of Acute Oxidative Stress. *Biochem. Pharmacol.* **2017,** *128,* 34–45.
73. Zacharias, Z.; Flouda, K.; Jepps, T. A.; Gammelgaard, B.; Schiesser, C. H.; Davies, M. J. Effects of a Novel Selenium Substituted-Sugar (1,4-Anhydro-4-Seleno-D-talitol,SeTal) on Human Coronary Artery Cell Lines and Mouse Aortic Rings. *Biochem. Pharmacol.* **2020,** *173,* 113631.
74. Carroll, L.; Pattison, D. I.; Fu, S.; Schiesser, C. H.; Davies, M. J.; Hawkins, C. L. Catalytic Oxidant Scavenging by Selenium-Containing Compounds: Reduction of Selenoxides and N-Chloramines by Thiols and Redox Enzymes. *Redox Biol.* **2017,** *12,* 872–882.
75. Borges, E. L.; Ignasiak, M. T.; Velichenko, Y.; Perin, G.; Hutton, C. A.; Davies, M. J.; Schiesser, C. H. Synthesis and Antioxidant Capacity of Novel Stable 5-Tellurofuranose Derivatives. *Chem. Commun.* **2018,** *54,* 2990–2993.
76. Soukup, G. A. *Nucleic Acids: General Properties;* eLS - American Cancer Society. John Wiley & Sons: New York, 2003.

77. Jordheim, L. P.; Durantel, D.; Zoulim, F.; Dumontet, C. Advances in the Development of Nucleoside and Nucleotide Analogues for Cancer and Viral Diseases. *Nat. Rev. Drug. Discov.* **2013,** *12,* 447–464.
78. Andries, O.; Mc Cafferty, S.; De Smedt, S. C.; Weiss, R.; Sanders, N. N.; Kitada, T. N (1)-Methylpseudouridine-Incorporated mRNA Outperforms Pseudouridine-Incorporated mRNA by Providing Enhanced Protein Expression and Reduced Immunogenicity in Mammalian Cell Lines and Mice. *J. Control. Rel.* **2015,** *217,* 337–344.
79. Taniike, H.; Inagaki, Y.; Matsuda, A.; Minakawa, N. Practical Synthesis of 4′-Selenopyrimidine Nucleosides using Hypervalent Iodine. *Tetrahedron.* **2011,** *67,* 7977–7982.
80. Yu, J.; Kim, J.-H.; Lee, H. W.; Alexander, V.; Ahn, H. C.; Choi, W. J.; Choi, J.; Jeong, L. S. New RNA Purine Building Blocks, 4′-Selenopurine Nucleosides: First Synthesis and Unusual Mixture of Sugar Puckerings. *Chem. Eur. J.* **2013,** *13,* 5528–5532.
81. Ishii, K.; Saito-Tarashima, N.; Ota, M.; Yamamoto, S.; Okamoto, Y.; Tanaka, Y.; Minakawa, N. Practical Synthesis of 4′-Selenopurine Nucleosides by Combining Chlorinated Purines and 'armed' 4-Selenosugar. *Tetrahedron* **2016,** *72,* 6589–6594.
82. Duarte, L. F. B.; Oliveira, R. L.; Rodrigues, K. C.; Voss, G. T.; Godoi, B.; Schumacher, R. F.; Perin, G.; Wilhelm, E. A.; Luchese, C.; Alves, D. Organoselenium Compounds from Purines: Synthesis of 6-Arylselanylpurines with Antioxidant and Anticholinesterase Activities and Memory Improvement Effect. *Bioorg. Med. Chem.* **2017,** *25,* 6718–6723.
83. Sorbera, L. A.; Castañer, J.; Martin, L.; Bayés, M. Tecadenoson: Antiarrhythmic, Adenosine A_1 Agonist. *Drugs Fut.* **2002,** *27,* 846.
84. Trivedi, B. K.; Bridges, A. J.; Pratt, W. C.; Priebe, S. R.; Bruns, R. F. N6-Bicycloalkyladenosines with Unusually High Potency and Selectivity for the Adenosine A1 Receptor. *J. Med. Chem.* **1989,** *32,* 8–11.
85. Ashton, T. D.; Aumann, K. M.; Baker, S. P.; Schiesser, C. H.; Scammells, P. J. Bioorg. Structure–Activity Relationships of Adenosines with Heterocyclic N^6-Substituents. *Med. Chem. Lett.* **2007,** *17,* 6779–6784.
86. Macdougall, P. E.; Aitken, H. M.; Scammells, P. J.; Kavanagh, Y.; Kyne, S. H.; Schiesser, C. H. 7-Selenabicyclo[2.2.1]heptane. *Chem. Commun.* **2012,** *48,* 9126–9128.
87. Du, J.; Cullen, J. J.; Buettner, G. R. Ascorbic Acid: Chemistry, Biology and the Treatment of Cancer. *Biochim. Biophys. Acta* **2012,** *1826,* 443–457.
88. Figueroa-Méndez, R.; Rivas-Arancibia, S. Vitamin C in Health and Disease: Its Role in the Metabolism of Cells and Redox State in the Brain. *Front. Physiol.* **2015,** *6,* 397–407.
89. Tanini, D.; Lupori, B.; Malevolti, G.; Ambrosi, M.; Nostro, P. L.; Capperucci, A. Direct Biocatalysed Synthesis of First Sulfur-, Selenium- and Tellurium Containing L-Ascorbyl Hybrid Derivatives with Radical Trapping and GPx-Like Properties. *Chem. Commun.* **2019,** *55,* 5705–5708.
90. Rinaudo, M. Chitin and Chitosan: Properties and Applications. *Prog. Polym. Sci.* **2006,** *31,* 603–632.
91. Gularte, M. S.; Anghinoni, J. M.; Abenante, L.; Voss, G. T.; Oliveira, R. O.; Vaucher, R. A.; Luchese, C.; Wilhelm, E. A.; Lenardão, E. J.; Fajardo, A. F. Synthesis of Chitosan Derivatives with Organoselenium and Organosulfur Compounds: Characterization, Antimicrobial Properties and Application as Biomaterials. *Carbohydr. Polym.* **2019,** *219,* 240–250.
92. Watson, J. D.; Crick, F. H. C. Molecular Structure of Nucleic Acids; A Structure for Deoxyribose Nucleic Acid. *Nature* **1953,** *171,* 737–738.
93. Nougayrède, J.-P.; Homburg, S.; Taieb, F.; Boury, M.; Brzuszkiewicz, E.; Gottschalk, G.; Buchrieser, C.; Hacker, J.; Dobrindt, U.; Oswald, E. *Escherichia coli* Induces DNA Double-Strand Breaks in Eukaryotic Cells. *Science* **2006,** *313,* 848–851.

94. De Clercq, E. The Discovery of Antiviral Agents: Ten Different Compounds, Ten Different Stories. *Med. Res. Rev.* **2008,** *28,* 929–953.
95. Schwarz, K.; Foltz, C. M. Selenium as an Integral Part of Factor 3 Against Dietary Necrotic Liver Degeneration. *J. Am. Chem. Soc.* **1957,** *79,* 3292–3293.
96. May, S. W.; Pollock, S. H. Selenium-Based Antihypertensives. *Drugs* **1998,** *56,* 959–964.
97. Mugesh, G.; du Mont, W.-W.; Sies, H. Chemistry of Biologically Important Synthetic Organoselenium Compounds. *Chem. Rev.* **2001,** *101,* 2125–2179.
98. Inoue, N.; Minakawa, N.; Matsuda, A. Synthesis and Properties of 4′-thioDNA: Unexpected RNA-like Behavior of 4′-thioDNA. *Nucleic Acids Res.* **2006,** *34,* 3476–3483.
99. Adiwidjaja, G.; Schulze, O.; Voss, J.; Wirsching, J. Preparation of Selenoanhydro- and Telluroanhydroglycofuranosides and Some Corresponding Nucleosides. *Carbohydr. Res.* **2000,** *325,* 107–119.
100. Moroder, H.; Kreutz, C.; Lang, K.; Serganov, A.; Micura, R. Synthesis, Oxidation Behavior, Crystallization and Structure of 2′-Methylseleno Guanosine Containing RNAs. *J. Am. Chem. Soc.* **2006,** *128,* 9909–9918.
101. Salon, J.; Sheng, J.; Gan, J.; Huang, Z. Synthesis and Crystal Structure of 2′-Se-Modified Guanosine Containing DNA. *J. Org. Chem.* **2010,** *75,* 637–641.
102. Sheng, J.; Jiang, J.; Salon, J.; Huang, Z. Synthesis of a 2′-Se-Thymidine Phosphoramidite and its Incorporation into Oligonucleotides for Crystal Structure Study. *Org. Lett.* **2007,** *9,* 749–752.
103. Jayakanthan, K.; Johnston, B. D.; Pinto, B. M. Stereoselective Synthesis of 4′-Selenonucleosides using the Pummerer Glycosylation Reaction. *Carbohydr. Res.* **2008,** *343,* 1790–1800.
104. Braga, A. L.; Severo, F. W. A.; Schwab, R. S.; Rodrigues, O. E. D.; Dornelles, L.; Braga, H. C.; Lüdtke, D. S. Synthesis of Selenium- and Tellurium-Containing Nucleosides Derived from Uridine. *Tetrahedron Lett.* **2009,** *50,* 3005–3007.
105. Maza, S.; López, Ó.; Martos, S.; Maya, I.; Fernández-Bolaños, J. G. Synthesis of the First Selenium-Containing Acyclic Nucleosides and Anomeric Spironucleosides from Carbohydrate Precursors. *Eur. J. Org. Chem.* **2009,** 5239–5246.

CHAPTER

9

Organoselenium reagents derived from natural compounds

Agata J. Pacuła-Miszewska and Jacek Ścianowski

Department of Organic Chemistry, Faculty of Chemistry, Nicolaus Copernicus University, Torun, Poland

9.1 Introduction

Nature creates an unlimited source of chemical inspiration. Structures that evolved through natural evolution are perfect in their individual way, enabling living organisms to grow, reproduce, develop, and even resist internal and external environmental factors. Natural compounds originated in the plant and animal kingdom can be classified as primary metabolites, molecules essential for survival like amino acids, sugars, fatty acids, and nucleic bases, and secondary metabolites, which affect the surroundings and not the producer and are not required for the organism to survive, including alkaloids and terpenes.[1,2] The applicability of natural compounds in chemical synthesis possesses several advantages including structural diversity, chirality transfer, and commercial availability. As reagents in chemical reactions, the utility of molecules derived from nature also fulfills the basic green chemistry rules as obtained from environment-friendly and recyclable sources.

The combination of a natural scaffold with a nonmetal selenium moiety sets an interesting perspective for greener and highly potent reagent design. Selenium itself is a part of the broad and versatile field of natural metabolites. For animals, it is an essential micronutrient incorporated in the structure of physiologically relevant molecules, including the 21st amino acid selenocysteine (Sec) (**1**), methylated Sec (MSC) (**2**), the diselenide of Sec selenocystine (SeCys) (**3**), selenomethionine (SeMet) (**4**), selenocystamine (**5**), and selenodiglutathione (**6**). As SeMet (**4**) serves as

Organochalcogen Compounds
DOI: https://doi.org/10.1016/B978-0-12-819449-2.00001-X

the main dietary source of selenium and, after incorporation into proteins, only plays a role in Se storage, the other Se derivatives facilitate important biochemical reactions. The glutathione adduct (**6**), formed from inorganic selenite and reduced glutathione in mammals, is a substrate for the dithiol–disulfide thioredoxin oxidoreductase system. The Se-cystamine (**5**) enables protein folding with oxidative –S–S– bond formation. Sec (**1**), with its metabolic derivatives **2** and **3**, can be classified as an important redox biocatalyst. It is an essential structural element of selenoproteins that takes part in reduction–oxidation reactions and in the form of free selenol keeps the redox homeostasis and the cells thiolsat at physiological values. The –SeH moiety of Sec is stable because of a specific chemical environment of the surrounding amino acids building up the final protein structure.[3,4]

Besides the mentioned forms of Sec (**1–3**), SeMet (**4**), selenocystamine (**5**), and selenodiglutathione (**6**), the group of naturally occurring organoselenium molecules is extended by simple compounds found in plants like dimethyl selenide (**7**), dimethyl diselenide (**8**), methylselenol (**9**), dimethyl selenyl sulfide (**10**) and dimethyl selenone (**11**) formed from inorganic selenium species absorbed from soil (Scheme 9.1).[5]

Looking at this group of compounds from the point of chemical synthesis, the catalytically active selenocysteine molecule (**1**) particularly stands out and highlights the potential of constructing efficient reagents from natural scaffolds with incorporated Se moieties. In terms of applicability, Sec (**1**) and corresponding derivatives **2** and **3** are commonly used in mechanistic investigations to evaluate the role of proteins and enzymes in selenium metabolism, the activity of specific Se metabolities, and as standard samples in tests evaluating the ability of organoselenium compounds to mimic the activity of important selenoenzymes like glutathione peroxidase (GPx), thioredoxin reductase (Trx), and iodothyronine deiodinase (ID). Several publications demonstrate their various applicability. MSC (**2**) has been proven to possess anticancer and antioxidants activity, stimulate wound healing,[6] act as a building

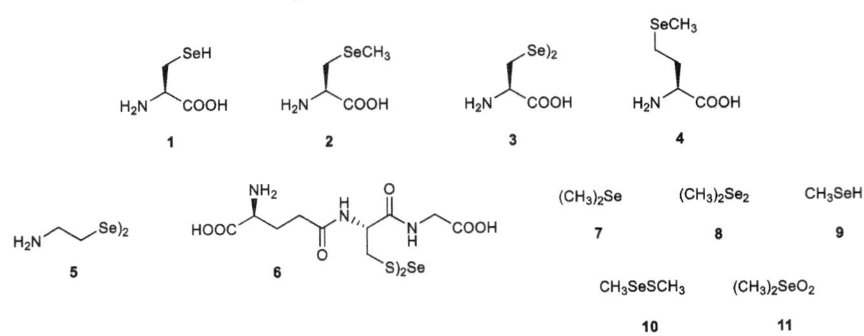

SCHEME 9.1 Natural organoselenium compounds **1–11**.

block in the synthesis of peptides,[7] and initiate cross-linking of proteins.[8] Furthermore, selenodiglutathione (**6**) has been efficiently used for fixing misfolded S–S protein bonds,[9] as substrate for Trx[10] and GPx-mimetic.[11]

The significant enzymatic activity of Sec is explicitly connected with the uniqueness of the Se atom. The advantages of selenium can be well presented in comparison to sulfur. The lower redox potential and weaker covalent Se–Se and Se–H bonds, compared to analog sulfur ones, are the main constituents of the higher reactivity of organoselenium molecules. In comparison to sulfur, selenium as a nucleophile increases the reaction rates 10^2–10^3 and as an electrophile even 10^4 orders higher.[12] Additionally, after facile introduction in the compounds structure, by selenenylation or selenocyclization reaction,[13] the obtained Se derivative can be easily manipulated and transformed to: (1) alkene, through oxidation and selenooxide elimination[14]; (2) nucleophilic substitution precursor, by oxidation to selenone[15]; (3) a radical precursor through the homolytic cleavage of Se–C bond.[16] Combining these features with the advantages of natural, usually chiral, scaffolds opens up broad application possibilities. Although the group of known naturally derived organoselenium compounds is broad, the application of these structures as reagents in chemical syntheses is still limited. This chapter presents the state of knowledge in the field with emphasis on molecules applied as reagents.

9.2 Organoselenium compounds derived from natural products

Until now, numerous examples of organoselenium compounds, structurally based on natural products, have been presented and found several applications in various fields of chemistry including medicinal chemistry, agriculture, analytical chemistry, and organic syntheses. Herein, this broad group of derivatives is presented and divided according to: (a) the general plan of the synthesis: modification of known organoselenium metabolites (*Section 9.2.1*); exchange of sulfur for selenium in the structure of natural organosulfur compounds (*Section 9.2.2*); incorporation of the Se moiety in the structure of non-chalcogen natural scaffolds (*Sections 9.2.3* and *9.2.4*), (b) the purpose of the synthesis and possible utility of the Se derivative: compounds as final products (*Sections 9.2.1, 9.2.2, and 9.2.3*) or intermediates (*Section 9.2.4*).

9.2.1 Modification of natural selenoamino acids

Selenoamino acids dominate in the field of known natural Se-metabolities. Several transformations have been conducted to improve their synthetic utility, pharmacological potential, or induce different biological activities.[17] Braga and coworkers used *L*-serine methyl ester as a substrate for the synthesis of *N*-Boc-protected selenocysteine derivative **12**, which was efficiently transformed to modified selenolanthionine (**13a**) and selenocystine (**13b**). The compounds were successfully investigated as substrates for the synthesis of selenium-containing peptides.[18] The Sec (**1**) was also modified by conjugating the selenol group with an alkyl, aryl or benzyl substituents. Derivatives **14** were evaluated as human cytochrome P450 inhibitors.[19,20] Other aromatic Sec derivatives **15** were tested as anticancer agents. In the rat kidneys, the compounds underwent transformation, by lyase enzymes, to the active chemopreventive selenol form.[21] The bioavailability of selenium from the unnatural amino acids was evaluated on the basis of the release from 3,3′-dimethylselenocystine (**16**). Se accumulated in liver and blood of rats and influenced the concentration of proline in blood serum.[22] To increase the anticancer activity as well, selenomethionie (**2**) was modified and obtained in the form of trifluoromethylselenide (**17**). In the presence of the enzyme methionine γ-lyase, it was proven to exhibit cytotoxic activity toward colon cancer[23] (Scheme 9.2).

9.2.2 Se-modification of natural organosulfur metabolites

Sulfur is a widespread element in nature broadly incorporated in the structure of natural metabolites including amino acids (methionine, cysteine), proteins, hormones, flavonoids, and vitamins. Although sulfur and selenium share the same group in the periodic table, they express different chemical behaviors. Since the biological importance of selenium was discovered, several sulfur-bearing natural compounds were modified to incorporate selenium instead of sulfur and evaluate its influence on the activity of the molecules. In 1965, Gunther and Moutner exchanged S for Se in the structure of coenzyme A (**18**), in which the presence of sulfur is essential to perform several metabolic activities including acyl group transfer and methylene group activation. Surprisingly, the obtained selenocoenzyme A diselenide (**19**) exhibited no catalytic or inhibitory activity in phosphoacetyltransferaze enzyme

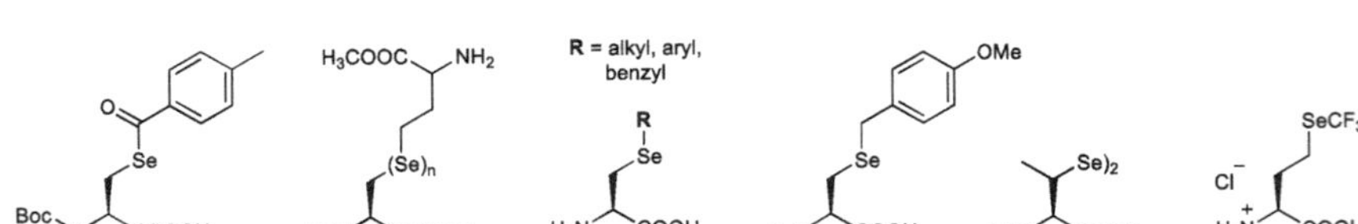

12 13a, n=1 13b, n=2 14 15 16 17

SCHEME 9.2 Modified selenoamino acids **12–17**.

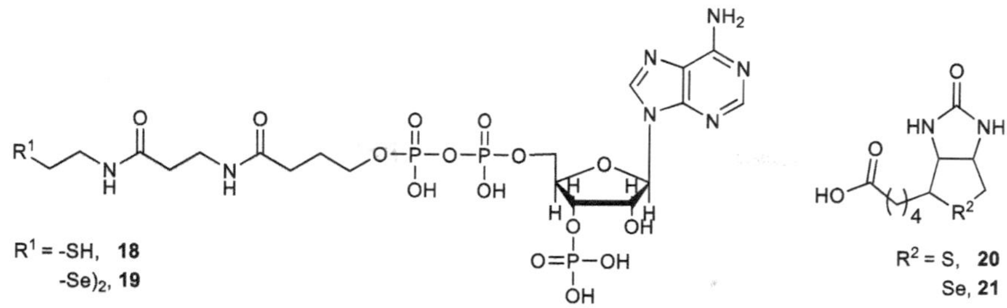

SCHEME 9.3 CoA **18** and Se-CoA **19**; Biotin **20** and Se-biotin **21**.

system and moderate activity (lower than for normal CoA) toward pigeon liver system.[24] Next, selenium analog **21** of biotin (vitamin B7, **20**) was synthetized and observed to be an efficient growth factor that enhanced the growth of biotin-dependent microorganisms[25] (Scheme 9.3).

A Se modification of insulin was also conducted. The disulfide bridge R–S–S–R in the compounds structure was exchanged to a diselenide one R–Se–Se–R. The modification enabled to obtain a long-lasting analog of the protein and a possible new drug in diabetes therapy.[26]

9.2.3 Semisynthetic Se- modification of other natural scaffolds

Apart from the sulfur–selenium exchange, several natural Se derivatives were obtained by semisynthetic modification of the primary or secondary metabolites. The C5–selenium substitution of nucleotides is a well-established procedure and led to the synthesis of an important class of derivatives.[27] Bardos et al. obtained C5–selenium analogs of uracil (**22**), 2′-deoxyuridine (**23**), and 2′-deoxyuridylic acid (**24**) in the form of diselenides. All compounds were tested as antibacterial and anticancer agents and exhibited inhibition toward *Lacticaseibacillus casei* thymidylate synthetase and antiproliferative potential on leukemia cell lines.[28] Selenenylated derivatives of uridine nucleosides, in the form of diselenide (**25**), ethoxyselenide (**26**), and seleninic acid sodium salt (**27**), were also synthetized by Knapp and coworkers. The antimalarial and anticancer activity assays, assessing the ability of compounds to inhibit enzymes responsible for the biosynthesis of pyrimidine nucleotides, revealed a high inhibitory potential.[29] Moreover, Schizani et al. synthetized 5-phenylselenyl pyrimidine derivatives **28**, which influenced the pyrimidine metabolism and in consequence exhibited anti-HIV and anticancer properties.[30,31] Pyrimidine was also functionalized with a selenocyanic moiety (**29**). The compound could be efficiently used as a DNA-labeling agent.[32] Additionally, a different modification based on the Se substitution on the ribosyl moiety was used for the synthesis of interesting

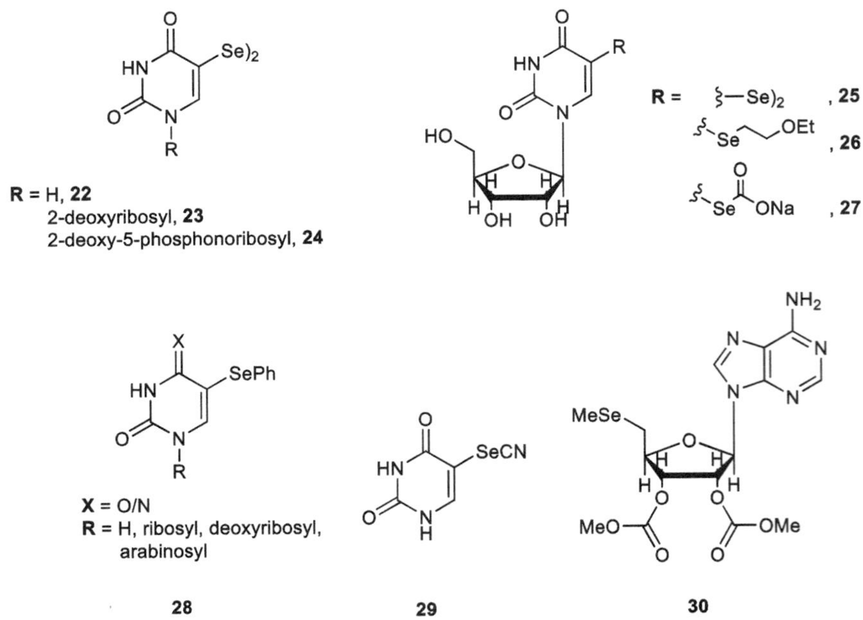

SCHEME 9.4 Se derivatives of nucleotides **22–30**.

insulin mimetic (**30**). The adenosine, 5′-Se-methyl-5′-seleno-, 2′,3′-diacetate was able to reduce hyperglycemia in insulin-resistant diabetes[33] (Scheme 9.4).

Selenosteroids can be assigned as the second wide group of nonnatural selenium metabolites. In the early 1980s, Sadek and coworkers published a series of estradiol derivatives for the synthesis of radiopharmaceuticals. The phenylselenide (**31**), obtained from 17-spirosteroid through epoxide opening with sodium phenyl selenide, could bind to estrogen receptors and be used in diagnostics of breast tumor.[34,35] A similar procedure was used by Braga et al. for the stereoselective synthesis of selenosteroids from cholesterol. The compounds **32** and **33** were obtained through the ring-opening of cholesterol epoxide with nucleophilic aromatic or aliphatic selenolates and further successfully tested as pro-oxidants.[36] Steroidal phenyl selenides were also synthetized by Santi and coworkers by treating the epoxide scaffold with PhSeZnCl. The obtained compound (**34**) exhibited promising antibacterial potential.[37] Interestingly, Zhang et al. also incorporated the –SePh moiety at the C2 position of the steroidal scaffold but by a Ru (II)-catalyzed C–H activation/selenenylation procedure.[38] The stereoselective steroid epoxide-opening was also efficiently performed by nucleophiles obtained from selenosugars leading to the formation of a C6–Se-substituted product (**35**).[39] Different selenocarbohydrate steroidal derivatives were obtained by Luboradzki and coworkers. First, synthesized homobetulin selenocyanate (**36**) was reduced to a free selenol, which was further glycosylated to the final homoselenosaponin (**37**). Surprisingly, only the selenocyanates showed cytotoxic potential, whereas the homoselenosaponins were inactive.[40] Various derivatives were also synthesized based on the structure of steroids diosgenin, hecogenin, and smilagenin, including diselenides, selenocyanates, isoselenocyanates, selenoureas, benzoselenazolones, and selenosemicarbazones.[41] The aglycon of the saponin dioscin—diosgenin—was modified by several research groups to obtain: selenides as cytotoxic agents (**38**),[42] selenocyanates with better antitumor activity than the known drug Paclitaxel (**39**),[43] and selenoureas with enhanced anticancer activity (**40**)[44] (Scheme 9.5).

9.2.4 Naturally derived Se compounds as intermediates in total synthesis

As the selenenylation reaction is a useful synthetic tool used to create new carbon–carbon and carbon–heteroatom (N, O) bonds, several organoselenium compounds are formed as intermediates in the total synthesis of diverse natural products. The selenium moiety is easily incorporated and further eliminated to obtain the final compound. In most total synthesis protocols, the used organoselenium reagent is generated from diphenyl diselenide. The introduction of the –SePh moiety can be performed in several ways. For example, in the synthesis of a natural antiinflammatory agent cyklobakuchiol A (**43**), the phenylselenenylation of (1*S*)-pinanone (**41**) was performed in the presence of SeO_2.[45] To obtain the antitumor alkaloid lycorine (**46**), the selenium

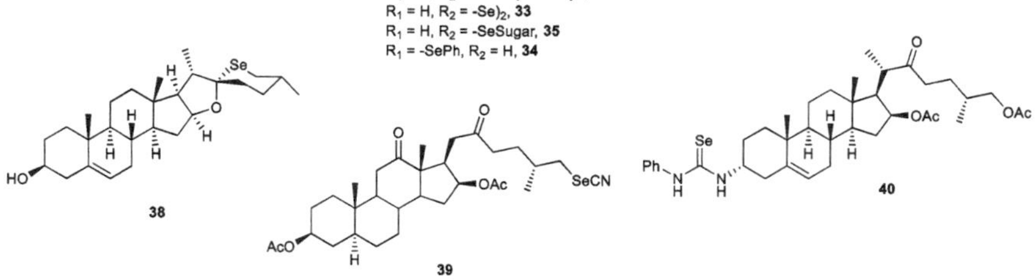

SCHEME 9.5 Steroidal Se derivatives **31–40**.

substituent was introduced by epoxide-ring opening by PhSeNa.[46] In both examples, the phenylselenyl group was further oxidized to selenoxide and underwent elimination with the formation of an alkene. On the contrary, in the case of the Se intermediate formed in the synthesis of fumiquinazoline C (**49**), heating of the compound **48** in a solution of CH_3CN/AcOH (25:1) initiated four sequential reactions that enabled the benzeneselenol elimination without previous oxidation to selenoxide[47,48] (Scheme 9.6).

The elimination of the −SePh can also be performed by a radical reaction as in the synthesis of geissoschizine, a precursor of various alkaloids, including strychnine. The acyl selenide (**51**), formed by the treatment of **50** with isobutyl chloroformate, and then PhSeNa, was eliminated by a radical decarbonylation yielding the deformylgeissoschizine (**52**)[49,50] (Scheme 9.7).

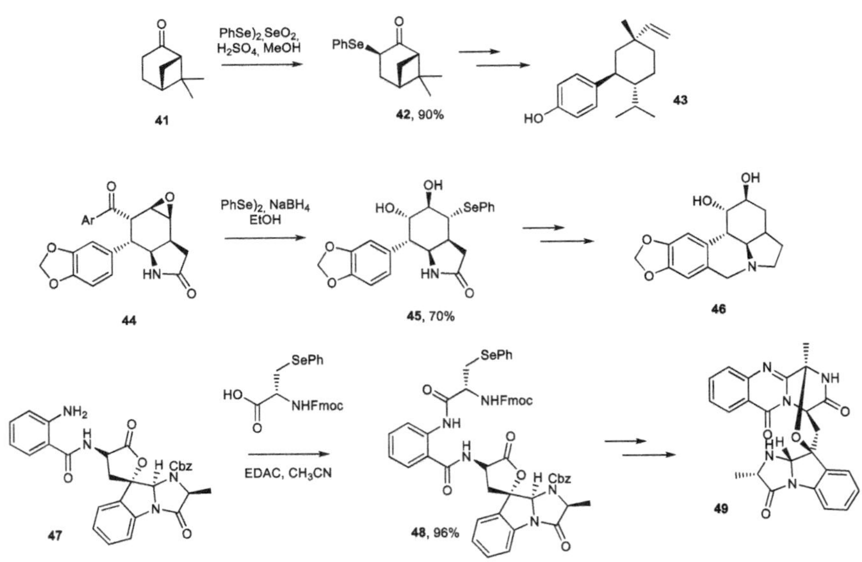

SCHEME 9.6 Examples of total syntheses with the presence of SePh intermediate **42**, **45**, and **46**.

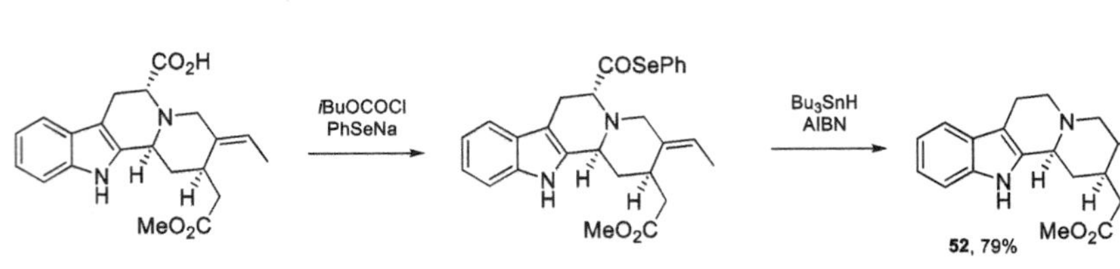

SCHEME 9.7 The radical decarbonylation in the synthesis of geissoschizine.

As it has been discussed, several natural scaffolds have been utilized to synthetize organoselenium molecules with interesting applications. Taking into account the synthetic potential of the presented compounds, especially in chiral transformations, it is surprising that only few have been used as reagents or catalysts in chemical reactions. The main part of this review summarizes the state of knowledge in the field and thoroughly describes the synthesis and applications of known Se reagents derived from natural products.

9.3 Naturally derived organoselenium compounds as reagents

Natural metabolites used to build up scaffolds for the Se-based reagents are limited to amino acids, sugars, and terpenes. In the case of amino acids and carbohydrates, the known derivatives were applied as acyl or glycosyl group transfer agents for the synthesis of peptides or oligosaccharides. In 2015, a series of seleno-*o*-benzaldehydes (**55**), obtained from Fmoc-protected amino acids (**53**) and diselenide (**54**), were used for peptide ligation. The amide bond of compound **57** was formed through an intramolecular reaction between the nitrogen, connected to the aldehyde group, and the carbonyl adjacent to the selenium atom (structure **56**).[51] Other examples of amino acid Se reagents used in peptide ligation or protein surface modification include β-selenol phenylalanine derivatives,[52] proline diselenide,[53] and Se-allylselenocysteine.[54]

Recently, Temperini and coworkers applied amino acid selenoesters (**58**) as acyl group transfer agents in peptide synthesis. Compounds **58** were formed from carboxylic acids and diphenyl diselenide, which after formation of the final peptide (**60**), was isolated as the only byproduct[55] (Scheme 9.8).

The phenylselenyl functionalization was also the dominating moiety in the structure of known selenoglycoside-based reagents. The

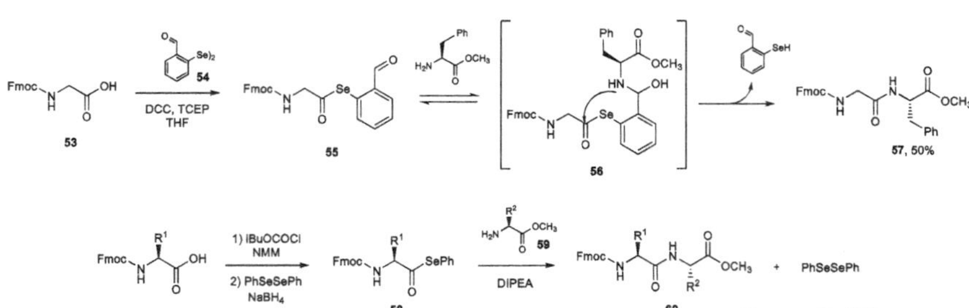

58	59	Yield
N-Fmoc-*L*-Phe-SePh	Ala	90%
N-Fmoc-*L*-Val-SePh	Phe	79%
N-Fmoc-*L*-Leu-SePh	Val	96%

SCHEME 9.8 Peptide synthesis using organoselenium reagents **55** and **58**.

derivatives were efficiently used in *O*-glycosylation reaction: as glycosyl donors in the synthesis of oligosaccharides (compound **61**)[56,57]; *O*-glycosylation promoted by iodonium ions (**62**)[58,59]; as glycosyl donors in a light-promoted procedure **62**[60]; and in the form of 2-azidoselenogalactosides for highly stereoselective synthesis of galactopyranosides (**63–65**).[61] Apart from the RSePh reagents, Szilagyi et al. used also isoselenuronium bromides (**66**) or chlorides (**67**) as glucoselenyl-transfer reagents for the synthesis of selenoglycosides[62] (Scheme 9.9).

In 2016, Ludtke and Wessjohann presented an interesting conjugation of steroids and carbohydrates through a selenide linkage (compound **68**). This procedure was based on ring-opening of steroidal epoxides by nucleophiles generated from Se-pyranosides and furanosides[39] (Scheme 9.10).

Nevertheless, the largest group of metabolites used to create reagents based on selenium are terpenes. This class of compounds possesses several synthetic advantages including structural diversity, availability in optically pure form, facile substituent introduction, and further chemical transformation. The majority of terpenyl organoselenium reagents

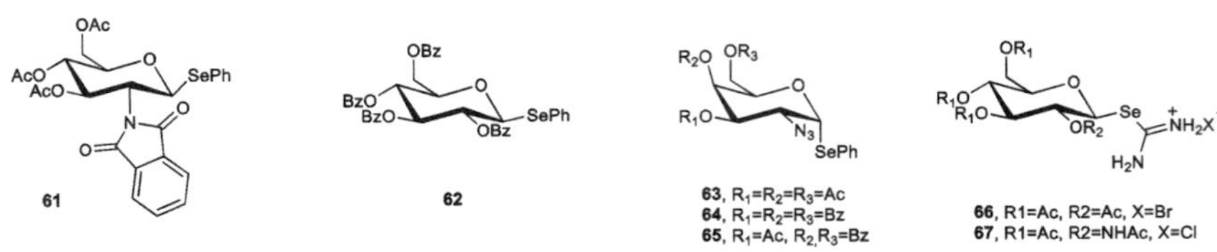

SCHEME 9.9 Selenoglycosides **61–67** used in glycosylation reaction.

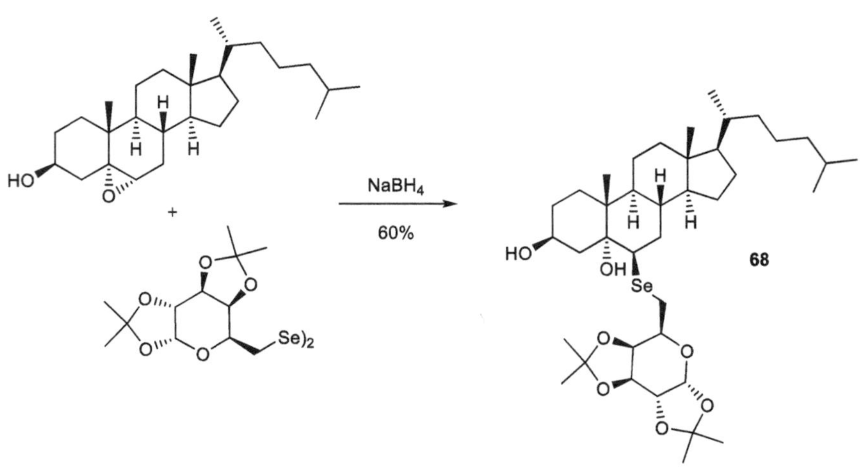

SCHEME 9.10 Synthesis of glyco-conjugated Se-steroid (**68**).

are built-up from the bicyclic camphane skeleton, functionalized on C2, C3, C9, or C10 position. However, other terpene scaffolds from *p*-menthane, carane, and pinane group have also been efficiently used in the synthesis of organoselenium reagents (Scheme 9.11).[63]

In 1972, when Sharpless presented the first example of [2,3]-sigmatropic rearrangement of an allylic Se(IV) species (**69**) to form a chiral allylic alcohol linalool (**70**),[64–66] a new application route opened up for organoselenium compounds to be used as precursors of new chiral molecules including alcohols and amines (Scheme 9.12).

Because of these findings, in the 1990s, Koizumi and coworkers synthetized a series of C_{10}-camphor derivatives – allylic chloroselenuranes (**72**)—obtained through the oxidation of allylic selenides (**71**). After treatment with lithium *N*-protected amides, carbon nucleophiles or aqueous solution of $NaHCO_3$, the compounds were transformed to chiral allylic selenimines (**73**), selenonium ylides (**74**), and selenoxides (**75**), respectively. All allylic Se-species **73–75** also underwent facile [2,3]-sigmatropic rearrangements to form chiral allylic amines (**76**), selenides (**77**) or alcohols (**78**)[67–69] (Scheme 9.13).

Other terpenyl selenides including acyclic (neryl, geranyl), monocyclic ((+)-carvotanacetyl, (−)-carvyl, perillyl), and bicyclic (pinane, carane) structures, were also efficiently applied as chiral precursors undergoing the same [2,3]-rearrangement.[70–74] The overall methodology is presented for (+)-10-phenylseleno-2-pinene (**79**), used as an example.

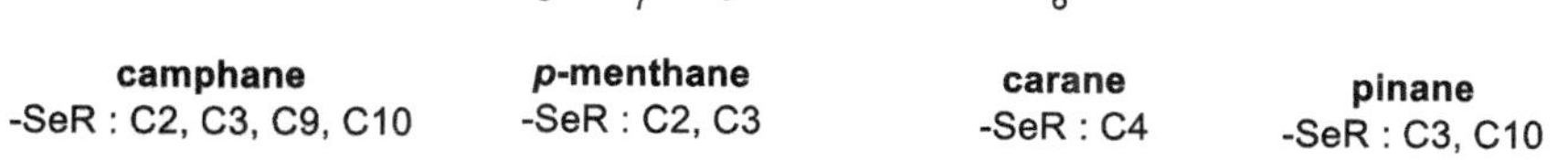

SCHEME 9.11 Various Se-functionalization of the terpenyl scaffold.

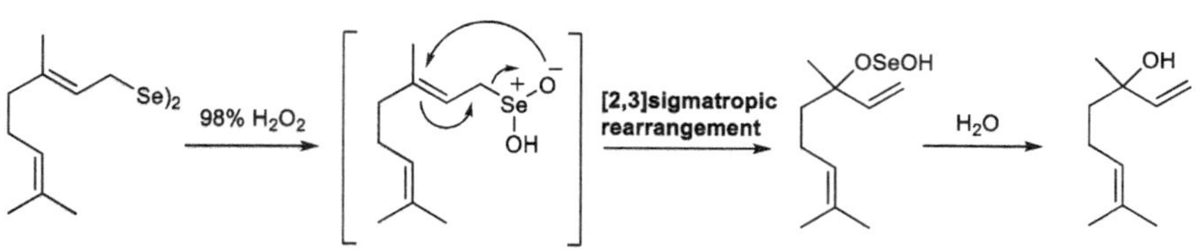

69 70

SCHEME 9.12 Synthesis of linalool (70) from geranyl diselenide.

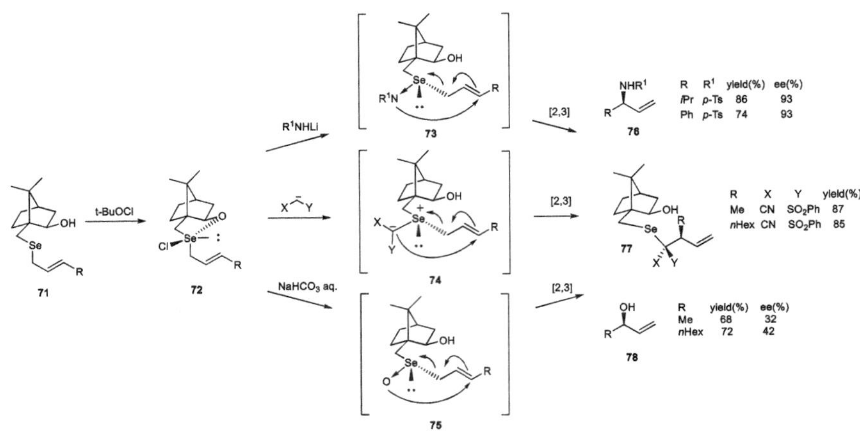

SCHEME 9.13 Allylic chloroselenouranes (**72**) as precursors of chiral compounds **76–78**.

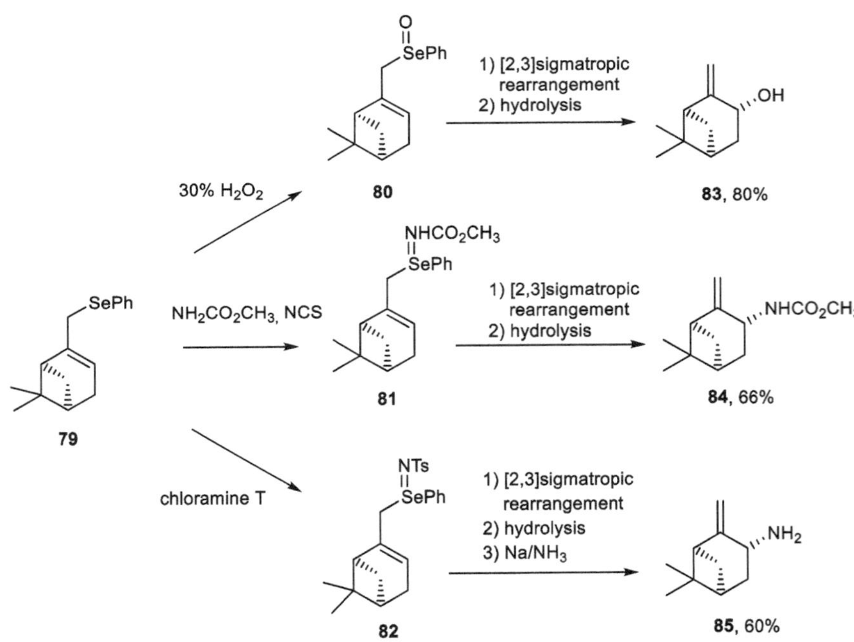

SCHEME 9.14 Application of (+)-10-phenylseleno-2-pinene (**79**).

Treatment of phenylselenide (**79**) with 30% H_2O_2, methylcarbamate and N-chlorosuccinimide, or chloramine-T resulted in the synthesis of selenoxide (**80**) and selenoimides (**81**) and (**82**), which were further rearranged to corresponding allyl alcohol (**83**), methylcarbamate (**84**), and, after additional reduction, amine (**85**) (Scheme 9.14).

In 2013, Ścianowski et al. obtained a series of allylic selenides from *p*-menthane, carane, and pinane selenols, and they also used them in the synthesis of chiral allylic alcohols.[75] The best enantiomeric excess was observed for caranylselenol (**86**). The rearrangement of allylic selenide (**87**) led to the final alcohol (**88**) with 34% enantiomeric excess (ee) (Scheme 9.15).

Next, Tiecco et al. obtained terpene selenols from C_3-camphor and menthyl diselenides by the treatment with zinc in acidic conditions. The compounds were used as nucleophiles to obtain corresponding methyl selenides.[76]

Additionally, the *p*-menthane, carane, and pinane selenols served as substrates for the synthesis of tritylated selenides. The chirality transfer from the terpene moiety, through the selenium atom, to the chiral trityl propeller was evaluated according to density functional theory structure and electronic circular dichroism spectra calculations.[77]

Other synthetic utility of terpene-derived organoselenium compounds covers their application as electrophilic reagents in selenenylation and selenocyclization reaction. The most exploited derivatives are those based on camphor with the Se-functionalization on the C3 carbon atom, especially the diselenide (**89**) presented by Back.[78,79] Compound **89** in the form of triflate (**90**) was efficiently used in

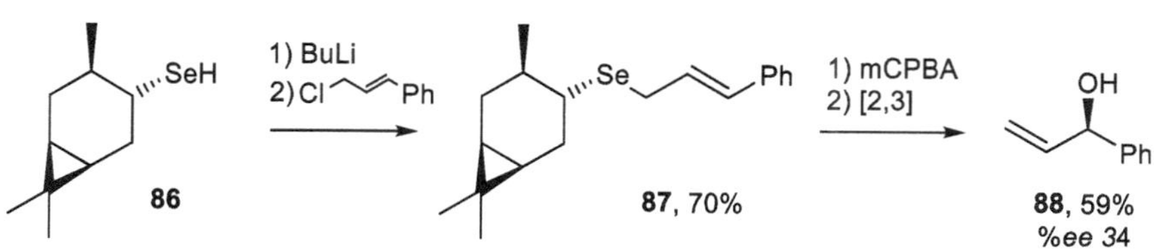

SCHEME 9.15 Synthesis of allylic alcohols using terpene selenol (**86**).

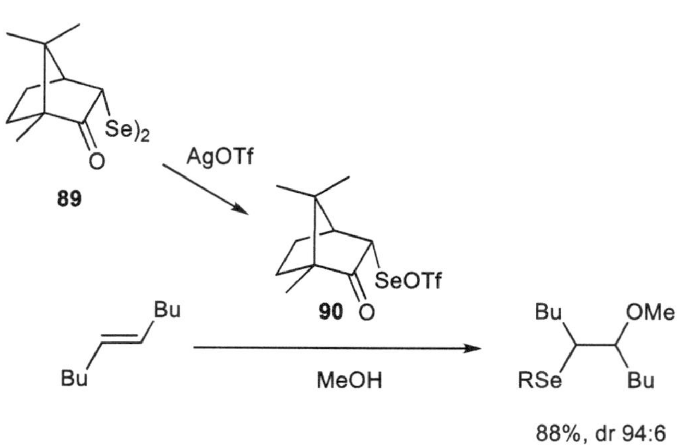

SCHEME 9.16 Methoxyselenenylation performed with camphor diselenide (**89**).

methoxyselenenylation of *trans*-dec-5-ene, with good diastereomeric excess 94:6 (Scheme 9.15) (Scheme 9.16).

On the other hand, compound **89**, with a chloride counter ion, used in the selenocyclization reaction gave almost no asymmetric induction in comparison to the additionally modified electrophilic reagents **91–94** (Scheme 9.17).[80–84]

Tiecco et al. have also broadly utilized the camphor diselenide (**89**) but in the form of selenenyl sulfate (**95**). The ability of facile introduction and elimination from the structure of the final product was efficiently tested in various synthetic procedures. In the synthesis of allylic ethers (**96**), the selenium moiety was eliminated by excess of ammonium persulfate.[85,86] To obtain oxazolines (**97**), (**97b**), and thiazolines (**98**), (**98b**), after amidoselenenylation, the deselenylation was performed through selenium activation by PhSeOTf or PhSeCl[87,88] (Scheme 9.18).

The same selenenylation–deselenylation sequence with the presence of compound (**95**) was also further implicated in the synthesis of perhydrofuro[2,3-b]furans[89] and 1,6-dioxaspiro[4,4]nones.[90] Further conversion of diselenide (**89**) to a nucleophilic (*R*)-camphorselenoacetone and methyl (*R*)-camphorselenoacetate enabled the application in TiCl4-mediated aldol condensation.[91,92]

Besides the camphor skeleton, different terpene scaffolds were used by Ścianowski and coworkers to also construct electrophilic organoselenium reagents. A series of diselenides derived from *p*-menthane, carane, and pinane was synthetized, transformed to corresponding selenyl triflates, and then used in asymmetric methoxyselenenylation of styrene[93–98] and selenocyclization of *o*-allylphenol.[99,100] The best results

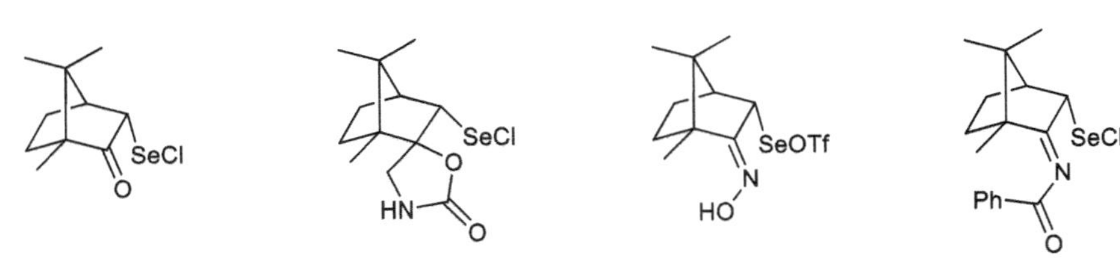

91 **92** **93** **94**

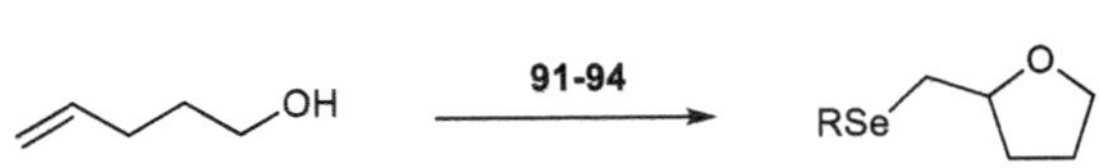

	yield (%)	dr
91	57	58:42
92	87	84:16
93	55	60:40
94	63	67:33

SCHEME 9.17 Selenocyclization reaction using camphor C3-modified reagents **91–94**.

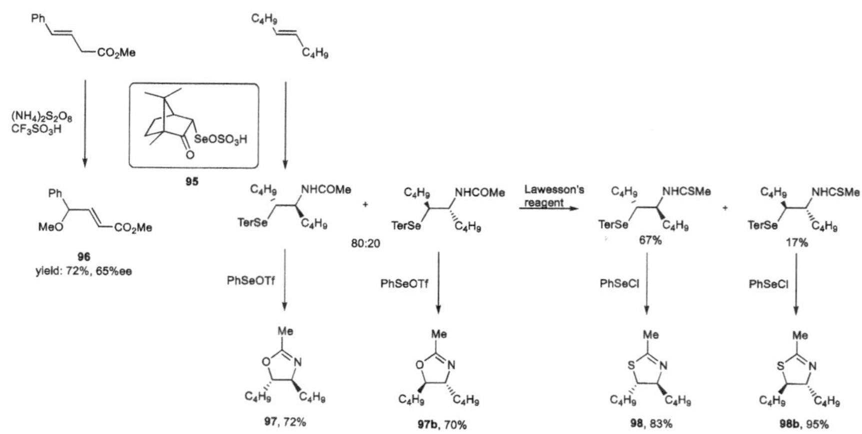

SCHEME 9.18 Examples of selenenylation–deselenylation procedure performed with reagent **95**.

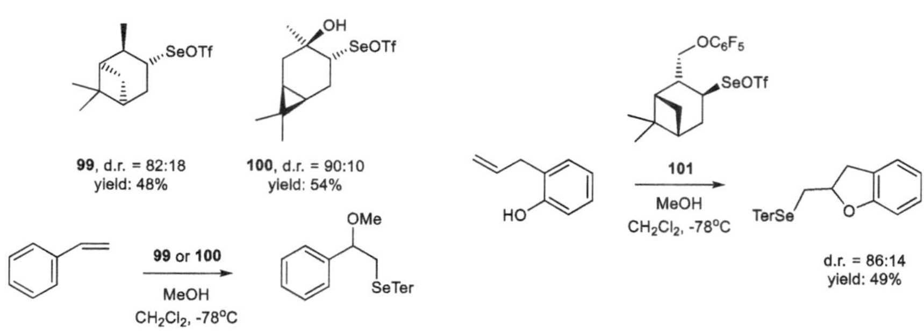

SCHEME 9.19 Selected results obtained for methoxyselenenylation of styene and selenocyclization of *o*-allylphenol.

were obtained for bicyclic structures derived from pinane (**99**), hydroxycarane (**100**), and myrtenol (**101**) (Scheme 9.19).

Organoselenium reagents based on monoterpenes were also efficiently used in epoxidation and cyclopropanation reactions. Huang et al. applied C_3–camphor selenonium salts in the synthesis of chiral epoxides. In case of salt **102**, the final diaryl oxirane was obtained with the best diastereomeric and enantiomeric ratios[101] (Scheme 9.20).

Next, the authors performed the first stereoselective cyclopropanation using C_3-functionalized camphor selenonium ylides (**103**). The results were promising with the best diastereomeric ratio 99:1[102] (Scheme 9.21).

Further examples of Se-catalyzed epoxide synthesis were described by Katanoka and coworkers. The C_2–camphor selenide (**104**) was successfully applied as catalyst in the Darzens reaction. The best ee was

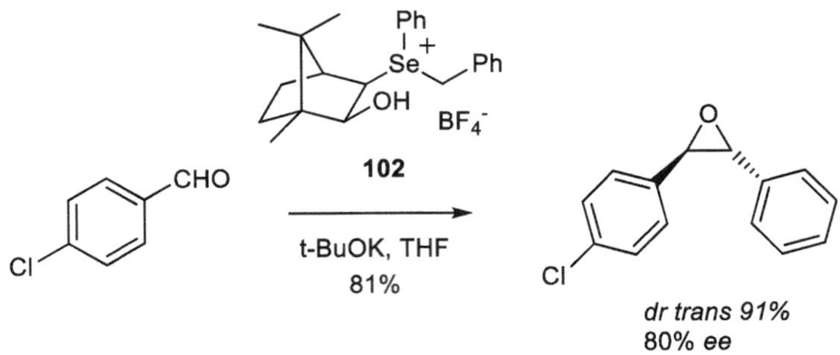

SCHEME 9.20 Selenonium salt (102) in asymmetric epoxidation.

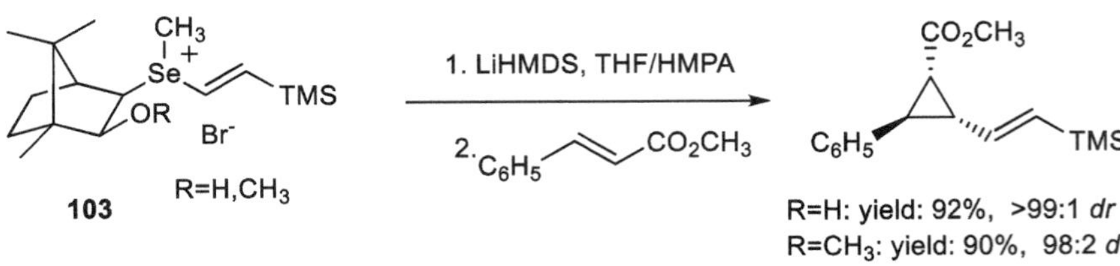

SCHEME 9.21 Cyclopropanation via selenonium ylides (103).

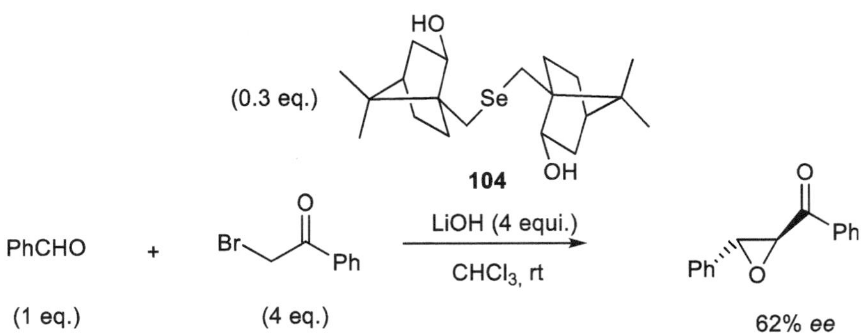

SCHEME 9.22 Asymmetric Darzens reaction catalyzed by selenide **104**.

observed when benzaldehyde was used as substrate yielding the final epoxide with 62% ee[103] (Scheme 9.22).

In 2014–2015, Ścianowski et al. used other terpene scaffolds to obtain salts applicable in asymmetric cyclopropanation and epoxidation. A series of selenonium salts derived from (−)-menthol and (+)-limonene were transformed to corresponding ylides and applied in cyclopropanation of vinylphosphonate. Excellent enantioselectivity was obtained for compound **105** with ratio for *cis* and *trans* enantiomers 98:2 and 99:1, respectively (Scheme 9.23).[104]

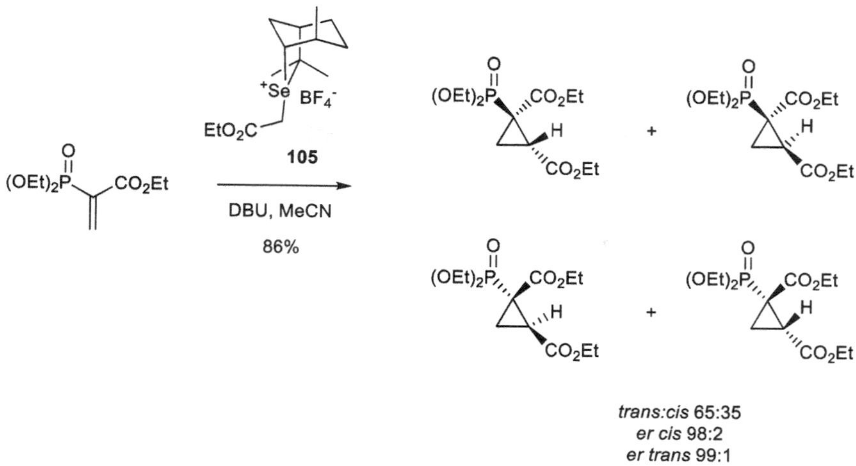

SCHEME 9.23 Selenonium salt (**105**) in asymmetric cyclopropanation.

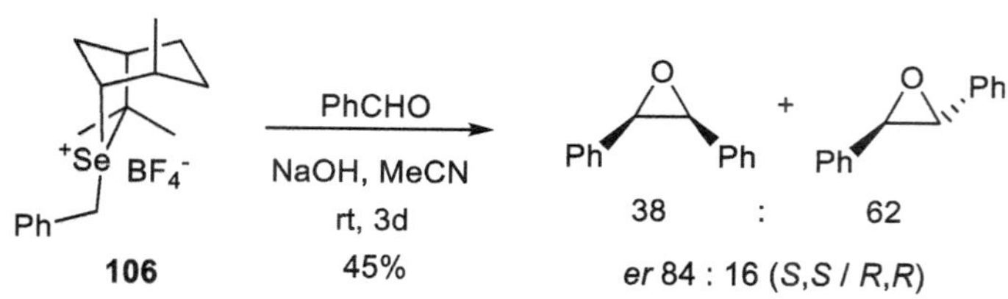

SCHEME 9.24 Selenonium salt (**106**) in asymmetric epoxidation.

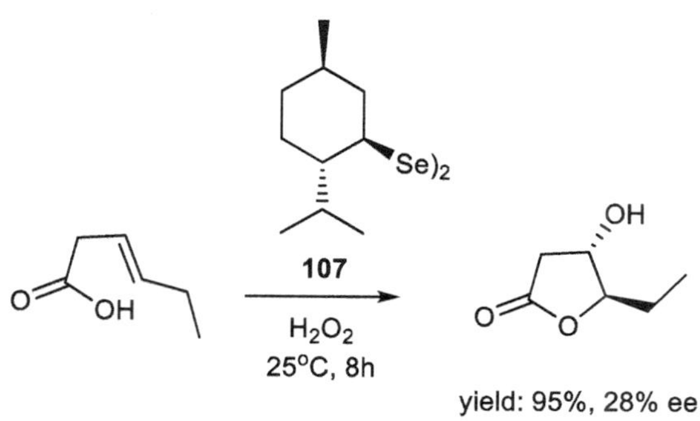

SCHEME 9.25 Diselenide (**107**) in oxacyclization of alkenoic acids.

The salts used in asymmetric epoxidation were derived from *p*-menthane, carane, pinane, and (+)-limonene (**106**) (presented in Scheme 9.24) and gave moderate enantiomeric ratios.[105]

The application of terpene-derived Se reagents in a different type of reactions was recently presented by Santi and coworkers. Dimenthyl diselenide (**107**) was used as catalyst in oxacyclization of alkenoic acids. The final product was obtained with high yield but low enantioselectivity[106] (Scheme 9.25).

Next, pinene selenide (**108**), obtained through selenium dioxide oxidation of the corresponding olefin, was used as a catalyst in allylic chlorination of alkenes, including asymmetric chlorination of terpenes. The reaction proceeded with excellent yields in mild conditions[107] (Scheme 9.26).

Furthermore, in 2016, Ścianowski et al. synthesized a series of diaryldiselenides o-substituted by O(S, Se)-caranyl moieties. The diselenides bearing o-(S-caranyl)- and o-(S-isocaranyl)- linkages were transformed to methylselenides (**109**) and (**110**) and tested as catalysts in the

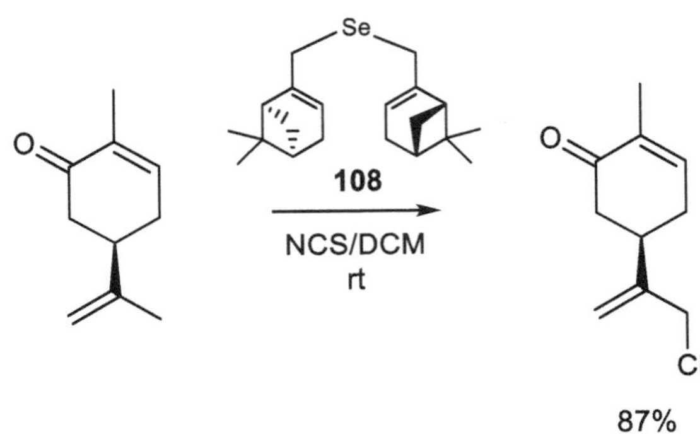

SCHEME 9.26 Catalytic allylic chlorination using selenide (**108**).

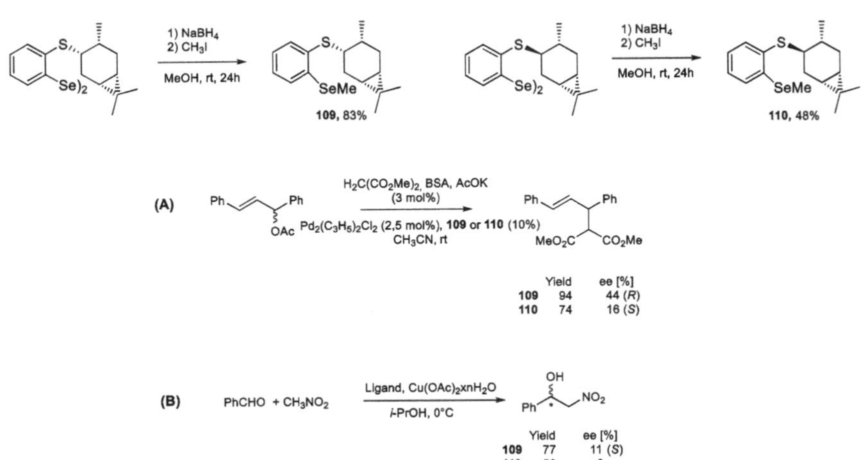

SCHEME 9.27 Catalytic Tsuji–Trost allylic alkylation (A) and Henry reaction (B) using selenides **107**.

Tsuji–Trost allylic alkylation (A) and Henry reaction (B). The Pd-catalyzed alkylation proceeded with moderate enantioselectivities[108] (Scheme 9.27).

9.4 Conclusions

Natural metabolites are continuously serving as role models for chemists when designing compounds with particular structure and meaningful synthetic features. Since the discovery of selenium importance, both in medicine and chemical synthesis, the field of natural organoselenium molecules has been extended by compounds obtained through modification of the original Se metabolite or insertion of the Se functionality in the structure of non-selenium natural derivatives, for example, exchange of sulfur with selenium in natural organosulfur molecules or functionalization with –SePh moiety for further synthetic transformation in natural compound synthesis. Currently, the group of known naturally derived oganoselenium compounds includes amino acids, nucleotides, sugars, steroids, and terpenes. However, only few (selected amino acids, sugars, and terpenes) have been used as reagents in chemical synthesis, whereas the Se-terpenyl derivatives have been efficiently applied in various chemical transformations: synthesis of alcohols and amines through [2,3]-sigmatropic rearrangement; as electrophilic species in selenenylation and selenocyclization reaction; cyclopropnation; epoxidation; and allylic chlorination of alkenes, the utility of selenoamino acids and Se-glycosides is still limited to acyl or glycosyl group transfer in the synthesis of peptides or oligosaccharides. The predicted applicability of the presented group of compounds in chemical syntheses, significant because of the structural diversity and chirality along with the reactivity of the Se-functionality, exposes the scientific potential in the field and the research paths that are still not explored.

References

1. Croteau, R.; Kutchan, T. M.; Lewis, N. G. In *Natural Products (Secondary Metabolites), Biochemistry & Molecular Biology of Plants;* Buchanan, B., Gruissem, W., Jones, R., Eds.; American Society of Plant Physiologists, 2000; pp 1250–1318.
2. Drew, S. W.; Demain, A. L. Effect of Primary Metabolites on Secondary Metabolism. *Ann. Rev. Microbiol.* **1977,** *31,* 343–356.
3. Pacuła, A. J.; Mangiavacchi, F.; Sancineto, L.; Lenardao, E. J.; Ścianowski, J.; Santi, C. An Update on "Selenium Containing Compounds from Poison to Drug Candidates: A Review on the GPx-like Activity". *Curr. Chem. Biol.* **2015,** *9,* 97–112.
4. Weekley, C. M.; Harris, H. H. Which Form Is That? The Importance of Selenium Speciation and Metabolism in the Prevention and Treatment of Disease. *Chem. Soc. Rev.* **2013,** *42,* 8870–8894.

5. Achibat, H.; AlOmari, N. A.; Messina, F.; Sancineto, L.; Khouili, M.; Santi, C. Organoselenium Compounds as Phytochemicals from the Natural Kingdom. *Nat. Prod. Commun.* **2015,** *11,* 1885–1892.
6. Dongsoo Kim, D.; Ku, B.; Choi, E.-M. Se-Methylselenocysteine Stimulates Migration and Antioxidant Response in HaCaT Keratinocytes: Implications for Wound Healing. *J. Trace Elem. Med. Biol.* **2020,** *58,* 126426–126433.
7. Yin, Y.; Fei, Q.; Liu, W.; Li, Z.; Suga, H.; Wu, Ch Chemical and Ribosomal Synthesis of Topologically Controlled Bicyclic and Tricyclic Peptide Scaffolds Primed by Selenoether Formation. *Angew. Chem. Int. (Ed.)* **2019,** *58,* 4880–4885.
8. Dantas de Araujo, A.; Perry, S. R.; Fairlie, D. P. Chemically Diverse Helix-Constrained Peptides Using Selenocysteine Crosslinking. *Org. Lett.* **2018,** *20,* 1453–1456.
9. Shimodaira, S.; Asano, Y.; Arai, K.; Iwaoka, M. Selenoglutathione Diselenide: Unique Redox Reactions in the GPxLike Catalytic Cycle and Repairing of Disulfide Bonds in Scrambled Protein. *Biochemistry* **2017,** *56,* 5644–5653.
10. Bjornstedt, M.; Kumar, S.; Holmgren, A. Selenodiglutathione Is a Highly Efficient Oxidant of Reduced Thioredoxin and a Substrate for Mammalian Thioredoxin Reductase. *J. Of. Biol. Chem.* **1992,** *267,* 8030–8034.
11. Yoshida, S.; Kumakura, F.; Komatsu, I.; Arai, K.; Onuma, Y.; Hojo, H.; Singh, B. G.; Priyadarsini, K. I.; Iwaoka, M. Antioxidative Glutathione Peroxidase Activity of Selenoglutathione. *Angew. Chem. Int. (Ed.)* **2011,** *50,* 2125–2128.
12. Steinmann, D.; Nauser, T.; Koppenol, W. H. Selenium and Sulfur in Exchange Reactions: A Comparative Study. *J. Org. Chem.* **2010,** *75,* 6696–6699.
13. Ścianowski, J.; Rafiński, Z. In *Organoselenium Chemistry: Between Synthesis and Biochemistry;* Santi, C., Ed.; Bentham, 2014; pp 8–60.
14. Jones, D. N.; Mundy, D.; Whitehouse, R. D. Steroidal Selenooxides Diastrereomeric at Selenium; syn-Elimination, Absolute Configuration, and Optical Rotatory Dispersion Characteristic. *J. Chem. Soc. Chem. Commun.* **1970,** 86–87.
15. Drabowicz, J.; Lewkowski, J.; Ścianowski, J. In *Organoselenium Chemistry: Synthesis and Reactions;* Wirth, T., Ed.; Wiley-VCH: Weinheim, 2012; pp 191–256.
16. Bowman, W. R. In *Organoselenium Chemistry: Synthesis and Reactions;* Wirth, T., Ed.; Wiley-VCH: Weinheim, 2012; pp 111–146.
17. Wessjohann, L. A.; Schneider, A. Synthesis of Selenocysteine and Its Derivatives with an Emphasis on Selenenylsulfide (SeS) Formation. *Chem. & Biodiv* **2008,** *5,* 375–388.
18. Braga, A. L.; Wessjohann, L. A.; Taube, P. S.; Galetto, F. Z.; de Andrade, F. M. Straightforward Method for the Synthesis of Selenocysteine and Selenocystine Derivatives from L-Serine Methyl Ester. *Synthesis* **2010,** *18,* 3131–3137.
19. Venhorst, J.; Rooseboom, M.; Vermeulen, N. P. E.; Commandeur, J. N. M. Studies on the Inhibition of Human Cytochromes P450 by Selenocysteine Se-conjugates. *Xenobiotica* **2003,** *33,* 57–72.
20. Rooseboom, M.; Vermeulen, N. P. E.; Durgut, F.; Commandeur, J. N. M. Comparative Study on the Bioactivation Mechanisms and Cytotoxicity of Te-Phenyl-L-tellurocysteine, Se-Phenyl-L-selenocysteine, and S-Phenyl-L-cysteine. *Chem. Res. Toxicol.* **2002,** *15,* 1610–1618.
21. Andreadou, I.; Menge, W. M. P. B.; Commandeur, J. N. M.; Worthington, E. A.; Vermeulen, N. P. E. Synthesis of Novel Se-Substituted Selenocysteine Derivatives as Potential Kidney Selective Prodrugs of Biologically Active Selenol Compounds: Evaluation of Kinetics of â-Elimination Reactions in Rat Renal Cytosol. *J. Med. Chem.* **1996,** *39,* 2040–2046.
22. Poluboyarinov, P. A.; Golubkina, N. A.; Aniskov, A. A.; Moiseeva, I. J.; Glebova, N. N.; Shvets, V. I. The Synthesis and Biological Activity of 3,3′-Dimethyl-L-Selenocystine, a New Selenocystine Derivative. *Russ. J. Bioorg. Chem.* **2019,** *45,* 241–247.

23. Block, E.; Booker, S. J.; Flores-Penalba, S.; George, G. N.; Gundala, S.; Landgraf, B. J.; Liu, J.; Lodge, S. N.; Pushie, M. J.; Rozovsky, S. Trifluoroselenomethionine: A New Unnatural Amino Acid. *Chem. Bio. Chem* **2016,** *17*, 1738–1751.
24. Giinther, W. H.; Mautner, H. G. The Synthesis of Selenocoenzyme A12. *J. Am. Chem. Soc.* **1965,** *87*, 2708–2716.
25. Piffeteau, A.; Gaudry, M.; Marquet, M. Biological Properties of Selenobiotin. *Biochem. Biophys. Res. Comm.* **1976,** *73*, 773–778.
26. Arai, K.; Takei, T.; Okumura, M.; Watanabe, S.; Amagai, Y.; Asahina, Y.; Moroder, L.; Hojo, H.; Inaba, K.; Iwaoka, M. Preparation of Selenoinsulin as aLong-Lasting Insulin Analogue. *Angew. Chem. Int. (Ed.)* **2017,** *56*, 5522–5526.
27. Li, X.-D.; Gao, Y.-D.; Sun, Y.-J.; Jin, X.-Y.; Wang, D.; Liu, L.; Cheng, L. A NaI/H2O2-Mediated Sulfenylation and Selenylation of Unprotected Uracil and its Derivatives. *Org. Lett.* **2019,** *21*, 6643–6647.
28. Choi, S.; Kalman, T. I.; Bardos, T. J. Synthesis of 5-Selenium-Substituted Uracil Derivatives. Inhibition of Thymidylate Synthetase by 5-Hydroseleno-2-deoxyuridylate. *J. Med. Chem.* **1979,** *22*, 618–621.
29. Abdo, M.; Zhang, Y.; Schramm, V. L.; Knapp, S. Electrophilic Aromatic Selenylation: New OPRT Inhibitors. *Org. Lett.* **2010,** *12*, 2982–2985.
30. Goudgaon, N. M.; Naguib, F. N. M.; el Kouni, M. H.; Schinazi, R. F. Phenylselenenyl- and Phenylthio-Substituted Pyrimidines as Inhibitors of Dihydrouracil Dehydrogenase and Uridine Phosphorylase. *J. Med. Chem.* **1993,** *36*, 4250–4254.
31. Schinazi, R. J.; Arbiser, J.; Ja, J.; Lee, S.; Kalman, T. I.; Prusoffl, W. H. Synthesis and Biological Activity of 5-Phenylselenenyl-Substituted Pyrimidine Nucleosides. *J. Med. Chem.* **1986,** *29*, 1293–1295.
32. Sosnowska, M.; Makurat, S.; Zdrowowicz, M.; Rak, J. 5-Selenocyanatouracil: A Potential Hypoxic Radiosensitizer. Electron Attachment Induced Formation of Selenium Centered Radical. *J. Phys. Chem. B.* **2017,** *121*, 6139–6147.
33. Lan, Z.-J.; Lei, Z.; Yiannikouris, A.; Reddy, T.; Reddy, Y.; Li, X.; Kincaid, H.; Eastridge, K.; Gadberry, H.; Power, C. Non-Peptidyl Small Molecule, Adenosine, 5′-Se-Methyl-5′-Seleno-, 2′,3′-Diacetate, Activates Insulin Receptor and Attenuates Hyperglycemia in Type 2 Diabetic Leprdb/db Mice. *Cell. Mol. Life Sci.* **2019**. Available from: https://doi.org/10.1007/s00018-019-03249-4.
34. Sadek, S. A.; Shaw, S. M.; Kessler, W. V.; Wolf, G. C. Selenosteroids as Potential Estrogen-Receptor Scanning Agents1. *J. Org. Chem.* **1981,** *46*, 3259–3262.
35. Sadek, S. A.; Pento, T.; Basmadjiant, G. P. Estrogenic and Antiestrogenic Activity of Novel Selenosteroids. *J. Pharm. Sci.* **1984,** *73*, 416–418.
36. Rodrigues, O. A. D.; de Souza, D.; Soares, L. C.; Dornelles, L.; Burrow, R. A.; Appelt, H. R.; Alves, C. F.; Alves, D.; Braga, A. L. Stereoselective Synthesis of Selenosteroids. *Tetrahedron Lett.* **2010,** *51*, 2237–2240.
37. Jastrzebska, I.; Mellea, S.; Salerno, V.; Grzes, P. A.; Siergiejczyk, L.; Niemirowicz-Laskowska, K.; Bucki, R.; Monti, B.; Santi, C. PhSeZnCl in the Synthesis of Steroidal β-Hydroxy-Phenylselenides Having Antibacterial Activity. *Int. J. Mol. Sci.* **2019,** *20*, 2121–2133.
38. Shu, S.; Fan, Z.; Yao, Q.; Zhang, A. Ru(II)-Catalyzed Direct C(sp2) – H Activation/Selenylation of Arenes with Selenyl Chlorides. *J. Org. Chem.* **2016,** *81*, 5263–5269.
39. Affeldt, R.; Santos, F.; da Silva, R.; Rodrigues, O.; Wessjohann, L.; Ludtke, D. Stereoselective Glycoconjugation of Steroids with Selenocarbohydrates. *RSC Adv.* **2016,** *6*, 93905–93914.
40. Sidoryk, K.; Rárová, L.; Oklešťková, J.; Pakulski, Z.; Strnad, M.; Cmocha, P.; Luboradzkie, R. Synthesis of 28a-Homoselenolupanes and 28a-Homoselenolupane Saponins. *Org. Biomol. Chem.* **2016,** *14*, 10238–10248.

41. Fuentes-Aguilar, A.; Romero-Hernández, L. L.; Arenas-González, A.; Merino-Montiel, P.; Montiel-Smith, S.; Meza-Reyes, S.; Vega-Báez, J. L.; Plata, G.; Padrón, J. M.; López, O. New Selenosteroids as Antiproliferative Agents. *Org. Biomol. Chem.* **2017,** *15,* 5041–5054.
42. Chen, P.; Wang, P.; Song, N.; Li, M. Convergent Synthesis and Cytotoxic Activities of 26-Thio- and Selenodioscin. *Steroids* **2013,** *78,* 959–966.
43. Fernández-Herreraa, M.; Sandoval-Ramíreza, J.; Sánchez-Sánchez, J.; López-Muñoz, H.; Escobar-Sánchez, H. M. Probing the Selective Antitumor Activity of 22-Oxo-26selenocyanocholestane Derivatives. *Eur. J. Med. Chem.* **2014,** *74,* 451–460.
44. Romero-Hernandez, L. L.; Merino-Montiel, P.; Montiel-Smith, S.; Meza-Reyes, S.; Vega-Baez, J.-L.; Abasolo, I.; Schwartz, S., Jr.; Lopez, O.; Fernandez-Bolanos, J. G. Diosgenin-based Thio(Seleno)Ureas and Triazolyl Glycoconjugates as Hybrid Drugs. Antioxidant and Antiproliferative Profile. *Eur. J. Med. Chem.* **2015,** *99,* 67–81.
45. Kawashima, H.; Kaneko, Y.; Sakai, M.; Kobayashi, Y. Synthesis of Cyclobakuchiols A, B, and C by Using Conformation Controlled Stereoselective Reactions. *Chem. Eur. J.* **2014,** *20,* 272–278.
46. Shin, H.-S.; Jung, Y.-G.; Cho, H.-K.; Park, Y.-G.; Cho.; Ch, -G. Total Synthesis of (±)-Lycorine from the *Endo*-Cycloadduct of 3,5Dibromo-2-pyrone and (E)-β-Borylstyrene. *Org. Lett.* **2014,** *16,* 5718–5720.
47. Snider, B. B.; Zeng, H. Total Synthesis of (-)-Fumiquinazolines A, B, C, E, H, and I. Approaches to the Synthesis of Fiscalin A. *J. Org. Chem.* **2003,** *68,* 545–563.
48. Snider, B. B.; Zeng, H. Total Syntheses of (–)-Fumiquinazolines C, E, and H. *Org. Lett.* **2002,** *4,* 1087–1090.
49. Deiters, A.; Chen, K.; Eary, C. T.; Martin, S. F. Biomimetic Entry to the Sarpagan Family of Indole Alkaloids: Total Synthesis of (+)-Geissoschizine and (+)-N -Methylvellosimine. *J. Am. Chem. Soc.* **2003,** *125,* 4541–4550.
50. Jarret, M.; Turpin, V.; Tap, A.; Gallard, J.; Kouklovsky, C.; Poupon, E.; Vincent, G.; Evanno, E. Bioinspired Oxidative Cyclization of the Geissoschizine Skeletonfor Enantioselective Total Synthesis of Mavacuran Alkaloids. *Angew. Chem. Int. (Ed.)* **2019,** *58,* 9861–9865.
51. Raj, M.; Wu, H.; Blosser, S. L.; Vittoria, M. A.; Arora, P. S. Aldehyde Capture Ligation for Synthesis of Native Peptide Bonds. *J. Am. Chem. Soc.* **2015,** *137,* 6932–6940.
52. Malins, L. R.; Payne, R. J. *Org. Lett.* **2012,** *14,* 3142–3145.
53. Townsend, S. D.; Tan, Z.; Dong, S.; Shang, S.; Brailsford, J. A.; Danishefsky, S. J. *J. Am. Chem. Soc.* **2012,** *134,* 3912–3916.
54. Lin, Y. A.; Chalker, J. M.; Davis, B. G. *J. Am. Chem. Soc.* **2010,** *132,* 16805–16811.
55. Temperini, A.; Piazzolla, F.; Minuti, L.; Curini, M.; Siciliano, C. General, Mild, and Metal-Free Synthesis of Phenyl Selenoesters from Anhydrides and their Use in Peptide Synthesis. *J. Org. Chem.* **2017,** *82,* 4588–4603.
56. Mehta, S.; Pinto, B. M. Novel Glycosidation Methodology. The Use of Phenyl Selenoglycosides as Glycosyl Donors and Acceptors in Oligosaccharide Synthesis. *J. Org. Chem.* **1993,** *58,* 3269–3276.
57. Mehta, S.; Pinto, B. M. Phenylselenoglycosides as Novel, Versatile Glycosyl Donors. Selective Activation over Thioglycosides. *Tetrahedron Lett.* **1991,** *32,* 4435–4438.
58. Van Well, R. M.; Karkkainen, T. S.; Ravindranathan Karths, K. P.; Field, R. A. Contrasting Reactivity of Thioglucosides and Selenoglucoside Donors Towards Promoters: Implications for Glycosylation Stereocontrol. *Car. Res.* **2006,** *341,* 1391–1397.
59. Meloncelli, P. J.; Martin, A. D.; Lowary, T. L. Glycosyl Iodides. History and Recent Advances. *Car. Res.* **2009,** *344,* 1110–1122.
60. Spell, M.; Wang, X.; Wahba, A. E.; Conner, E.; Ragains, J. An α-Selective, Visible Light Photocatalytic Glycosylation of Alcohols with Selenoglycosides. *Car. Res.* **2013,** *369,* 42–47.

61. Khatuntseva, E. A.; Sherman, A. A.; Tsvetkov, Y. E.; Nifantiev, N. E. Phenyl 2-azido-2-Deoxy-1-Selenogalactosides: A Single Type of Glycosyl Donor for the Highly Stereoselective Synthesis of α- and α-2-azido-2-Deoxy-D-Dalactopyranosides. *Tetrahedron Lett.* **2016,** *57,* 708–711.
62. Kumar, A. A.; Illyes, T. Z.; Kover, K. E.; Szilagyi, L. Convenient Syntheses of 1,2-Trans Selenoglycosides Using Isoselenuranium Salts as Glycosylselenenyl Transfer Reagents. *Car. Res.* **2012,** *360,* 8–18.
63. Pacuła, A. J.; Ścianowski, J. Terpenes As Green Starting Materials for New Organoselenium and Organotellurium Compounds. *Current Green. Chem.* **2016,** *3,* 36–50.
64. Sharpless, K. B.; Lauer, R. F. Selenium Dioxide Oxidation of Olefins. Evidence for the Intermediacy of Allylseleninic Acids. *J. Am. Chem. Soc.* **1972,** *94* (20), 7154–7155.
65. Sharpless, K. B.; Lauer, R. F. Facile Thermal Rearrangements of Allyl Selenides and Diselenides. [1,3] and [2,3] Shifts. *J. Org. Chem.* **1972,** *37* (24), 3973–3974.
66. Sharpless, K. B.; Young, M. W.; Lauer, R. F. Reactions of Selenooxides: Thermal Syn-Elimination and H218O Exchange. *Tetrahedron Lett.* **1973,** *22,* 1979–1982.
67. Kurose, N.; Takahashi, T.; Koizumi, T. Asymmetric [2,3]Sigmatropic Rearrangement of Chiral Allylic Selenimides. *J. Org. Chem.* **1996,** *61,* 2932–2933.
68. Kurose, N.; Takahashi, T.; Koizumi, T. Asymmetric [2,3]Sigmatropic Rearrangement of Chiral Selenonium Ylides. *J. Org. Chem.* **1997,** *62,* 4562–4563.
69. Kurose, N.; Takahashi, T.; Koizumi, T. First Synthesis of Optically Pure Selenuranes and Stereoselective Alkaline Hydrolysis. their Application to Asymmetric [2,3]sigmatropic Rearrangement of Allylic Selenoxides. *Tetrahedron: Asymmetry* **1997,** *53* (36), 12115–12129.
70. Uzarewicz, A.; Ścianowski, J.; Bąkowska-Janiszewska, J. Synthesis and Reactions of Organic Compounds with a Nitrogen Atom. Part XV. Reactions of (+)-3-Chloro-2(10)-Pinene and (-)-10-Chloro-2-Pinene with Phenyltelluro- and Phenylselenosodium. *Pol. J. Chem.* **1999,** *73,* 1791–1796.
71. Uzarewicz, A.; Ścianowski, J.; Bąkowska-Janiszewska, J. Reaction of (+)-3-Carene and (+)-2-Carene with T-Butyl Hypochlorite or N-Chlorosuccinimide in the Presence of Free Radical Catalysts. *Pol. J. Chem.* **2000,** *74,* 777–783.
72. Uzarewicz, A.; Ścianowski, J.; Bąkowska-Janiszewska, J. Synthesis and Reactions of Organic Compounds with a Nitrogen Atom. Part XVI. Reactions of (-)-4-Chloro-3(10)-Carene and (+)-10-Chloro-3-Carene with Phenyltelluro- and Phenylselenosodium. *Pol. J. Chem.* **2000,** *74,* 1079–1083.
73. Uzarewicz, A.; Ścianowski, J.; Bąkowska-Janiszewska, J. Synthesis and Reactions of Organic Compounds with a Nitrogen Atom. Part XVII. Reactions of Acyclic and Monocyclic Chlorides with Phenyltelluro- and Phenylselenosodium. *Pol. J. Chem.* **2001,** *74,* 649–656.
74. Ścianowski, J.; Tafelska-Kaczmarek, A. Syntheses and Reactions of Allyl Phenylselenides. A Convenient Method of the Allyl Methylcarbamates Synthesis. *Pol. J. Chem.* **2007,** *81,* 529–534.
75. Ścianowski, J.; Rafalski, J.; Banach, A.; Czaplewska, J.; Komoszyńska, A. Synthesis and Reactions of Optically Active Selenols Derived from Monoterpenes. *Tetrahedron: Asymmetry* **2013,** *24,* 1089–1096.
76. Claudio Santi, C.; Santoro, S.; Testaferri, L.; Tiecco, M. A Simple Zinc-Mediated Preparation of Selenols. *Synlett* **2008,** *10,* 1471–1474.
77. Skowronek, P.; Ścianowski, J.; Pacuła, A. J.; Gawronski, J. Chirality Transfer Through Sulfur or Selenium to Chiral Propellers. *RSC Adv.* **2015,** *5,* 69441–69444.
78. Back, T. G.; Dyck, B. P.; Parvez, M. Unexpected Formation of 1,3-Diselenetanes from the Reaction of Camphor Enolate with Selenium. *J. Chem. Soc., Chem. Comm.* **1994,** 515–516.

79. Back, T. G.; Dyck, B. P.; Parvez, M. 1,3-Diselenetanes and 1,3-Dithietanes Derived from Camphor. Formation, Structure, Stereochemistry, and Oxidation to Selenoxide and Sulfoxide Products. *J. Org. Chem.* **1995,** *60*, 703–710.
80. Back, T. G.; Dyck, B. P. Asymmetric Cyclization of Unsaturatedalcohols and Carboxylic Acids with Camphor-Based Selenium Electrophiles. *Chem. Comm.* **1996,** 2567–2568.
81. Back, T. G.; Nan, S. Asymmetric Methoxyselenenylations with Camphor-Based Selenium Electrophiles. *J. Chem. Soc., Perkin Trans. 1* **1998,** 3123–3124.
82. Back, T. G.; Dyck, B. P.; Nan, S. Asymmetric Electrophilic Methoxyselenenylations and Cyclizations with 3-Camphorseleno Derivatives. *Tetrahedron* **1999,** *55*, 3191–3208.
83. Back, T. G.; Moussa, Z. New Chiral Auxiliaries for Highly Stereoselective Asymmetric Methoxyselenenylations. *Org. Lett.* **2000,** *2* (19), 3007–3009.
84. Back, T. G.; Moussa, Z.; Parvez, M. Asymmetric Methoxyselenenylations and Cyclizations with 3-Camphorseleno Electrophiles Containing Oxime Substituents at C-2. Formation of an Unusual Oxaselenazole from an Oxime-Substituted Selenenyl Bromide. *J. Org. Chem.* **2002,** *67*, 499–509.
85. Tiecco, M.; Testaferri, L.; Santi, C.; Marini, F.; Bagnoli, L.; Temperini, A. Asymmetric Selenomethoxylation of Alkenes with Camphorselenenyl Sulfate. *Tetrahedron Lett.* **1998,** *39*, 2809–2812.
86. Tiecco, M.; Testaferri, L.; Marini, F.; Santi, C.; Bagnoli, L.; Temperini, A. Asymmetric Oxyselenenylation-Deselenylation Reactions of Alkenes Induced by Camphor Diselenide and Ammonium Persulfate. A Convenient One-Pot Synthesis of Enantiomerically Enriched Allylic Alcohols and Ethers. *Tetrahedron: Asymmetry* **1999,** *10*, 747–757.
87. Tiecco, M.; Testaferri, L.; Santi, C.; Tomassini, C.; Marini, F.; Bagnoli, L.; Temperini, A. Asymmetric Amidoselenenylation of Alkenes Promoted by Camphorselenenyl Sulfate: A Useful Synthetic Route to Enantiopure Oxazolines. *Eur. J. Org. Chem.* **2000,** 3451–3457.
88. Tiecco, M.; Testaferri, L.; Santi, C.; Tomassini, C.; Marini, F.; Bagnoli, L.; Temperini, A. Asymmetric Synthesis of Thioamido Selenides. A Simple Synthetic Route to Enantiopure Thiazolines. *Tetrahedron: Asymmetry* **2002,** *13*, 429–435.
89. Tiecco, M.; Testaferri, L.; Bagnoli, L.; Scarponi, C.; Purgatorio, V.; Temperini, A.; Marini, F.; Santi, C. Synthesis of Enantiomerically Pure Perhydrofuro[2,3-b]furans. *Tetrahedron: Asymmetry* **2005,** *16*, 2429–2435.
90. Tiecco, M.; Testaferri, L.; Bagnoli, L.; Scarponi, C.; Temperini, A.; Marini, F.; Santi, C. Organoselenium Mediated Asymmetric Cyclizations. Synthesis of Enantiomerically Pure 1,6-dioxaspiro[4.4]nonanes. *Tetrahedron: Asymmetry* **2006,** *17*, 2768–2774.
91. Tiecco, M.; Testaferri, L.; Marini, F.; Sternativo, S.; Santi, C.; Bagnoli, L.; Temperini, A. Asymmetric Aldol Reactions from Titanium Enolates of α-Seleno Ketones and Esters. *Tetrahedron: Asymmetry* **2004,** *15*, 783–791.
92. Tiecco, M.; Testaferri, L.; Marini, F.; Sternativo, S.; Santi, C.; Bagnoli, L.; Temperini, A. Conjugated Additions of Selenium Containing Enolates to Enones—Enantioselective Synthesis of δ-Oxo-A-Seleno Esters and their Facile Transformations. *Eur. J. Org. Chem.* **2005,** 543–551.
93. Ścianowski, J.; Rafiński, Z.; Wojtczak, A. Syntheses and Reactions of New Optically Active Terpene Diselenides. *Eur. J. Org. Chem.* **2006,** 3216–3225.
94. Ścianowski, J. Convenient Route to Dialkyl Diselenides from Alkyl Tosylates. Synthesis Of Di(Cis-Myrtanyl) Diselenide. *Tetrahedron Lett.* **2005,** 3331–3334.
95. Rafiński, Z.; Ścianowski, J. Synthesis and Reactions of Enentiomerically Pure Dialkyl Diselenides from the P-Menthane Group. *Tetrahedron: Asymmetry* **2008,** *19*, 1237–1244.
96. Ścianowski, J.; Rafiński, Z.; Wojtczak, A.; Burczyński, K. Synthesis and Reactions of β–Hydroxyselenides and β-Hydroxyselenides. *Tetrahedron: Asymmetry* **2009,** *20*, 2871–2879.

97. Ścianowski, J.; Szumera, J.; Kleman, P.; Pacuła, A. J. Synthesis and Reactions of Terpenyl Diselenides Functionalized with Phenyl and Naphthyl Groups. *Tetrahedron: Asymm* **2016,** *27,* 238–245.
98. Ścianowski, J.; Szumera, J.; Pacuła, A. J.; Rafiński, Z. Reactivity of Dipinanyl Diselenides Functionalized at the C-10-Position with -CH2O(Se)Ph, -OH and -OCPh3 Substituents. *Arkivoc* **2017,** *II,* 272–284.
99. Rafiński, Z.; Ścianowski, J.; Wojtczak, A. Synthesis and Reactions of Optically Active Dialkyl Diselenides from Pinane Group. *Lett. Org. Chem.* **2009,** *6,* 321–328.
100. Ścianowski, J.; Rafiński, Z.; Szuniewicz, A.; Wojtczak, A. New Chiral Selenium Electrophiles Derived from Functionalized Terpenes. *Tertrahedron* **2009,** *65,* 10162–10174.
101. Li, X.-L.; Wang, H.-Y.; Huang, Z.-Z. New Camphor-Derived Selenonium Ylides: Enantioselective Synthesis of Chiral Epoxides. *Aust. J. Chem.* **2005,** *58,* 749–752.
102. Wang, H.-Y.; Yang, F.; Li, X.-L.; Yan, X.-M.; Huang, Z.-Z. First Example of Highly Stereoselective Synthesis of 1,2,3-Trisubstituted Cyclopropanes via Chiral Selenonium Ylides. *Chem. Eur. J.* **2009,** *15,* 3784–3789.
103. Watanabe, S.; Hasebe, R.; Ouchi, J.; Nagasawa, H.; Kataoka, T. Enantioselective Darzens Reaction Using Organoselenide-Lithium Hydroxide Complex. *Tertahedron Lett.* **2010,** *51,* 5778–5780.
104. Midura, W. H.; Ścianowski, J.; Banach, A.; Zając, A. Asymmetric Synthesis of Cyclopropyl Phosphonates using Chiral Terpenyl Sulfonium and Selenonium Ylides. *Tetrahedron: Asymm* **2014,** *25,* 1488–1493.
105. Banach, A.; Ścianowski, J.; Uzarewicz-Baig, M.; Wojtczak, A. Terpenyl Selenides: Synthesis and Application in Asymmetric Epoxidation. *Eur. J. Org. Chem.* **2015,** *16,* 3477–3485.
106. Luca Sancineto, L.; Mangiavacchi, F.; Tidei, C.; Bagnoli, L.; Marini, F.; Gioiello, A.; Scianowski, J.; Santi, C. Selenium-Catalyzed Oxacyclization of Alkenoic Acids and Alkenols. *Asian J. Org. Chem.* **2017,** *6,* 988–992.
107. Boualy, B.; El Houssame, S.; Sancineto, L.; Santi, C.; Ait Ali, M.; Stoeckli-Evans, H.; El Firdoussi, L. *N. J. Chem.* **2016,** *40,* 3395–3399.
108. Ścianowski, J.; Pacuła, A. J.; Zielińska-Błajet, M.; Wojtczak, A. New Diphenyl Diselenides o-Substituted by an O(S,Se)-Caranyl Skeleton—Synthesis and Application in Asymmetric Reactions. *N. J. Chem.* **2016,** *40,* 6697–6705.

Index

Note: Page numbers followed by "*f*" and "*t*" refer to figures and tables, respectively.

B

D

E

F

G

FSC
www.fsc.org
MIX
Papier | Fördert gute Waldnutzung
FSC® C083411

Kolibri 360 GmbH
Ferdinand-Jühlke-Straße 7
99095 Erfurt, Deutschland
produktsicherheit@kolibri360.de